Electromechanics And Electric Machines

Electromechanics And Electric Machines

S. A. NASAR
Department of Electrical Engineering
University of Kentucky
Lexington, Kentucky

And

L. E. UNNEWEHR
Scientific Research Laboratories
Ford Motor Company
Dearborn, Michigan

John Wiley & Sons
New York □ Chichester □ Brisbane □ Toronto

Library of Congress Cataloging in Publication Data:

Nasar, S. A.
Electromechanics and electric machines.

Includes bibliographical references and index.
1. Electric machinery. 2. Electromechanical devices.
I. Unnewehr, L. E., 1925– joint author. II. Title.

TK2000.N37 621.31′04′2 78-8967
ISBN 0-471-03536-X

Printed in the United States of America

10 9 8 7 6 5 4 3 2

To my uncle, Professor Majnoon Gorakhpuri, who through his own immeasurable literary contributions, profoundly influenced me. With deepest regards and affection,
S. A. Nasar

To Jean for her love and understanding,
L. E. Unnewehr

PREFACE

The purpose of this book is to develop the fundamental principles of electromechanics and electric machines and to guide the reader in the efficient applications of the governing principles of electromechanical devices that make up our modern technological society. It is unnecessary to recount the widespread use of both static and dynamic machines in this society, for they are a part of every aspect of our daily lives, from the starting of our cars, through operating numerous household appliances and providing most of the mechanical work in industry, to supplying an important energy conversion process in the production of electrical energy. These electromechanical devices also perform a significant role as actuators and transducers in control systems of all types, such as converting position or velocity to an electric signal as timing devices and as components in instrument and display systems. In short, electromechanical devices are a part of many of the systems with which engineers and technologists from many disciplines are concerned today, so that a knowledge of the basic concepts of torque, force, induced voltage, and electromagnetic fields that describe these devices is essential.

As the title of this book implies, several areas of background study are important for its understanding. The principal subject matter in which the reader should be prepared are electric circuit theory and basic mechanics. A course background in these subjects equivalent to an engineering college junior-level is recommended. Magnetic circuit theory, which is derived both from electromagnetic field theory and from analogies to static electric circuit theory, is also important. A review of magnetic circuit theory is supplied in Chapter 2 for those readers who are unfamiliar with the concepts and terms used in this theory.

With an emphasis upon fundamental concepts and principles, the aim of this book is to provide an understanding of many types of problems involving electromechanical devices. As noted above, the most frequent contact with an electromechanical device by today's engineer is as a component in a system. The differential equations and equivalent circuits for representing many types of devices and machines in systems models are developed for use in systems analysis, performance prediction, and so forth. Guidance on handling the nonlinearities associated with most machines and techniques for linear approximations are also given, and the theory and application of machines as control devices is presented.

The design of the common types of rotating machines and transformers – induction motors, many types of dc commutator motors, and appliance motors, and so forth—has become a mature science as a result of the outstanding developments in rotating machine theory during the first half of the twentieth century. Much of this design is now handled by routine computer design programs based upon this theory, and the typical reader of this book is not likely to need this highly specialized field of knowledge. However, there is a growing interest in many of the nonstandard configurations of rotating, linear, and static electromechanical devices as a result of recent aerospace activities, renewed interest in electric transportation, and the development of more sophisticated control systems. The design and analysis of these nonstandard devices generally require a return to basic principles, for which this book is suitable.

Finally, there are two significant emphases of the milieu in which this book was written that have contributed to its approach: economics and energy conservation. Both are vital in the proper application of electromechanical devices, and they are closely interrelated. There is potential for reducing the energy consumption of many systems by increasing the efficiency (or reducing the losses) of the electromechanical devices in the system. This may be accomplished by increasing the size of the device in some cases, by improving the devices cooling system, by using different materials in the device such as higher quality magnetic materials or bearings, or, in some cases, by employing an entirely different device that will perform the required function. All these alternatives will generally necessitate additional cost in the device and perhaps in other components of the total system. The usable lifetime and reliability of the device will also influence its overall cost. In the design and application of most electromechanical devices, there are many trade-offs available that permit choices between efficiency and cost. As the need to conserve the world's energy and materials resources becomes more urgent, the engineer must be cognizant of these choices. This book attempts to point out these trade-offs wherever possible. In some cases, problems at the end of the chapters are used for this purpose.

To briefly review the contents of the book, Chapter 1 presents a general introduction. The ac and dc operations of magnetic circuits, as well as magnetic circuits with permanent magnets, are considered in sufficient detail in Chapter 2. In a one-semester course, selected sections in this chapter may be skipped by the instructor. Transformers are discussed in Chapter 3, of which the first four sections can be covered in a one-semester course. Chapter 4, which deals with incremental motion electromechanical systems, such as transducers, provides the foundation for the formulation of the energy conversion process and its dynamics. Descriptive details and steady-state characteristics of dc machines are presented in Chapter 5. It is not recommended that the entire chapter be covered in a first course—much of the descriptive material may be given as reading assignments, followed by class discussions. The pertinent topics must be judiciously chosen by the instructor. Chapters 6 and 7 deal with induction and synchronous machines. These chapters are fairly short and, except for Sections 6.1, 7.5, and 7.6, the remainder of the two chapters are recommended for a first course. Solid-state control of electric motors is given in Chapter 8, which is fairly extensive. The choice of topics for presentation in a first course is left to the instructor. Chapter 9 outlines the general theory of electric machine dynamics. Although this chapter is not recommended for a first course, students should be made aware of its contents by at least scanning it.

Standard International (SI) units are used throughout the book. However, where other units are in common usage, these are also given. Unit conversion, wire table, and certain computer applications are included in the appendixes.

S. A. Nasar

Lexington, Kentucky

L. E. Unnewehr

Dearborn, Michigan

CONTENTS

Electromechanics And Electric Machines

Chapter 1
Introduction

This is a book about the principles of electromechanics and their application in the design and analysis of electric motors, generators, transformers, and other electromechanical devices. The subject matter contained in this book encompasses a wide variety of disciplines, from advanced mathematics and computer science through many phases of electrical and mechanical engineering to the physical and material sciences. The devices and systems to which the principles of electromechanics are applied are likewise diverse in size, in the type of construction, in materials used in construction, in rotational speed and electrical frequency, and in ultimate application.

The importance of electromechanical devices in almost every aspect of life does not need to be emphasized. The number of electric motors in the average U. S. residence today is probably a minimum of 10 and can easily exceed 50. There are a minimum of 5 rotating electric machines on even the most Spartan compact automobile, and this number is increasing steadily as emission and fuel economy systems are added. In an aircraft, there are many more. Electromechanical devices are involved in every industrial and manufacturing process of a technological society. Many rotating machines have been on the moon and play an important role in most aerospace systems. It may come as a surprise to the reader, but more people travel by means of electrical propulsion each day—in elevators and horizontal people movers—than by any other mode of propulsion. The recent electrical blackouts in several major U. S. cities are a reminder of the almost total dependence of activity in urban areas upon electric machines.

This book, therefore, deals with a vast and significant topic. An understanding of the principles of electromechanics is important for all who wish to extend the usefulness of electrical technology in order to ameliorate the problems of energy, pollution, and poverty that presently face mankind. It is hoped that the reader will keep sight of the long-range potential usefulness of electromechanical devices and electric machines while using this book, even though he may at times become bogged down in the details of theory and analysis in some of the later chapters.

From the brief listing of the applications of electric machines in the preceding paragraph, it is obvious that many portions of this industry are mature and are meeting the needs of society with relatively little need for research or advanced development. For example, the motor used in the garbage disposal of a modern home is probably designed by a relatively simple computer program and manufactured in a totally automated process. The hundreds of millions of clock motors manufactured each year as well are almost totally standardized in both design and manufacture. The same can also be said of many types of industrial motors.

This is only a part of the picture, however, as one looks at the state of electromechanical development today. Even this apparently placid, static state-of-the-art in the manufacture of conventional motors may be changing drastically in the near future. In a recent effort to improve the efficiency of small induction motors used in homes, offices, business, and industrial plants, it was estimated by the Federal Energy Administration that from one to two million barrels of oil per day could be saved in the United States by improving the efficiency of these motors by 20%.[1] This particular effort makes use of changes in capacitor size and winding connections in single-phase induction motors.

Efforts are also continually being made to improve efficiency of such motors by means of improved materials and design, while keeping in mind availability of materials, adverse environmental effects in the manufacture and use of materials (especially insulation materials), and the *energy cost* of manufacturing these materials. For example, aluminum has many desirable electrical characteristics for electromechanical applications and is one of the most abundant metals on the earth but it is very costly in terms of energy use to process from raw materials.

Besides the changes in the manufacture and operation of conventional machines that are beginning to occur out of necessity for energy and environmental reasons, there are many exciting applications for new machine configurations, for unusual operation of old configurations, for sophisticated electronic control of all types of machines, and for improved understanding of theory and design techniques to achieve more economical and energy efficient machine designs. Many of the newer applications involve a new look at some old machines, such as redesigning commutator motors or operating a conventional squirrel-cage induction motor from a transistor inverter in order to develop an economically competitive electric car. Others involve the design of totally new motor configurations, such as the brushless dc motors being developed for aerospace, automotive, and industrial applications.

Electronic control of electric machines has been in use almost from the dawn of the electronic era, beginning with the relatively crude mercury-arc rectifier-controlled motors. However, with the advent of and the present-day rapid development of solid-state power devices, integrated circuits, and low-cost computer modules, the range, quality, and precision of motor control has become practically unlimited.

The integration of electromechanical devices and electronic circuits has only just begun. The environment has always offered challenges to the design and operation of electric machines. For example, effective and reliable electrical insulation was one of the nastiest problems for early machine designers. Recently, rotating machines and other electromechanical devices have had to be developed for an environment, including various types of nuclear radiation for nuclear power generating plants and for several space vehicles. Extremes of reliability in these environments have also been required in the space applications. Finally, as new sources of energy come into economic viability, electromechanical energy converters will be needed with characteristics to match such energy sources as solar converters, windmills, various nuclear configurations, coal-to-oil conversion processes, hydrogen systems, and so forth.

In this chapter, besides acquainting the reader with the exciting possibilities for advanced development of electromechanical devices, we shall discuss some of the basic concepts common to most electromechanical devices, review the methods of analysis that will be presented in subsequent chapters, and shall introduce the major classifications of rotating machines.

It is obvious that one book cannot do justice to the many topics that have been mentioned in this introduction thus far. The purpose of this book is to present the fundamentals of electromechanics and to apply these fundamentals to a few basic configurations of electromechanical devices. Some design guides and "rules of thumb" useful in analysis are given, but there is much more to the story than can be related in these few pages. Therefore, we trust that full use will be made by the reader of the references given with each chapter. Of course, there is much material beyond these references, and it is the hope of the authors that this book will stimulate further investigation into this "beyond."

1.1 TYPES OF ROTATING MACHINES

There are four principal classes of rotating machines: dc commutator, induction, synchronous, and polyphase commutator machines. There are several other types of machines that do not fit conveniently into any of these classifications. Some of the latter include stepper motors, which are, in general, synchronous machines operated in a digital manner; torque motors, which are either dc commutator or brushless synchronous machines operated in the torque (zero or low-speed) mode; homopolar machines, which are a variation of the Faraday disc generator principle and which are used to supply low-voltage, high-current for plating loads;[2] and electrostatic machines, which fall into a different category of theory and practice from the electromagnetic machines to be discussed in this book.[3]

1. *DC commutator machines.* These are commonly referred to as just "dc machines," and are distinguished by the mechanical switching device known as the commutator. They are widely used in traction and industrial applications and are discussed in Chapter 5.
2. *Induction machines.* The induction motor is the so-called "work-horse" of industry, but it is also the principal appliance motor in homes and offices. It is simple, rugged, durable, and long-lived, which accounts for its widespread acceptability in almost all aspects of technology. It can be operated as a generator and is so used in various aerospace and hydroelectric applications. Induction motors, because of their simple rotor structure, can operate at very high speeds. Figure 1.1 pictures an aerospace induction motor which operates at speeds near 64,000 rpm when driven from a source of 3200 Hz. (See chapter 6.)
3. *Synchronous machines.* The synchronous machine is probably the most diversified machine configuration, and it is often difficult to recognize the many variations that this class of machines can take on. The term *synchronous* refers to the relationship between the speed and frequency in this class of machine, which is given by (see Chapter 7)

$$rpm = 120\frac{f}{p} \tag{1.1}$$

where

rpm = machine speed in rev/min

f = frequency of applied source in Hz

p = number of poles on the machine

A synchronous machine operates only at a synchronous speed, while induction machines, often termed *asynchronous* machines, operate at speeds somewhat below synchronous speeds. A wide variety of synchro-

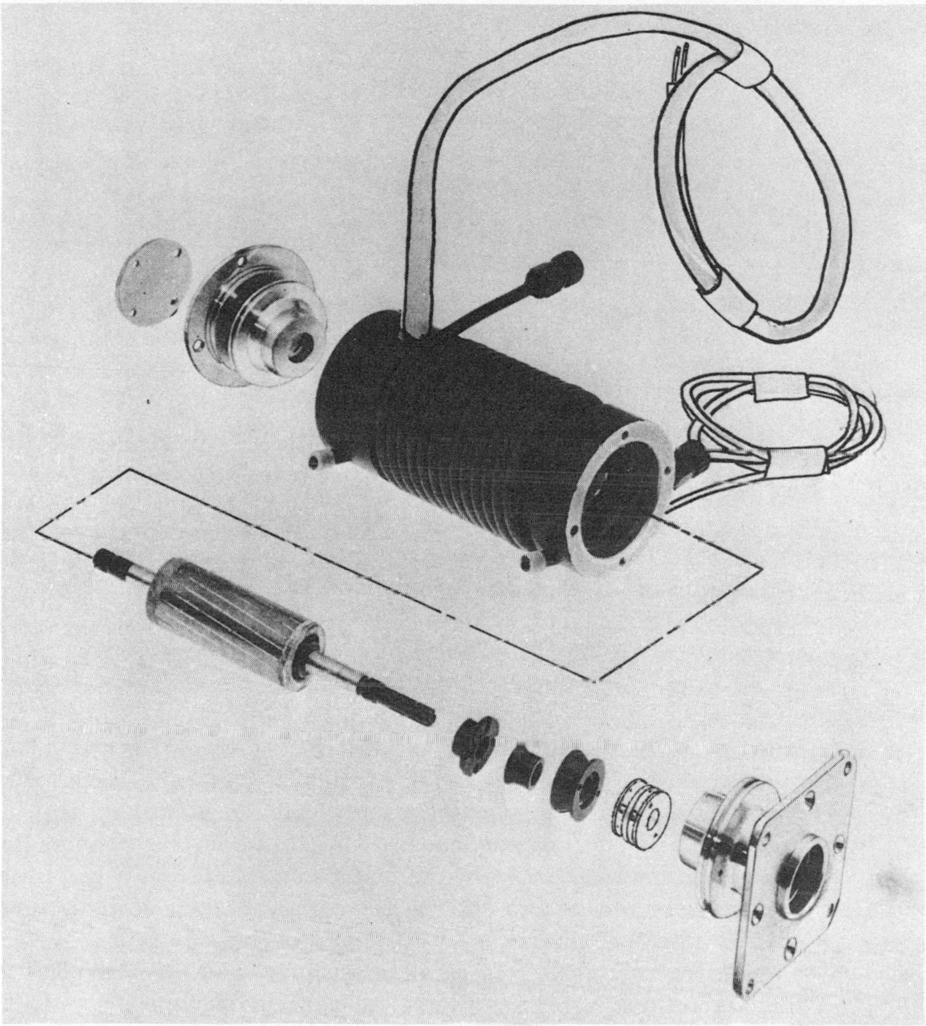

Figure 1.1. Exploded view of a five-hp, five lb., 3200-Hz, 62,000 rpm induction motor for aerospace applications. (Courtesy The Garrett Corporation.)

nous machines are in common use today:

a. *Conventional.* This is the standard synchronous machine (discussed in Chapter 7) which requires a dc source of excitation in its rotor. It is the machine used in most central-station electrical generating plants (as a generator), and in many motor applications for pumps, compressors, and so forth. A cut-away of a central-station generator is shown in Figure 1.2.

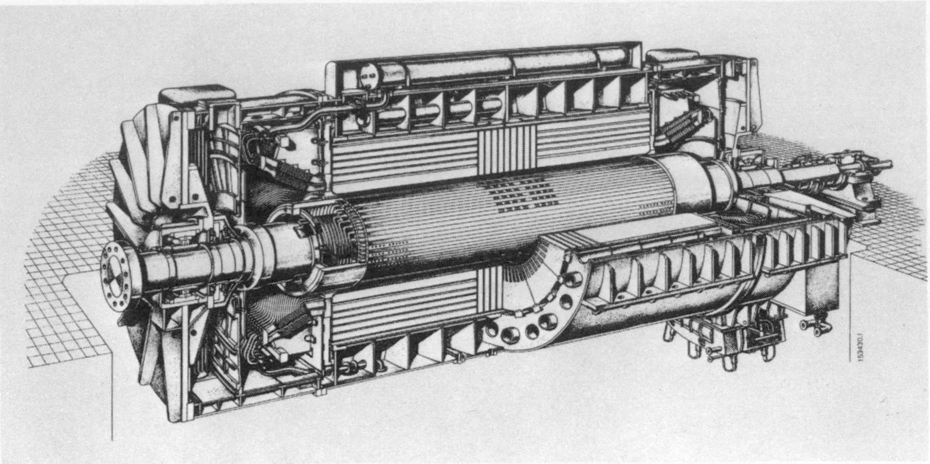

Figure 1.2. Cut-away of a water-cooled turbine generator. (Courtesy Brown, Boveri Company.)

b. *Reluctance.* This is a conventional machine without the dc field excitation and is discussed in Chapters 4 and 7. It is one of the simplest machine configurations and has recently been used in applications conventionally supplied by induction motors. In very small power ratings, it is used for electric clocks, timers, and recording applications.
c. *Hysteresis.* This configuration, like the reluctance configuration, requires only one electrical input (singly excited). The rotor of a hysteresis motor is a solid cylinder constructed of permanent magnet materials. Hysteresis motors are used in electric clocks, phonograph turntables, and other constant-speed applications. Recently, hysteresis motors have been used in applications requiring larger power output, such as centrifuge drives.
d. *Rotating rectifier.* This configuration, in performance, is identical to the conventional synchronous machine, except that the field excitation is supplied from an ac auxiliary generator and from rectifiers located on the rotating member.
e. *Inductor and flux-switch.* These are brushless synchronous motors and generators and have been applied to many aerospace and traction applications. Like the reluctance configuration, the inductor and flux-switch configurations operate on a variable-reluctance principle. The varying reluctance, as a function of rotor position, is achieved by means of the rotor design. Figure 1.3 illustrates the rotor construction of an inductor alternator; Figure 1.4 illustrates a fully assembled inductor machine.
f. *Lundell.* This configuration is used in automotive alternators and,

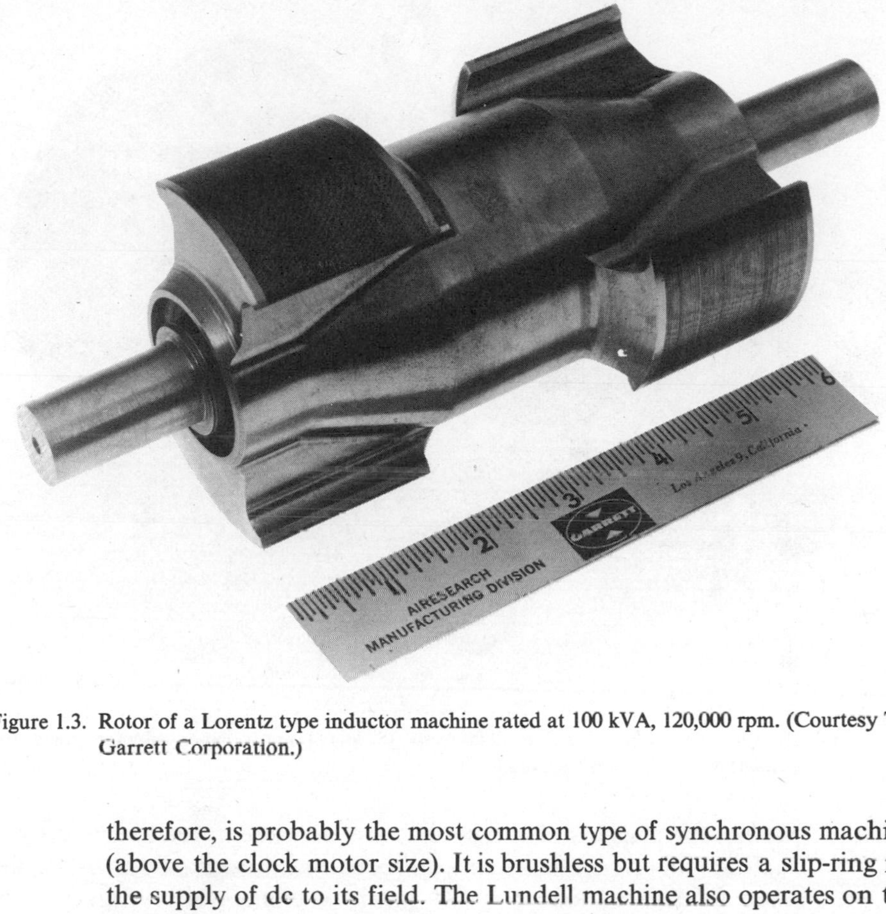

Figure 1.3. Rotor of a Lorentz type inductor machine rated at 100 kVA, 120,000 rpm. (Courtesy The Garrett Corporation.)

therefore, is probably the most common type of synchronous machine (above the clock motor size). It is brushless but requires a slip-ring for the supply of dc to its field. The Lundell machine also operates on the principle of varying magnetic reluctance due to the rotor construction.

g. *Beckey-Robinson and Nadyne Rice.* These are brushless synchrononous machines that depend upon rotor magnetic structure (varying reluctance) for operation. They have been used in many aerospace applications.

h. *Permanent magnet.* This is a conventional synchronous machine in which the field excitation is supplied by a permanent magnet (PM) instead of by a source of electrical energy. It has the potential for very high energy efficiency since there are no field losses and can, in general, be constructed at a low cost. An example of a rotor of a permanent magnet alternator used in an aerospace application is shown in Figure 1.5.

High efficiency is a characteristic of PM machines. However, to achieve relatively large power levels, PM machines require the use of

Figure 1.4. A fully assembled 520-kVA 3900-rpm, 12-pole Lorentz type inductor alternator. (Courtesy The Garrett Corporation.)

permanent magnets of a type that are, at present, relatively costly—such as cobalt-platinum, and the cobalt-rare earth alloys. Also, the fixed field excitation of a PM machine eliminates one element of control that is a principal advantage of synchronous machines over induction machines—the field control.

4. *Polyphase commutator machines.* A common form of induction machines is known as the wound-rotor or slip-ring induction machine. In this variation, the squirrel-cage bars and shorting rings are replaced by conventional windings that, essentially, duplicate the windings on the stator. The terminals of these windings are brought out for external connection through slip rings or a "polyphase commutator." Exciting these rotor windings in various manners results in a wide variety of machine characteristics. The excitation must be at a prescribed voltage and frequency which are related to the speed difference between the rotor and stator of the machine. The excitation is injected through the commutator to the rotor circuit.

Figure 1.5. Rotor for an eight-pole permanent magnet machine rated 60 kVA, 30,000 rpm, illustrating a machine used in aerospace applications where high efficiency and high specific power (watt per kg) are required. (Courtesy The Garrett Corporation.)

It is possible to obtain variable-speed characteristics over a wide range with polyphase commutator machines, the first truly variable-speed ac motors—which are still used for this application, especially in Europe. The emergence of low-cost electronic devices for control of stator voltage and frequency in cage induction, dc commutator, and synchronous motors has lessened the importance of polyphase commutator machines.

1.2 EFFICIENCY, ENERGY, AND LOSSES

An important factor in the applications of electromechanical devices of all types is the efficiency of the device. Efficiency can have different meanings in different types of physical systems. In fact, it can have a fairly general meaning that is used in everyday conversation, which is "how well a specific job is done." In mechanical systems, use is made of thermal efficiency and mechanical efficiency, which describe the efficiency of two phases of a given process and also "ideal" efficiencies. In the electrical systems that will be discussed in this book, efficiency is always used with only one meaning:

$$\eta = \frac{\text{output power or energy}}{\text{input power or energy}} \tag{1.2}$$

This can also be expressed in terms of mechanical and electrical losses in either energy or power terms as

$$\eta = \frac{\text{output}}{\text{output} + \text{losses}} = \frac{\text{input} - \text{losses}}{\text{input}} \tag{1.3}$$

The SI units of power are watts, abbreviated, W; SI units of energy are joules, J, or wattseconds, Ws, or watthours, Wh.

The energy use or efficiency of an electric machine is becoming increasingly significant and is one of the more important design criteria today. Therefore, it is important to know how to calculate the numerator and denominator of the above equations. In electromechanical devices, either the numerator or denominator of the above equations is a mechanical power or energy. Mechanical power of a rotating machine can be expressed as

$$P_m = T_{av}\Omega_{av} \tag{1.4}$$

where

$$T_{av} = \text{shaft torque in newton-meters}$$

$$\Omega_{av} = \text{shaft speed in radians/second}$$

On the electrical side of a machine, power is expressed as

$$P_e = VI\cos\Theta \qquad \text{(sinusoidal)} \tag{1.5}$$

or

$$P_e = V_{av}I_{av} \qquad \text{(DC or pulse)}$$

where

V = terminal voltage in volts

I = terminal current in amperes

Θ = power factor angle

In the above equations and throughout this book, rms parameters are designated by uppercase, unsubscripted symbols; time-averaged parameters are designated by uppercase symbols and subscript, "*av*". Power calculated by these equations is *average power*. It is also quite common to have instantaneous quantities on the right-hand side of these equations, in which lowercase symbols would be used and the power on the left would be referred to as *instantaneous power*. The use of both average and instantaneous power is common in the analysis of electromechanical systems. Energy W is the time integral of power; that is,

$$W = \int p \, dt \tag{1.6}$$

and the SI units of energy are joules or wattseconds.

It is often important to calculate efficiency in terms of energy rather than power, since the efficiency of most electromechanical devices varies over a wide range as a function of both speed and mechanical torque or load. Machines that operate under widely different conditions of speed and torque, such as traction motors, or which are operated over a "duty cycle" of varying torque and speed levels during a certain time period, are evaluated in terms of energy efficiency. The general expression for either power or energy efficiency is the same, as shown in (1.2).

1.3 RMS AND AVERAGE VALUES

Electromechanical devices often operate with input and/or output parameters that have irregular waveforms. The use of the word, "irregular," implies that one must define what is regular. Regular waveforms in electromechanical systems are steady dc or sinusoidal ac, and many of the characteristics of the materials used in electromechanical devices are defined in terms of these waveforms. However, certain parameters in electromechanical devices, such as the exciting current in transformers and induction motors, do not fit into these regular waveforms even when the applied signal is regular. With the use of electronic control, many other parameters become irregular. Therefore, it is useful to know how to calculate rms and average values of irregular waveforms in the analysis of electromechanical systems. The calculation of rms values can be found in most electrical circuit texts, but is repeated here for convenience:

$$A=\left[\frac{1}{\tau}\int_0^\tau a^2\,dt\right]^{\frac{1}{2}} \tag{1.7}$$

where

a = instantaneous value of the parameter

τ = time period over which rms value is to be calculated

t = time, s

The average value of a parameter must be more carefully defined. For periodic functions, the term, "average" refers to the *half-wave average*, defined as

$$A_{HWA}=\frac{2}{\tau}\int_0^{\tau/2} a\,dt \tag{1.8}$$

where $\tau/2$ = one-half period of a periodic function. However, it is also common to refer to the *time-average* of a parameter, defined over an arbitrary period of time, T_0, rather than a half-period:

$$A_{av}=\frac{1}{T_0}\int_0^{T_0} a\,dt \tag{1.9}$$

These concepts are particularly important in the analysis of machines excited from electronic control systems, as discussed in Chapter 8.

1.4. METHODS OF ANALYSIS

An electromechanical device is in itself a "system," since it usually consists of several electric and magnetic circuits. A number of analytical methods are useful in the analysis of electromechanical devices:

1. **Mathematical systems analysis using conventional differential equations or state equations**: This approach is used in Chapters 4 and 9, and in certain portions of other chapters.
2. **Magnetic field theory**: This is developed in Chapter 2 and will be applied to many configurations in subsequent chapters.
3. **Electrical circuit theory**: Most electromechanical devices can be represented by electrical equivalent circuits. This approach is most useful in the simulation of device performance by means of computer modeling techniques.

4. **Design and application approach**: This is a quasi-mathematical technique used in the design and application of electromechanical devices. It makes use of many of the results of the more rigorous methods described above but this method also introduces many "rules of thumb" based upon experience, geometric considerations, and thermal analysis.

All of these approaches are valid and will be introduced in subsequent portions of this book.

There are many features of the various rotating machine configurations described in Section 1.1 that are common. Of this commonality, the basic electromechanical principles are probably the most obvious, and it has long been the desire of many analysts to express this in mathematical form.[4,5,6] This approach, known as *generalized machine theory*, is a beautiful description of the mathematical and geometric form of electromechanical devices and of the commonality among diverse configurations of rotating machines. It is also most valuable in developing mathematical models for computer simulation of the dynamic characteristics of rotating machines.

The limitations of generalized machine theory are the need for relatively advanced mathematical methods for its full utilization and the fact that this highly mathematical approach cannot describe many of the physical, structural, and magnetic features of the actual electromagnetic device that are so important in their design and application. Chapter 9 of this book presents an introduction to this important body of theory.

The analysis of electromagnetic devices by means of circuit theory has also been found useful, especially in computer simulation of electrical systems containing electromechanical devices. It will be seen in subsequent chapters that most devices can be represented by relatively simple electrical and mechanical networks, although there are always limiting restrictions on the range of operating parameters over which these networks are valid. The common "black box" approach can be used with most electromagnetic devices just as it is so widely applied in electronic circuit analysis. This is particularly true of the transformer, which has no mechanical "ports" on its black box. Transformers are often represented as a simple mutual inductance, as developed in Chapter 2.

The danger of both generalized machine theory and circuit theory methods of analysis is that the physical restrictions that occur in all electromagnetic devices can seldom be properly introduced. All such devices "saturate" in one way or another, that is, they are limited in operating range by thermal, magnetic, or structural maximum, or by satisfactory commutation. Also, there can be major differences in the required model representation as a function of the device loading even within the above limits. Finally, most electromagnetic devices are frequency-sensitive, and most models are valid only over a limited range of frequencies. Therefore, to achieve a good understanding of electromagnetic devices requires a study of the *physics* of the device as well as its mathematical description. This approach, which we shall pursue in this book is outlined in the following section.

1.5 METHOD OF APPROACH

Chapters 2 and 4 deal with the general theory related to most electromechanical systems and devices. The remaining chapters deal with the major configurations of rotating machines and transformers. In these chapters, the following general approach is used.

1. **Configuration description**. The basic configuration will be described; applications and characteristics are presented in general terms.
2. **Description of physical and structural features**. Some guides in the design of devices will be given; references for further design studies are cited.
3. **Development of electromechanical principles**. Principles applicable to the specific configuration are derived and applied.
4. **Development of physical,** magnetic, and circuit models.
5. **Application of principles and models** in the analysis of the major performance characteristics. These include power, torque, efficiency, voltage and speed regulation, transient performance, and so forth.
6. **Discussion of limitations** on performance in practical devices.
7. **Analytical and computer simulation techniques** useful in the design and analysis of the configuration.
8. **Description of various practical machines** that embody the basic configuration.

1.6 RATINGS AND LIMITATIONS ON ELECTROMECHANICAL DEVICES

It has been noted above that all electromechanical devices—as well as all physical systems, including the human body—have limitations in performance. This limitation has been referred to as saturation, since it implies that there is an upper limit beyond which a given parameter cannot increase. This is most closely descriptive of the saturation that occurs in ferromagnetic materials (discussed in Chapter 2) and in electronic amplifiers.

In electromechanical devices, there are other limitations that do not generally conform to this well-known characteristic and that result in physical damage to the device. These are thermal, rotational, and commutation limits. The simplest and most obvious of these is the rotational limitation, which is the rotational speed at which the rotational element starts to fly apart. The *speed rating* of a rotating device is always well below this structural and bearing design limit (usually 50% or less) and is determined by the frequency, voltage, and power at which the machine is to be operated.

The basic power limit on rotating machines and transformers is a thermal limit. The ratings of standard classes of machines and transformers are based upon this limitation, which is usually stated in terms of a temperature rise above ambient temperature. Machines and transformers, in contrast to electronic components, have a large thermal mass and a very long thermal time constant. Therefore, there is considerable *overload* capability (in excess of rated power) in most electromechanical devices.

Standard motors and transformers (used in appliances, industrial applications, etc.) are rated in terms of *continuous* or *duty-cycle* operation. For each such device, there are also overload or short-time ratings, which can be obtained from the manufacturer. In appliance motors, these are often little above the duty-cycle or continuous rating. In industrial dc motors, on the other hand, a one-minute rating of from 2.5 to 4 times the continuous rating is possible. The nameplate designation of ratings is discussed in Chapter 5.

The commutation limit on commutator machines is the most complex and least standardized limitation. This limit is a somewhat subjective limit and is related to the sparking and ring fire that occurs on the commutator. It is highly dependent upon the environment (such as atmospheric pressure, humidity, particle content of the air) as well as upon the speed, voltage, and relative level of machine excitation.

The commutation limit results in one of the few sources of pollution in rotating machines in that it ionizes air and produces ozone, and can cause severe electromagnetic radiation. It is likely that many of the readers of this book will be involved in methods to reduce these pollutants as new government regulations are imposed. The details of commutation are discussed in Chapter 5.

The *physical size* of standard rotating machines is designated by a *frame size*. These are specified by a standardizing agency known as the National Electrical Manufacturing Association (NEMA)[7]. Frame sizes for a given power rating have decreased steadily through the years as new materials become available to give improved thermal and magnetic characteristics.

The thermal class of rotating machines and transformers is set, basically, by the type of conductor used in its windings. Both aluminum and copper conductors are used in electromechanical devices, but copper magnet wire is by far the most common. Appendix II gives the wire table for American Wire Gauge (AWG) designations of wire size for standard magnet wire used in electromechanical applications.

1.7 ECONOMIC CONSIDERATIONS

Although economics has always been an important consideration in the use of transformers and rotating machines, this book will be devoted primarily to the theory and technical characteristics of these devices. In the application of

type produce forces or torques resulting from the presence of the magnetic force field. There is a class of electric machines, known as *electrostatic machines*, whose forces result from the presence of electric fields but these will not be treated in this book.

2. Fields are three-dimensional spacial phenomena, and their analyses and understanding require some capability to visualize in the abstract. It follows that geometric characteristics are important in the application of force fields to the production of useful forces. Rotary motion results from a rotary arrangement of magnetic fields in this class of machines. There is another class of machines and devices in which linear motion results from a linear arrangement of magnetic fields.

3. A truly three-dimensional analysis of a field becomes very complex and time-consuming and will tie up tremendous blocks of computer storage when used in computer analysis methods. Fortunately, three-dimensional analysis is seldom necessary because of the property of fields known as *symmetry*. Symmetry considerations allow one to resolve the three-dimensional problem into one of two or even one dimension(s) within a limited region of space, thus simplifying the analysis and conceptual difficulties. Much of the task of analyzing machines and electromechanical devices rests upon identifying the symmetry of its fields. Fortunately, this task has been accomplished for most of the standard configurations by early investigators, but any new configuration presents a very interesting challenge.

Tests for symmetry simplification revolve around answers to two questions: (a) what dimensional coordinate components of the field do *not* exist?, and (b) with which dimensional coordinates does the field *not* vary? These tests will be applied to examples later in this chapter. Probably the most "symmetrical" of electromagnetic devices is a transformer with toroidal (doughnut-shaped) core and distributed windings (i.e., windings wound uniformly around the circumference of the toroid). Envision taking a cross-sectional "slice" perpendicular to the core. No matter where this slice is taken around the circumference, one would expect the magnetic field relationships across the cross section of this slice to be the same, since there is no change in the geometry nor in the winding as one moves around the circumference of the toroid. Therefore, the magnetic field can be examined on the basis of this two-dimensional cross section of the slice.

4. The *form* of the mathematics used to describe a field depends upon the choice of dimensional coordinates. Most of us see, think, and even feel in terms of Cartesian (rectangular) coordinates. However, most rotating machines are best described by means of cylindrical coordinates.

Field equations for systems described by Cartesian coordinates are of the class known as linear, homogeneous, equations, with which most of us are reasonably familiar. In cylindrical systems, the equations result in a form known as Bessel equations. These are less familiar to the average reader, but the numerous tables of Bessel functions that now exist and the availability of computer routines to handle Bessel functions allow this form of mathematics to be handled with almost equal ease as that using Cartesian coordinates. The third standard set of coordinates, spherical coordinates, is applicable to relatively few configurations of electromechanical devices and will not be treated in this book.

Another set of coordinates is unique to the study of electric machinery and introduces a fourth dimensional concept, motion. It is a means of relating the electrical and magnetic quantities on a moving rotor of an electric machine to the stationary electrical circuit connected to the stator of the machine.

2.1 REVIEW OF ELECTROMAGNETIC FIELD THEORY

The analysis of electric machines and electromechanical devices appropriately begins with the adaptation of Maxwell's equations to the specific spacial symmetry and materials coefficients associated with this class of system. We shall start this analysis using vector notation, since this notation is extremely valuable in determining various directional parameters of induced voltage, force, torque, and so forth. Subsequently, after these directional considerations have been established, scalar notation will be used. In *all equations*, the Standard International (SI), sometimes termed "rationalized MKS," system of units is assumed. In magnetic systems, two other systems of units have been widely used by practitioners, the CGS system, and the English system. Relationships among these systems are given in Appendix I. Just to ensure understanding, the symbol, **K**, indicates a vector. Unless otherwise indicated, Cartesian coordinates are assumed. Maxwell's equations which govern the electromagnetic phenomena at any point in space are given as

$$\begin{aligned} \nabla\cdot\mathbf{B}&=0\\ \nabla\cdot\mathbf{D}&=\rho\\ \nabla\times\mathbf{H}&=\mathbf{J}+\frac{\partial\mathbf{D}}{\partial t}\\ \nabla\times\mathbf{E}&=-\frac{\partial\mathbf{B}}{\partial t}\end{aligned} \tag{2.1}$$

As noted above, for almost all electromagnetic systems, the charge distribution, ρ, and the electric field flux density, **D**, can be assumed to be negligibly small. By means of Stokes's theorem, it is possible to transform the third and fourth of Maxwell's equations into their "integral form," applicable over a region in space:

$$\oint\mathbf{H}\cdot d\mathbf{L}=I \tag{2.2}$$

$$\oint\mathbf{E}\cdot d\mathbf{L}=-\int_s\frac{\partial\mathbf{B}}{\partial t}\cdot d\mathbf{S} \tag{2.3}$$

These two equations are known, respectively, as Ampere's law and Faraday's law, named for the persons who first experimentally verified these relationships.

It is most important to observe the directional parameters described by the vector notation in these equations, for these are the basis of the "left" and "right-hand" rules that are of use in machine analysis.

A third significant field relationship is the Lorenz force equation,

$$d\mathbf{F} = I\,d\mathbf{L} \times \mathbf{B} \tag{2.4}$$

where I is the current flowing in a differential conductor of length, $d\mathbf{L}$. One simple application of (2.4) is to integrate this differential force over a volume in which the current flows in a conductor and the flux density, $\mathbf{B}$, is uniform:

$$\begin{aligned} \mathbf{F} &= \oint I\,d\mathbf{L} \times \mathbf{B} \\ &= I\mathbf{L} \times \mathbf{B} \\ &= BIL \sin \Theta \mathbf{a}_F \end{aligned} \tag{2.5}$$

where Θ is the angle between the direction of the conductor and the magnetic field. In many machine configurations, this angle is 90°, giving,

$$F = BLI \tag{2.6}$$

which is the well-known BLI rule used in machine analysis.

Of use in evaluating many nonstandard machine or device configurations is the Biot-Savart law:

$$d\mathbf{H} = \frac{I\,d\mathbf{L} \times \mathbf{a}_R}{4\pi R^2} \tag{2.7}$$

which describes the magnetic field intensity, $d\mathbf{H}$, at a point in space due to a differential current element of length, $d\mathbf{L}$, carrying a current, I, at a distance, R, in the direction (from the current element) $\mathbf{a}_R$. In integral form, the Biot-Savart law becomes

$$\mathbf{H} = \oint \frac{I\,d\mathbf{L} \times \mathbf{a}_R}{4\pi R^2} \tag{2.8}$$

2.2 MAGNETIC MATERIALS

In free space, $\mathbf{B}$ and $\mathbf{H}$ are related by the constant, μ_0, known as the *permeability of free space*:

$$\mathbf{B} = \mu_0 \mathbf{H} \tag{2.9}$$

and

$$\mu_0 = 4\pi \times 10^{-7} \text{ henry/m}$$

The value for μ_0 given above is that of the Standard International (SI) system of units; the SI unit of **B** is the tesla and **H** is in ampere/meter. Since it is still common for material characteristics to be given in CGS units and sometimes in English units, the units for (2.9) in these two systems of units are given in Appendix I.

Within a material, (2.9) must be modified to describe a different magnetic phenomenon than that occurring in free space:

$$\mathbf{B} = \mu \mathbf{H}$$
$$\mu = \mu_R \mu_0 \tag{2.10}$$

where μ is termed *permeability* and μ_R *relative permeability*, a nondimensional constant. Permeability in a material medium as defined by (2.10) must be further qualified as applicable only in regions of homogeneous (uniform quality) and *isotropic* (having the same properties in any direction) materials. In materials not having these characteristics, μ becomes a vector (rather a tensor). Finally, it should be noted that for some common materials, (2.10) is *nonlinear*, and μ varies with the magnitude of **B**. This results in several subdefinitions of permeability related to the nonlinear B-H characteristic of the material, which will be discussed below.

A material is classified according to the nature of its relative permeability, μ_R, which is actually related to the internal atomic structure of the material and will not be discussed further at this point. Most "nonmagnetic" materials are classified as either *paramagnetic*, for which μ_R is slightly greater than 1.0, or *diamagnetic*, in which μ_R is slightly less than 1.0. However, for all practical purposes, μ_R can be considered as equal to 1.0 for all of these materials.

There is one interesting case of diamagnetism that is becoming of interest in certain types of electromagnetic devices. This is "perfect diamagnetism" (Meissner effect) which occurs in certain types of materials known as *superconductors* at temperatures near absolute zero. In such materials $\mathbf{B} \to 0$ and μ_R is essentially zero; that is, no magnetic field can be established in the superconducting material. This phenomenon has several potential applications, for instance in several types of rotating machines and in a switching device.

Magnetic properties in matter are related to the existence of permanent magnetic dipoles within the matter. Such dipoles exist in paramagnetic materials but, as noted above, the resulting magnetism is so weak as to classify these materials as nonmagnetic. There are several further classifications of materials that exhibit greater degrees of magnetism, but only two classes are discussed in detail here: *ferromagnetic* and *ferrimagnetic* materials. Ferromagnetic materials are further subgrouped into *hard* and *soft* materials, this classification roughly corresponding to the physical hardness of the materials. Soft ferromagnetic materials include the element: iron, nickel, cobalt, and one rare earth element; most soft steels; and many alloys of the four elements. Hard ferromagnetic materials include the permanent magnet materials such as the alnicos, several alloys of cobalt with the rare earth elements, chromium steels, certain copper-nickel alloys, and many other metal alloys. Ferrimagnetic materials are the *ferrites* and are composed of iron oxides having the formula $MeO{\cdot}Fe_2O_3$, where *Me* represents a metallic ion. Ferrites are subgrouped into hard and soft ferrites, the former being the permanent magnetic ferrites, usually barium or strontium ferrite. Soft ferrites include the nickel-zinc and manganese-zinc ferrites and are used in microwave devices, delay lines, transformers and other generally high frequency applications. There is a third class of magnetic materials of growing importance made from powdered iron particles or other magnetic material suspended in a nonferrous matrix such as epoxy or plastic. Sometimes termed *superparamagnetic*, powdered iron parts are formed by compression or injection molding techniques and are widely used in electronics transformers and as cores for inductors. Permalloy (molybdenum-nickel-iron powder) is one of the earliest and best known of the powdered materials.

There are a number of magnetic properties of magnetic materials that are of importance in the study of electromagnetic systems: permeability at various levels of flux density, saturation flux density, H at various levels of flux density, temperature variation of permeability, hysteresis characteristic, electrical conductivity, Curie temperature, and loss coefficients. These parameters vary widely among the different types of materials so that this discussion will be kept very general. Because of the nonlinear characteristic of most magnetic materials, graphical techniques are generally quite valuable in describing their magnetic characteristics. The two graphical characteristics of most importance are known as the B-H curve or magnetization characteristic and the hysteresis loop. There are many well-known laboratory methods for obtaining these characteristics or for displaying them on an oscilliscope. Figure 2.1 shows a typical B-H characteristic. This characteristic can be obtained in two ways: the *virgin* B-H curve, obtained from a totally demagnetized sample; or the *normal* B-H curve, obtained as the tips of hysteresis loops of increasing magnitude. There are slight differences between the two methods, which are not of importance for our purposes. The B-H curve is the result of *domain* changes within the magnetic material. In ferromagnetic materials, the material is divided up into small regions or domains (approximately 10^{-2} to 10^{-5} cm in size) in each of which all dipole moments are spontaneously aligned. When the material is completely demagnetized, these domains have random orientation resulting in zero net flux density in any finite sample. As an external magnetizing force, H, is applied to the material, those domains that happen to be in line with the direction of H tend to grow, increasing B (region I in Figure 2.1). In region II as H is further

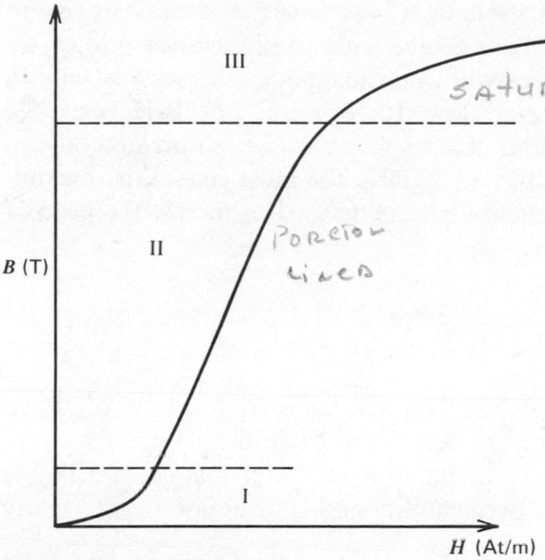

Figure 2.1. A typical *B-H* curve.

increased, the domain walls move rapidly until each crystal of the material is a single domain. In region III, the domains rotate in a direction until all domains are aligned with H. This results in *magnetic saturation* and the flux density within the material cannot increase beyond the *saturation density*, B_s. The small increase that occurs beyond this condition is due to the increase in the space occupied by the material according to the relationship $B = \mu_0 H$. It is often convenient to subtract out this component of "free space" flux density and observe only the flux density variation within the material. Such a curve is known as the *intrinsic magnetization* curve and is of use in the design of permanent magnet devices.

The three regions shown in Figure 2.1 are also of value in describing the nonlinear permeability characteristic. From (2.10) it is seen that permeability is the slope of the B-H curve. In the following discussion, relative permeability is assumed, that is, the factor μ_0, is factored out. The slope of the B-H curve is actually properly called *differential permeability*, or

$$\mu_d = \frac{1}{\mu_0} \frac{dB}{dH} \tag{2.11}$$

Initial permeability is defined as

$$\mu_i = \lim_{H \to 0} \frac{1}{\mu_0} \frac{B}{H} \tag{2.12}$$

and is seen to be the permeability in region I. This is of importance in many electronics applications where signal strength is low. It can also mislead one in measuring inductance of a magnetic core device with an inductance bridge, for the low signal strength in most bridges will often magnetize the sample only in region I where permeability is relatively low. In region II, the B-H curve for many materials is relatively straight and, if a magnetic device is operated only in this region, linear theory can be used. In all regions, the most general permeability term is known as *amplitude permeability* and is defined as merely the ratio of B to H at any point on the curve, or

$$\mu_a = \frac{1}{\mu_0}\frac{B}{H} \tag{2.13}$$

In general, permeability has to be defined on the basis of the type of signal exciting the magnetic material. There are additional definitions for pulsed and sinusoidal excitation, which will not be included here. *Maximum permeability* is the maximum value of the amplitude permeability and is of importance in many electronics applications.

The second graphical characteristic of interest is the hysteresis loop, a typical sample of which is shown in Figure 2.2. This is a *symmetrical hysteresis loop* obtained only after a number of reversals of the magnetizing force between plus and minus H_s. This characteristic illustrates several parameters of most magnetic materials, the most obvious being the property of hysteresis itself. The area within the loop is related to the energy required to reverse the magnetic domain walls as the magnetizing force is reversed. This is a nonreversible energy and results in an energy loss known as the *hysteresis loss*. This area varies with temperature and the frequency of reversal of H in a given material.

The second quadrant of the hysteresis loop is of much value in the analysis of devices containing permanent magnets. An example of this portion of hysteresis loop for Alnico V is shown in Figure 2.3. The intersection of the loop with the horizontal (H) axis is known as the *coercive force*, H_c, and is a measure of the magnet's ability to withstand demagnetization from external magnetic signals. Often shown on this curve is a second curve, which is the product of B and H plotted as a function of H, known as the *energy product*, and which is a measure of the energy stored in the permanent magnet. The value of B at the vertical axis is known as the *residual flux density*.

The *Curie temperature* or Curie point, T_c, is the critical temperature above which a ferromagnetic material becomes paramagnetic.

Up to this point, we have not indicated numerical values for these various parameters. The parameter values for several common magnetic materials are given in Table 2.1. Several B-H curves are given in Figure 2.4. It is of importance to note the typical values of relative permeability for good magnetic materials and to compare them with values of electrical conductivity for good

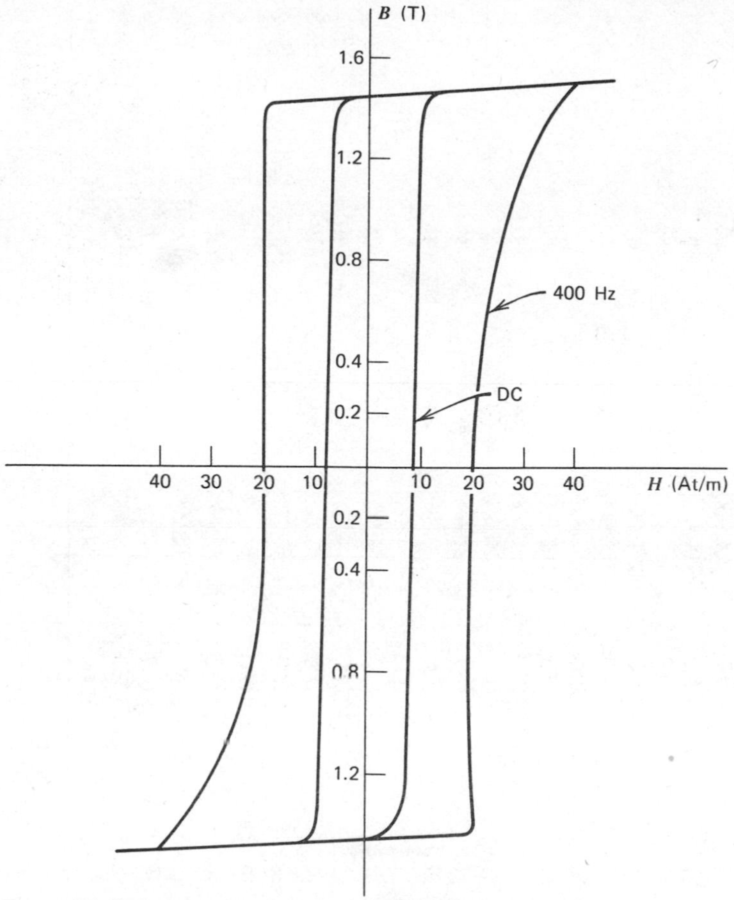

Figure 2.2. Deltamax tape-wound core 0.002-in. strip hysteresis loop.

electrical conductors. There are some magnetic materials, such as permalloy, supermendur, and other nickel alloys that have a *maximum* relative permeability of over 100,000, giving a ratio to the permeability of a nonmagnetic material, such as air or free space, of 10^5. High permeability of this magnitude can be realized only in a few materials and only over a very limited range of operation. The permeability ratio between good and poor magnetic materials over a typical working range of operation is more like 10^4 at best. However, the ratio of electrical conductivity of a good electrical conductor, such as copper, to a good insulator, such as polystyrene, is of the order of 10^{24}. This implies, correctly, that no material such as a good magnetic insulator exists except for the superconductors mentioned previously. This point will become more apparent as we discuss magnetic circuits.

Table 2.1 **Characteristics of Soft Magnetic Materials**

Trade Name	Principal Alloys	Saturation Flux Density (T)	H at B_{sat} (A/m)	Amplitude Permeability Max-μ_m	Coercive Force—H_c (A/m)	Electrical Resistivity (μohm-cm)	Curie Temperature (°C)
48 NI	48% Ni	1.25	80	200,000		65	
Monimax	48% Ni	1.35	6,360	100,000	4.0	65	398
High Perm 49	49% Ni	1.1	80			48	
Satmumetal	Ni,Cu	1.5	32	240,000		45	398
Permalloy (sheet)	Ni,Mo	0.8	400	100,000	1.6	55	454
Moly Permalloy (powder)	Ni,Mo	0.7	15,900	125			
Deltamax	50% Ni	1.4	25	200,000	8	45	499
M-19	Si	2.0	40,000	10,000	28	47	
Silectron	Si	1.95	8,000	20,000	40	50	732
Oriented-T	Si	1.6	175	30,000		47	
Oriented M-5	Si	2.0	11,900		26	48	746
Ingot Iron	None	2.15	55,000		80	10.7	
Supermendur	49% Co,V	2.4	15,900	80,000	8	26	
Vanadium Permendur	49% Co,V	2.3	12,700	4,900	92	40	932
Hyperco 27	27% Co	2.36	70,000	2,800	198	19	925
Flake Iron	Carbonal power	$\approx$0.8	5,200	5–130		10^5–10^{15}	
Ferrotron (powder)	Mo,Ni	(Linear)	(Linear)	5–25		10^{16}	
Ferrite	Mg,Zn	0.39	1,115	3,400	13	10^7	135
Ferrite	Mn,Zn	0.453	1,590	10,000	6.3	3×10^7	190
Ferrite	Ni,Zn	0.22	2,000	160	318	10^9	500
Ferrite	Ni,Al	0.28	6,360	400	143	—	500
Ferrite	Mg,Mn	0.37	2,000	4,000	30	1.8×10^8	210

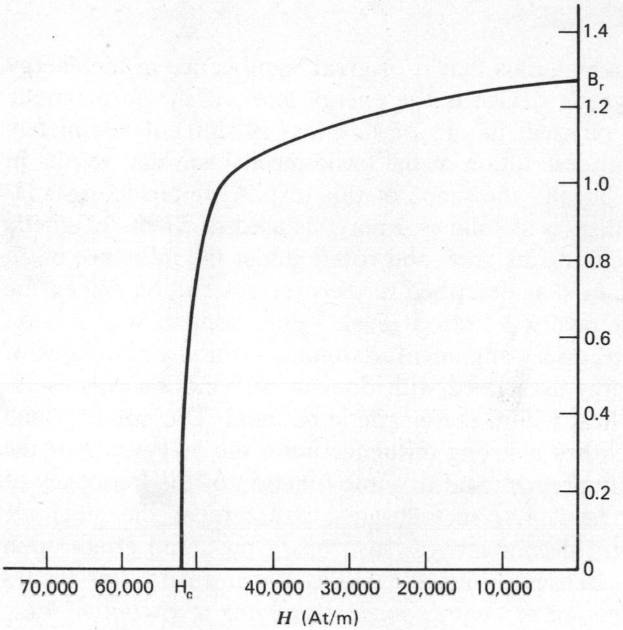

Figure 2.3. Demagnetization curve of Alnico V.

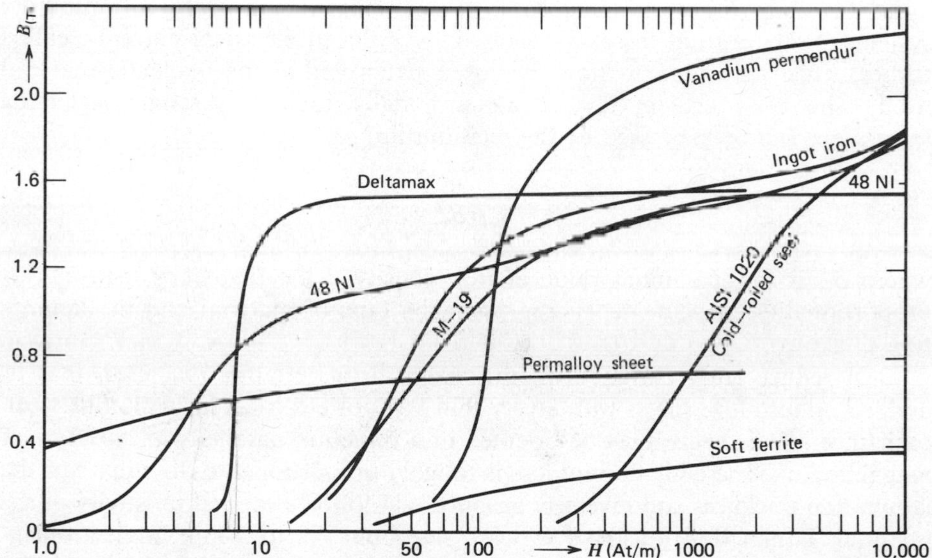

Figure 2.4. B-H curves of selected soft magnetic materials.

2.3 MAGNETIC LOSSES

A characteristic of magnetic materials that is of great significance in the energy efficiency of an electromagnetic device is the energy loss within the magnetic material itself. The actual physical nature of this loss is still not completely understood and a theoretical description of the basic mechanism that results in magnetic material losses is beyond the scope of this text. A simplistic explanation of this complex mechanism is as follows. Energy is used to effect "magnetic domain wall motion" as the domains grow and rotate under the influence of an externally applied magnetic field as described in the previous section. When the external field is reduced or reversed from a given value, domain wall motion again occurs to realize the necessary alignment of domains with the new value of the magnetic field. The energy associated with domain wall motion is irreversible and manifests itself as heat within the magnetic material. The *rate* at which the external field is changed has a strong influence upon the magnitude of the loss, and the loss is generally proportional to some function of the frequency of variation of the magnetic field. The metallurgical structure of the magnetic material, including its electrical conductivity, also has a profound effect upon the magnitude of the loss. In electric machines and transformers, this loss is generally termed the *core loss*, or sometimes *magnetizing loss* or *excitation loss*.

Traditionally, core loss has been divided up into two components: *hysteresis loss* and *eddy current loss*. The hysteresis loss component has been alluded to above and is generally held to be equal to the area of the low-frequency hysteresis loop times the frequency of the magnetizing force in sinusoidal systems. Eddy current losses are caused by induced electrical currents, called eddies, since they tend to flow in closed paths within the magnetic material itself. The eddy current loss in a sinusoidally excited material, neglecting saturation, can be expressed by the relationship,

$$P_e = k_e f^2 B_m^2 \text{ watts} \tag{2.14}$$

where B_m is the maximum value of flux density, f the frequency, and k_e is a proportionality constant depending upon the type of material and the lamination thickness.

To reduce eddy current loss, the magnetic material is *laminated*, that is, divided up into thin sheets with a very thin layer of electrical insulation between each sheet. The sheets must be oriented in a direction parallel with the flow of magnetic flux. The eddy current loss is roughly proportional to the square of the lamination thickness and inversely proportional to the electrical resistivity of the material. Lamination thickness varies from about 0.5 to 5 mm in electromagnetic devices used in power applications and from about 0.01 to 0.5 mm for devices used in electronics applications. Many magnetic cores used in electronics transformers and inductors are tape-wound from very thin strips of magnetic material. It should be recognized that laminating a magnetic part generally increases its volume. This increase may be appreciable, depending upon the method used to bond the laminations together. The ratio of the volume actually occupied by magnetic material to total volume of a magnetic part is known as the *stacking factor*. This factor is important in accurately calculating flux densities in magnetic parts. Table 2.2 gives typical stacking factors for the thinner lamination sizes.

Table 2.2

Lamination Thickness (mm)	Stacking Factor
0.0127	0.50
0.0254	0.75
0.0508	0.85
0.1–0.25	0.90
0.27–0.36	0.95

Stacking factor approaches 1.0 as the lamination thickness increases. In powdered iron and ferrite magnetic parts, there is an "equivalent stacking factor," which is approximately equal to the ratio of the volume of the magnetic particles to overall volume.

Whereas (2.14) and the statements made concerning hysteresis loss are good "rules of thumb" for evaluating variations of these loss components with various field parameters, they are quite inadequate for analytical predictions of absolute values of core loss. Therefore, core loss should be determined from experimental data. Most manufacturers of magnetic materials have obtained core loss data under the condition of sinusoidal excitation for most of their products. Figures 2.5 and 2.6 show measured core loss values for two common types of magnetic materials, M-15, a 3% silicon steel widely used in transformers and small motors, and 48NI, a nickel alloy used in many electronics applications, respectively. These data are obtained by a measurement known as the Epstein frame method on sheet samples of material. Figure 2.6 *b* shows measured core loss in a ferrite material.

As a word of caution, many electromagnetic devices are finding increasing application in circuits in which the voltages and currents have waveshapes that cannot be classified by any of the standard waveforms such as sine waves, steady dc, square waves, and so forth. In many such circuits, power levels are relatively large and, therefore, the measurement of power, losses, and efficiency is a significant factor in their design and application. The source of these nonstandard waveforms is frequently the switching action of semiconductors in systems that include inverters, cycloconverters, controlled rectifiers, and so forth. Core loss data from sinusoidal measurements are generally not adequate for such systems. In general, core loss measurements should be performed with the device excited from a source whose waveform is as close to that under which the device will be actually operated as possible. Power measurements under conditions of nonsinusoidal excitation generally require the use of Hall-effect, thermal, or electronic-multiplier types of wattmeters.

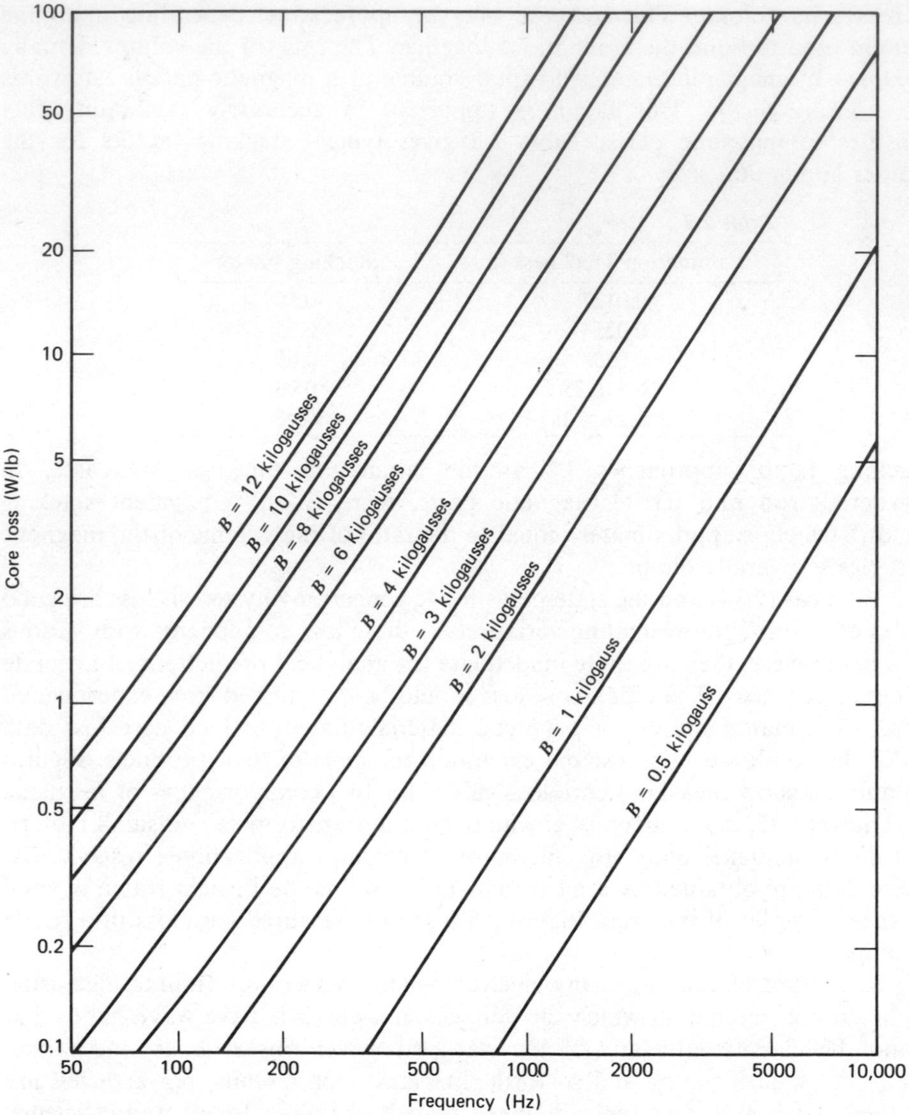

Figure 2.5. Core loss for nonoriented silicon steel, 0.019-in.-thick lamination. (Courtesy Armco Steel Corporation.)

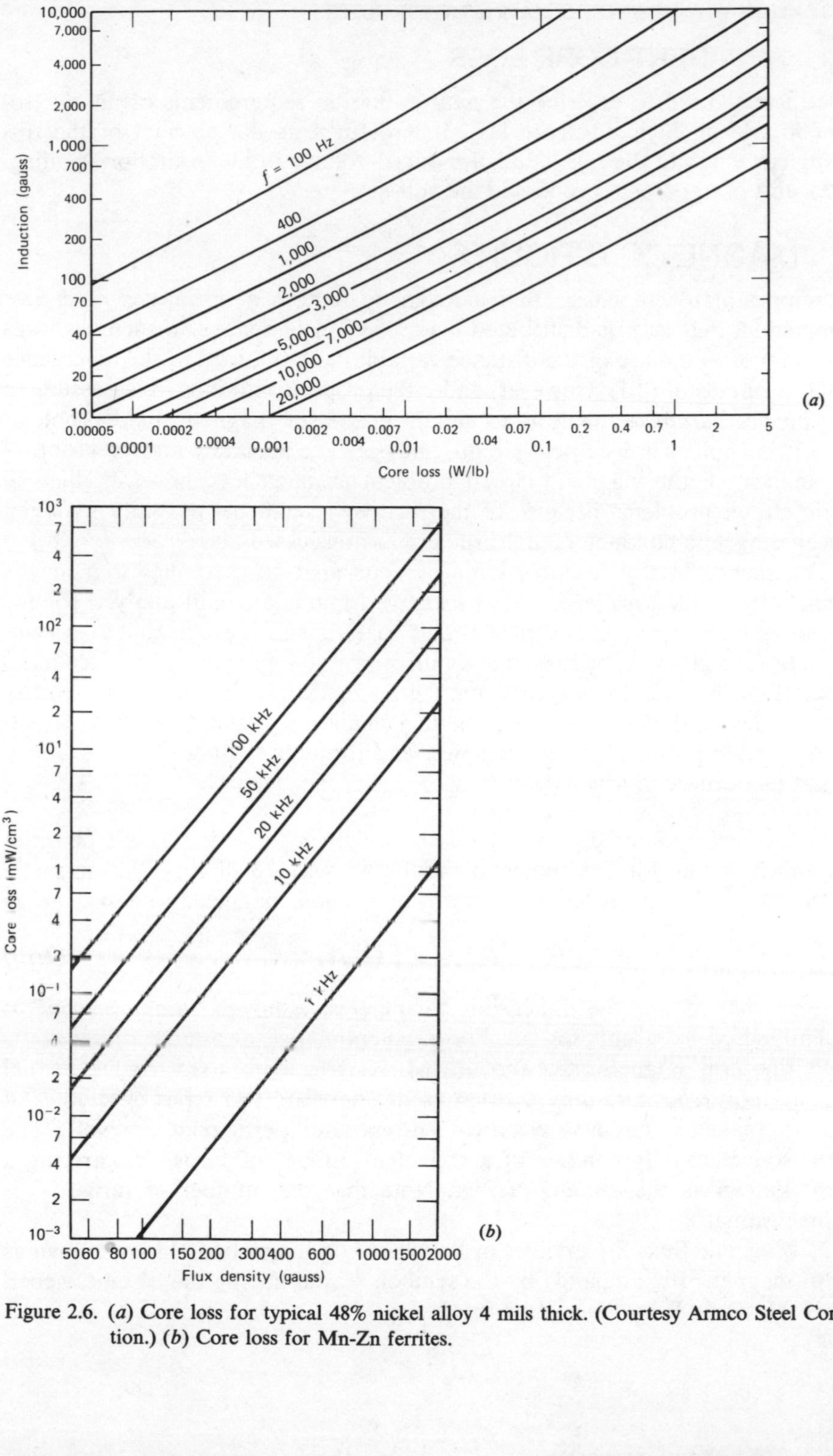

Figure 2.6. (*a*) Core loss for typical 48% nickel alloy 4 mils thick. (Courtesy Armco Steel Corporation.) (*b*) Core loss for Mn-Zn ferrites.

2.3.1 APPARENT CORE LOSS

This is a term used to describe the *total* excitation requirements of an electromagnetic system, including core loss. It is defined as the product of the rms exciting current with the rms value of induced voltage in the excitation winding. The SI unit of apparent core loss is the volt-ampere.

2.4 MAGNETIC CIRCUITS

It is important to emphasize that a magnetic field is a *distributed parameter* phenomenon, that is, it is distributed over a region of space. As such, rigorous analysis requires the use of the distance variables as contained in the divergence and curl symbols of (2.1). However, under the proper conditions, it is possible to apply *lumped parameter* analysis to certain classes of magnetic field problems just as it is applied in electrical circuit analysis. The accuracy and precision of such analysis in the magnetic circuit problem is much less, however, than in electric circuit problems because of the relatively small permeability variation between magnetic conductors and insulators as discussed above.

This section briefly describes lumped circuit analysis as applied to magnetic systems, often called *magnetic circuit analysis*. Magnetic circuit analysis follows the approach of simple dc electrical circuit analysis and is applicable to systems excited by dc signals or, by means of an incremental approach, to low-frequency ac excitation. Its usefulness is in sizing the magnetic components of an electromagnetic device during design stages, in calculating inductances, and in determining air-gap flux densities for power and torque calculations.

Let us begin with a few definitions.

1. Magnetic potential: For regions in which no electric current densities exist, which is true for the magnetic circuits we wish to discuss, the magnetic field intensity, **H**, can be defined in terms of a *scalar* magnetic potential, M, as

$$\mathbf{H} = \nabla M; \ M = \int \mathbf{H} \cdot d\mathbf{L} \tag{2.15}$$

It is seen that M has the dimension of amperes, although "ampere-turns" is frequently used as a unit for M. For a potential rise or source of magnetic energy, the term magnetmotive force (MMF) is frequently used. As a potential drop, the term *reluctance drop* is often used. There are two types of sources of MMF in magnetic circuits: electrical current and permanent magnets. The current source usually consists of a coil of a number of turns, N, carrying a current known as the *exciting current*. Note that the number of turns, N, is nondimensional.

2. Magnetic flux: Streamlines or flowlines in a magnetic field are known as lines of magnetic flux, denoted by the symbol, ϕ, and having the SI unit, weber. Flux is related to **B** by the surface integral,

$$\phi = \int_{\mathbf{s}} \mathbf{B} \cdot \mathbf{dS} \tag{2.16}$$

3. Reluctance: Reluctance is a component of magnetic impedance, somewhat analogous to resistance in electric circuits except that reluctance is not an energy loss component. It is defined by a relationship analogous to Ohm's law,

$$\phi = \frac{M}{R} \tag{2.17}$$

The SI unit of magnetic reluctance is henry^{-1}. In regions containing magnetic material that is homogeneous and isotropic and where the magnetic field is uniform, (2.17) gives further insight into the nature of reluctance. If we assume that the flux density has only one directional component, B, and is uniform over a cross section of area, A_m, taken perpendicular to the direction of B, then (2.16) becomes, $\phi = BA_m$. We also assume that H is nonvarying along the length, l_m, in the direction of B, and (2.17) becomes, with some rearranging,

$$R = \frac{M}{\phi} = \frac{Hl_m}{BA_m} = \frac{l_m}{\mu A_m} \tag{2.18}$$

which is similar to the expression for electrical resistance in a region with similarly uniform electrical properties.

4. Permeance: The permeance, P, is the reciprocal of reluctance and has the SI unit, henry. In electronics transformer analysis, the term, *induction factor*, A_L, is frequently used and is identical to what is here called permeance. Permeance and reluctance are both used to describe the geometric characteristics of a magnetic field, mainly for purposes of calculating inductances.

5. Leakage Flux: Between any two points at different magnetic potentials in space, a magnetic field exists, as shown by (2.15). In any practical magnetic circuit there are many points—or, more generally—planes at magnetic potentials different from each other. The magnetic field between these points can be represented by flow lines or lines of magnetic flux. Where these flux lines pass through regions of space—generally air spaces, electrical insulation, or structural members of the system—rather than along the main path of the circuit, they are termed *leakage flux lines*. In coupled circuits with two or more windings, the definition of leakage flux is specific, namely that flux which links one coil but not the other.

Leakage is a characteristic of all magnetic circuits and it can never be completely eliminated. At dc or very low ac excitation frequencies, magnetic shielding consisting of thin sheets of high permeability material can reduce leakage flux. This is done not by eliminating leakage but by establishing new levels of magnetic potential in the leakage paths to better direct the flux lines along the desired path. At higher frequencies of excitation, electrical shielding, such as aluminum foil, can reduce leakage flux by dissipating its energy as induced currents in the shield.

6. Fringing: Fringing is somewhat similar to leakage and is a term used to describe the spreading of flux lines in an air gap of a magnetic circuit. Figure 2.7

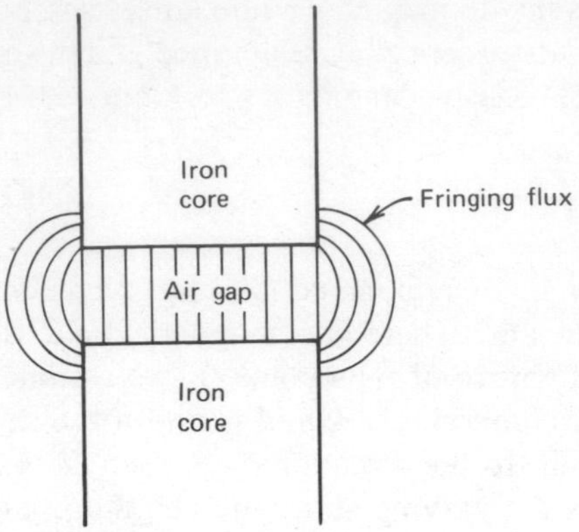

Figure 2.7. Flux fringing at an air gap.

illustrates fringing at a gap. Fringing results from lines of flux that appear along the sides and edges of the two magnetic members at each side of the gap, which are at different magnetic potentials. Fringing is almost impossible to calculate analytically, except in the simplest of configurations. Fringing has the effect of increasing the effective area of the air gap, which must be considered in the design of a magnetic circuit. The relative effect of fringing increases with the length of the air gap.

2.5 AMPERE'S LAW APPLIED TO A MAGNETIC CIRCUIT

According to (2.2) the integral around any closed path of the magnetic field intensity, **H**, equals the electrical current contained within that path. A word about directions in using this integral expression is in order here: Positive current is defined as flowing in the direction of the advance of a right-handed screw turned in the direction in which the closed path is traversed.

Let us apply Ampere's law to the simple magnetic circuit whose cross section is shown in Figure 2.8, consisting of a magnetic member of mean length, l_m, in series with an air gap of length, l_g, around which are wrapped three coils of turns N_1, N_2, and N_3, respectively. The path of magnetic flux, ϕ, is shown along the mean length of the magnetic member and across the air gap. Let the line integration proceed in a clockwise manner. Current directions are shown in the three coils. Note that for the directions shown, current direction is into the plane of the paper for the conductors enclosed by the integration path for coils 1 and 3 and out of the plane of the paper for coil 2. From the left-hand side of (2.2), we obtain

$$\oint \mathbf{H}\cdot d\mathbf{l} = \int_0^{l_m} \mathbf{H}_m \cdot d\mathbf{l} + \int_0^{l_g} \mathbf{H}_g \cdot d\mathbf{l} \tag{2.19}$$

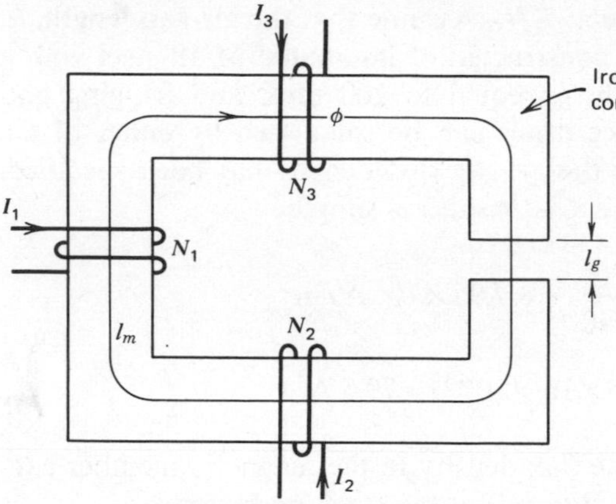

Figure 2.8. A composite magnetic circuit, with multiple excitation (mmf's).

If the magnetic material is linear, homogeneous, and isotropic, and if leakage flux is neglected, (2.19) becomes

$$\oint \mathbf{H} \cdot d\mathbf{l} = H_m l_m + H_g l_g = \phi(R_m + R_g) = M_m + M_g \tag{2.20}$$

where R_m and R_g are the reluctances of the magnetic member and gap, respectively, and M_m and M_g represent the magnetic potential or reluctance drop across these two members of the magnetic circuit. The right-hand side of (2.2) gives

$$I = N_1 I_1 + N_3 I_3 - N_2 N_2 \tag{2.21}$$

Combining (2.20) and (2.21) yields

$$N_1 I_1 + N_3 I_3 - N_2 I_2 - M_m - M_g = 0 \tag{2.22}$$

We may generalize Ampere's law on the basis of this simple example as stating that "the sum of the magnetic potentials around any closed path is equal to zero," which is analogous to Kirchoff's voltage relationship in electrical circuits. Note that this generalization follows even without the simplifying assumptions used to eliminate the integral form of (2.19).

Example 2.1

Let us illustrate the use of (2.22) by assuming some numerical values in the circuit of Figure 2.8 and solving the following problem: Determine the number of ampere-turns required to establish a flux density of 1 tesla in the air gap. It will serve no instructive purpose to have three coils, so we shall set I_2 and I_3

equal to zero and solve for the product, I_1N_1. Assume that the air-gap length, l_g, is 0.1 mm; the magnetic member is constructed of laminated M-19 steel with a stacking factor of 0.9 and of length, l_m, equal to 100 mm; and fringing and leakage are neglected. The reluctance drops can be calculated by either of the forms in the interior of (2.20); since the air-gap flux density has been specified, the form in terms of the magnetic field intensities is simple:

$$H_g = \frac{B_g}{\mu_0} = \frac{1.0}{4\pi \times 10^{-7}} = 7.95 \times 10^5 \text{ A/m}$$

$$M_g = H_g l_g = (7.95 \times 10^5)(10^{-4}) = 79.5 \text{ A}$$

If we neglect fringing and leakage, the flux density in the magnetic member can be assumed to be the air-gap density divided by the stacking factor, or

$$B_m = \frac{1.0}{0.9} = 1.11 \text{ tesla}$$

From the curve for M-19 steel in Figure 2.4, at this value of flux density,

$$H = 130$$

$$M_m = 130 \times 0.1 = 13$$

The required ampere-turns in the exciting coil,

$$N_1 I_1 = M_m + M_g = 92.5 \text{ A (ampere-turns)}$$

Example 2.2
Using the same configuration and numerical values as in Example 2.1, determine the required ampere-turns in the exciting coil for establishing a *flux* of 0.001 Wb in the air-gap. Let us solve this problem using reluctances, neglecting leakage but not fringing effects. We shall need to know the cross-sectional area of the magnetic member to determine reluctances; assume that $A_m = 16 \text{ cm}^2$ (gross).

Equation (2.18) can be used to determine the air-gap reluctance. Assume that fringing effects increase the effective gap area over the area of the steel surface facing it by 10%. The reluctance becomes

$$R_g = \frac{10^{-4}}{(4\pi \times 10^{-7})(1.1 \times 0.0016)} = 4.5 \times 10^4$$

If we neglect leakage, the same flux will exist in the magnetic member. The flux density in the magnetic material is

$$B_m = \frac{0.001}{0.9 \times 0.0016} = 0.695 \text{ tesla}$$

From the M-19 curve in Figure 2.4 the amplitude permeability is

$$\mu_a = \frac{B_m}{\mu_0 H_m} = \frac{0.695}{(4\pi \times 10^{-7})(54)} = 10{,}240 = \mu_R$$

The reluctance of the magnetic member is

$$R_m = \frac{l_m}{\mu_R \mu_0 A_m} = \frac{0.1}{10240(4\pi \times 10^{-7})(0.9 \times 0.0016)} = 0.54 \times 10^4$$

The required exciting ampere-turns are

$$N_1 I_1 = \phi(R_g + R_m) = 0.001(5.04 \times 10^4) = 50.4 \text{ A}$$

There are several conclusions to observe from these simple examples:

1. The magnetic core with air-gap is analogous to a simple series dc circuit as shown in Figure 2.9.
2. Because of symmetry about the plane of the paper of the system of Figure 2.8, two-dimensional representation of the magnetic field is acceptable.
3. The calculation of reluctances is a more cumbersome approach than the use of magnetic field intensities for determining reluctance drops.
4. The reader may wonder about solving the inverse problem to the above two examples, that is, given the exciting ampere-turns, determine the flux (or flux density) in the air-gap. A little thought will show that there is no direct analytical solution to this problem due to the nonlinear *B-H* characteristic of the magnetic material: The flux density must be known before the field intensity or reluctance of the magnetic member can be determined. There are several graphical techniques for solving this inverse problem. This type of problem is applicable to iterative computer techniques, although considerable computer storage space may be required, since the *B-H* characteristics of the magnetic material must be stored for repeated access during the iterative process.

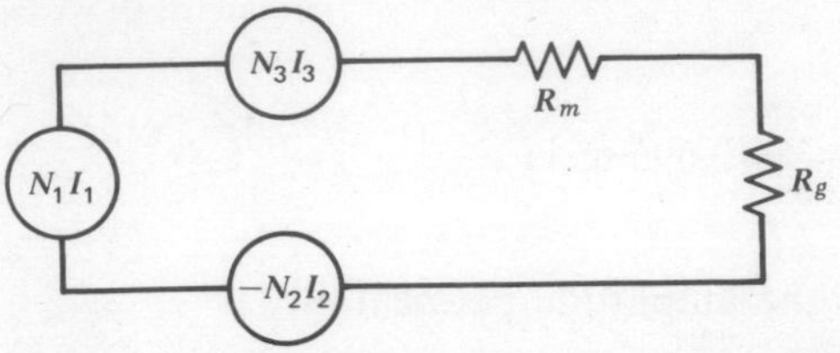

Figure 2.9. Approximate equivalent circuit for Figure 2.8.

2.6 LIMITATIONS OF MAGNETIC CIRCUIT APPROACH

The number of problems in practical magnetic circuits that can be solved by the approach outlined in the preceding two sections is rather limited, despite the similarity of this approach to simple dc electrical circuit theory. Discussion (4) after Example 2.2 has illustrated but one of the limitations. The purpose of introducing magnetic circuits is more to state some very fundamental principles and definitions which are necessary for an understanding of electromagnetic systems rather than as a problem solving technique. The limitations of magnetic circuit theory rest primarily on the nature of magnetic materials as contrasted with conductors, insulators, and dielectric materials. Most of these limitations have already been introduced as "assumptions" presented in the discussion of magnetic circuits. Let us assess the significance of these assumptions:

1. Homogeneous magnetic material. Most materials used in practical electromagnetic systems can be considered homogeneous over finite regions of space, allowing the use of the integral forms of Maxwell's equations and calculations of reluctances and permeances.
2. Isotropic magnetic materials. Many sheet steels and ferrites are *oriented* by means of the metallurgical process during their production. Oriented materials have a "favored" direction in their grain structure, giving superior magnetic properties when magnetized along this direction. These directional characteristics, when present, must always be considered in the analysis of a magnetic circuit, but the integral equations are still generally applicable.
3. Nonlinear characteristic. This is an inherent property of all ferro- and ferrimagnetic materials. This may appear to be a serious obstacle to any meaningful analytical approach to electromagnetic systems, but there are several reasons why this is not the case:
   a. As can be seen from the *B-H* curves shown in Figures 2.1 and 2.4, there is a considerable portion of the curve for most materials that can be approximated by a straight line. Many devices are operated at flux densities mainly in this range and the device can be considered essentially linear.

b. The nonlinear *B-H* characteristic of magnetic materials manifests itself in the relationship between flux and exciting current in electromagnetic systems; the relationship between flux and induced voltage is a *linear* relationship as given by Faraday's law, (2.3). It is possible to treat these nonlinear excitation characteristics separately in many systems, such as is done in the equivalent circuit approach to transformers and induction motors.

c. An inductance whose magnetic circuit is composed of a magnetic material is a nonlinear electrical circuit element, such as a coil wound on a magnetic toroid. With an air-gap in the magnetic toroid, however, the effect of the nonlinear magnetic material upon the inductance is lessened. Rotating machines and many other electromechanical devices have air-gaps in their magnetic circuits, permitting the basic theory of these devices to be described by means of linear equations. As has been pointed out, the paths for leakage fluxes are primarily through nonmagnetic regions of space, and the inductances associated with these fluxes—leakage inductances—are linear circuit elements.

4. Saturation. All engineering materials and devices exhibit a type of saturation, when output fails to increase with input, such as in the saturation of an electronic amplifier. Magnetic saturation, defined in Section 2.2, is a part of the nonlinear characteristic discussed in the previous paragraph but is generally treated as a separate characteristic of a magnetic circuit. Saturation is very useful in many electromagnetic devices, such as magnetic amplifiers and saturable reactors. The magnetic materials frequently used in these and other types of magnetic switching devices are called *square-loop* materials, since their hysteresis loops (Figure 2.2) can be approximated by a square or rectangle with sides parallel to the *B* and *H* axes.

   The assumption of this loop characteristic permits a straightforward mathematical approach to the analysis of magnetic circuits. In most rotating machines, saturation is a rather undesirable feature that must be carefully considered in the design and operation of machines. In the complex magnetic circuits of most rotating machines, saturation will generally occur in one portion of the circuit before it does in others. Operating a magnetic circuit into the saturated region of its *B-H* curve generally results in low energy efficiency and undue heating. The separation between the "unsaturated" and "saturated" portions of the *B-H* curve is known as the *knee of the B-H curve*, the region in which the curve makes the sharpest bend towards the horizontal. In all but the square-loop materials, the knee is a fairly gradual effect.

5. Leakage and fringing flux. This is a property of all magnetic circuits. It is best treated as a part of the generalized solution of magnetic field distribution in space, often called a boundary value problem. In many

rotating-machine magnetic circuits, boundaries between regions of space containing different types of magnetic materials (usually, a boundary between a ferromagnetic material and air) are often planes or cylindrical surfaces which, in a two-dimensional cross section, become straight lines or circles. Leakage inductances can frequently be calculated in such regions by calculating the reluctance or permeance of the region using fairly simple integral formulations. The spacial or geometric coefficients so obtained are known as permeance coefficients. An example of this approach is given in Section 2.10.

2.7 THE IDEAL MAGNETIC CIRCUIT

Using the assumptions discussed in the previous section, we can now define the ideal magnetic circuit: The ideal magnetic circuit is composed of magnetic materials that are homogeneous, isotropic, and linear and has *infinite permeability*. Air-gap fringing is also usually neglected. A magnetic material with infinite permeability is analogous to a perfect electrical conductor, that is, a conductor with infinite electrical conductivity, and the relationships between the **B** and **H** vectors are analogous to the relationships between **J** and **E** in the perfect conductor. These relationships are derived in most texts on field theory and will only be summarized here. Consider Figure 2.10, which shows the boundary between two regions of different magnetic permeability. Let region 1 be free space with $\mu_1 = \mu_0$; region 2 is the magnetic material with $\mu_2 \to \infty$. The boundary conditions, in terms of field components normal to the boundary, H_n and B_n, and tangential to the boundary, H_t and B_t can be shown to be (assuming zero current density on the boundary surface)

$$B_{1n} = B_{2n}$$

$$H_{1t} = H_{2t} \tag{2.23}$$

where the subscripts, 1 and 2, refer to the respective regions. Within the magnetic material, region 2,

$$H_{2t} = \frac{B_{2t}}{\mu_2} \to 0(\mu_2 \to \infty) \tag{2.24}$$

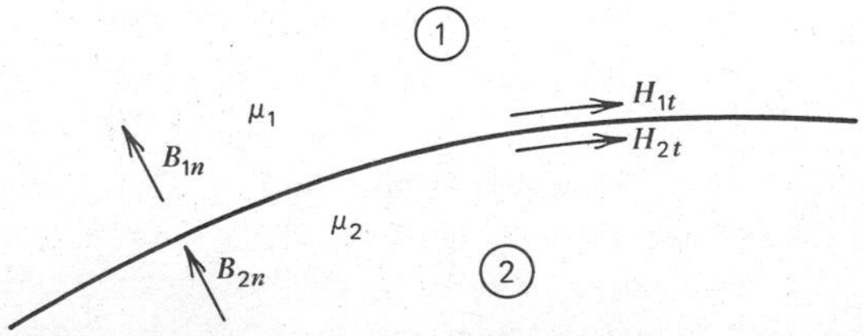

Figure 2.10. Magnetic materials interface.

showing that the tangential component of field intensity in the magnetic material is zero as the permeability approaches infinity. Therefore, from (2.23), the tangential component of the magnetic field in region 1 at the boundary is also zero. Also, from (2.15), it can be seen that M, the magnetic potential, is zero along a path parallel to the tangential field within the magnetic material. There are two important conclusions from this analysis:

1. Flow lines or lines of magnetic flux are perpendicular to the surface of a perfect magnetic conductor.
2. There is no potential difference, or reluctance drop, between points or planes at different locations within a perfect magnetic conductor.

These characteristics of an ideal magnetic circuit are frequently used in magnetic circuit analysis.

2.8 FARADAY'S LAW AND INDUCED VOLTAGE

To relate a time variation of the magnetic flux to an electric field variation around a closed path, we recall Faraday's law, (2.2) of Section 2.1, that is

$$\oint \mathbf{E}\cdot d\mathbf{l} = -\int_s \frac{\partial \mathbf{B}}{\partial t}\cdot d\mathbf{s} \qquad (2.2)$$

To determine the direction of the electric field vector, **E**, in this expression, we use the fingers of the right hand to indicate the positive direction about the closed path. The thumb will then indicate the direction of $d\mathbf{s}$; a flux density **B** in the direction of $d\mathbf{s}$, *increasing with time*, results in a direction for **E** *opposite to* the positive direction about the closed path. The left-hand side of (2.2) is termed "emf," or induced voltage, e. The right-hand side has been defined as the magnetic flux in (2.16). Using these definitions, the scalar form of Faraday's law is

$$e = -\frac{d\phi}{dt} \qquad (2.25)$$

We define *flux linkages*, λ, as the product of a number of turns, N, and the flux, ϕ, linking N, or

$$\lambda = N\phi \qquad (2.26)$$

If the line integral in (2.2) is made about N closed paths representing N series turns, (2.25) becomes

$$e = -\frac{d\lambda}{dt} \qquad (2.27)$$

Faraday's law is applicable to the case where the time variation of voltage results from a time-varying flux linking a stationary coil, as in a transformer; or from a coil or conductor physically moving through a static flux; or from any combination of the two situations. For the case of a conductor moving in a magnetic field, the induced voltage is often termed "motional emf" and is defined as

$$\text{motional emf} = \oint (\mathbf{U} \times \mathbf{B}) \cdot d\mathbf{l} \tag{2.28}$$

where $\mathbf{U}$ is the velocity vector of the conductor. A special application of this relationship is useful in rotating machine analysis. Assume a conductor of length, l, moving perpendicular to the direction of a uniform magnetic field, B, at constant velocity, U; (2.28) becomes

$$\text{motional emf} = BlU \tag{2.29}$$

The direction associated with motional emf can be worked through the integral and vector processes on the right-hand side of (2.28) in a process similar to that described under (2.2) but a simple rule, known as the *right-hand rule*, is much easier in cases where (2.29) is applicable: Extend the thumb, first, and second fingers of the right hand so that they are mutually perpendicular to each other. If the thumb represents the direction of U and the first finger the direction of B, the second finger represents the direction of the emf along l.

2.9 ENERGY RELATIONS IN A MAGNETIC FIELD

The potential energy in a magnetic field is defined throughout space by the volume integral,

$$W = \frac{1}{2}\int_{\text{vol}} \mathbf{B} \cdot \mathbf{H}\, dv = \frac{1}{2}\int_{\text{vol}} \mu H^2 dv = \frac{1}{2}\int_{\text{vol}} \frac{B^2}{\mu} dv \tag{2.30}$$

Example 2.3

Determine the potential magnetic energy in the air-gap and magnetic material of the magnetic circuit of Example 2.2. Under the assumptions used in Example 2.2, the field distribution was uniform in both the gap and the magnetic material, facilitating the use of (2.30). In the gap, $B_g = 0.001/(1.1 \times 0.0016) = 0.57$ tesla.

$$W = \frac{1}{2}\left(\frac{B_g^2}{\mu_0}\right)(\text{vol}) = \frac{1}{2}\left(\frac{0.57^2}{4\pi \times 10^{-7}}\right)(1.1 \times 0.0016 \times 10^{-4})$$

$$= 0.0228 \text{ J}$$

In the magnetic material, $B_m = 0.695$ tesla.

$$H_m = 54 \text{ A/m (from Figure 2.4)}$$

$$W = \frac{1}{2} B_m \times H_m \times \text{vol} = \frac{1}{2} \times 0.695 \times 54 \times (0.9 \times 0.0016) \times 0.1$$

$$= 0.0027 \text{ J}$$

It is evident that most of the energy is required to establish the flux in the air-gap.

2.10 INDUCTANCE

Inductance is one of the three circuit constants in electrical circuit theory and is defined as flux linkage per ampere,

$$L = \frac{\lambda}{i} = \frac{N\phi}{i} \tag{2.31}$$

Consider the magnetic toroid around which are wound "n" distinct coils electrically isolated from each other, as shown in Figure 2.11. The coils are linked magnetically by the flux, ϕ, some portion of which links each of the coils. A number of inductances can be defined for this system:

$$L_{km} = \frac{\text{flux linking the } k\text{th coil due to the current in the } m\text{th coil}}{\text{current in the } m\text{th coil}}$$

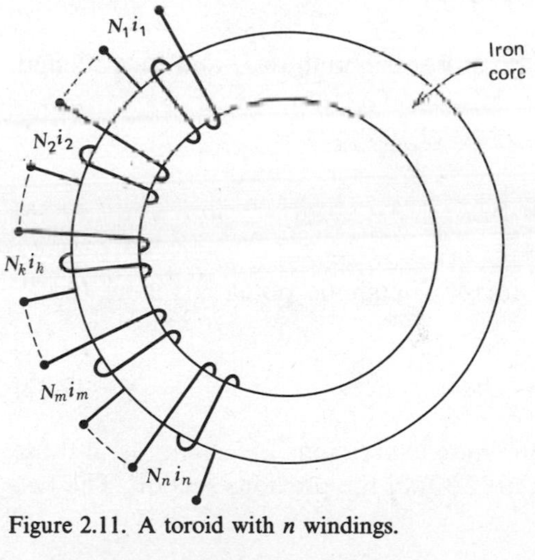

Figure 2.11. A toroid with n windings.

Mathematically, this can be stated as

$$L_{km} = \frac{N_k (k\phi_m)}{i_m} \tag{2.32}$$

where k is the portion of the flux due to coil m that links coil k and is known as the *coupling coefficient*. By definition, its maximum value is 1.0. A value of k less than 1.0 is attributable to leakage flux in the regions between the location of coil k and coil m. The reader probably recognizes that when the two subscripts in (2.32) are identical, the inductance is termed *self-inductance*; when different, the inductance is termed *mutual inductance* between coils k and m. Mutual inductances are symmetrical, that is,

$$L_{km} = L_{mk} \tag{2.33}$$

Inductance can be related to the magnetic parameters derived earlier in this chapter. In (2.32), ϕ_m can be replaced, using (2.17), by the magnetic potential of coil m, M_m, divided by the reluctance of the magnetic circuit, R; the magnetic potential of coil m, however, is $N_m I_m$. Making these substitutions in (2.32) gives

$$L_{km} = \frac{kN_k N_m}{R} = kN_k N_m P \tag{2.34}$$

where P is the permeance, the reciprocal of the reluctance. In a simple magnetic circuit such as the toroid of Figure 2.11, the reluctance, from the (2.18) can be substituted into (2.34), giving

$$L_{km} = \frac{kN_k N_m A_t \mu_t}{l_t} \tag{2.35}$$

where μ_t, A_t, and l_t are the permeability, cross-sectional area, and mean length of the toroid, respectively.

Stored energy can be expressed in terms of inductance:

$$W = \frac{1}{2} Li^2 \tag{2.36}$$

By substituting for L from (2.31) and for Ni (magnetic potential) from (2.17), (2.36) can be expressed as

$$W = \frac{1}{2} R\phi^2 \tag{2.37}$$

At this point, it is interesting to compare these expressions for energy with those in terms of the field quantities as given in (2.30) in the previous section. The two forms are equivalent.

Example 2.4
Calculate the self-inductance of coil 1 in Example 2.2 (Figure 2.8). To obtain a numerical value for inductance, the number of turns in the coil must be specified. Let us assume 10 turns.

$$L_{11}=\frac{N_1\phi_1}{I_1}=\frac{(10)(0.001)}{5.04}=0.00198 \text{ henry}$$

If we should assume 100 turns,

$$L_{11}=\frac{(100)(0.001)}{0.504}=0.198 \text{ henry}$$

Note that the inductance varies with the square of the turns.

Example 2.5
Calculate the mutual inductance between coils 1 and 2 in Example 2.2, assuming 10 turns in coil 1 and 20 turns in coil 2. Since leakage was neglected in Example 2.2, $k=1.0$; that is, all of the flux produced by coil 1 links coil 2.

$$L_{21}=L_{12}=\frac{(20)(0.001)}{5.04}=0.00398 \text{ henry}$$

Example 2.6
Calculate the inductance of coil 1 of the toroid in Figure 2.11, assuming material is 48% Ni alloy and that the toroid mean circumference is 0.1 m and cross-sectional area is 0.0016 m^2, with a stacking factor of 0.9; $N_1=10$.

From the *B-H* curve for 48% Ni alloy in Figure 2.4, we find that the *absolute* amplitude permeability in the linear range is 0.115. Using (2.35) with $k=1.0$, we obtain

$$L_{11}=\frac{(10^2)(0.0016)(0.115)}{0.1}=0.184\ H$$

It is seen that the inductance of this toroid is much larger than that of the circuit of Figure 2.8 used in Example 2.5, even though the mean length and area of the two cores and the winding turns are equal. This is attributable to the effect of the gap in the circuit of Figure 2.8 and also to the higher permeability of the magnetic material in the toroid.

Example 2.7
Determine the inductance of the armature slot, the cross section of which is shown in Figure 2.12, assuming that the lower portion of the slot of height y_2 is

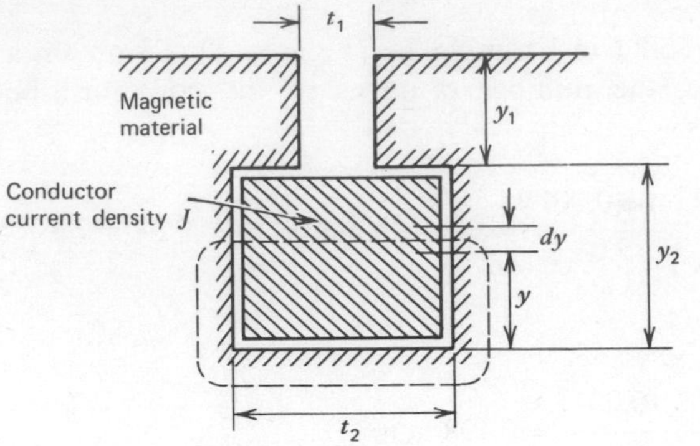

Figure 2.12. A slot cross section.

filled with conductors carrying a current density, J ampere/m^2, perpendicular to the plane of the paper. Also assume that the steel magnetic material surrounding the slot has infinite permeability. This example calculates an expression for the "slot inductance," an important component of the leakage inductance of many types of rotating machines, and will also illustrate the use of permeance coefficients and the concept of *partial flux linkages*. Furthermore, it will illustrate the use of symmetry, for Figure 2.12 represents the cross section of a portion of an armature of length, l_a in the direction perpendicular to the plane of the paper. The two-dimensional analysis is possible, since only one component of the current density, J, is present. The assumption of infinite permeability of the magnetic material on the sides and bottom of the slot means that flux lines leave the sides perpendicular. Therefore, the flux lines can be considered as horizontal components.

Partial flux linkage is the term used to describe flux lines that link only a portion of an electrical conductor or only a portion of the turns of a coil. In this example, a flux line across the slot at any height below y_2 links only the current below the flux line. (The path of the flux line is closed through the magnetic material as shown.) It should be noted that the assumption of horizontal flux lines is not exactly correct in a practical configuration, but an analytical solution would be impossible without this assumption.

Consider the differential flux in the "strip" at a distance, y, from the slot bottom; the strip has height, dy; width, l_a (into the paper); and a length, t_2. The magnetic potential enclosed by this "strip" is

$$M_y = J t_2 y$$

The permeance of the "strip" is

$$dP_y = \frac{\mu_0 l_a \, dy}{t_2}$$

The flux through the "strip" is, from (2.17),

$$d\phi = M_y\, dP_y = \mu_0 J l_a y\, dy$$

The total flux across the gap is

$$\phi_2 = \mu_0 J l_a \int_0^{y_2} y\, dy = \mu_0 J l_a \frac{y_2^2}{2}$$

The total magnetic potential of the slot is

$$M_s = J y_2 t_2$$

The permeance of the lower portion of the slot is

$$P_2 = \frac{\phi_2}{M_s} = \frac{\mu_0 l_a y_2}{2t_2}$$

The permeance of the upper portion of the slot is

$$P_1 = \mu_0 y_1 \frac{l_a}{t_1}$$

The slot inductance is

$$L_s = P_1 + P_2 = \mu_0\left(\frac{y_1 l_a}{t_1} + \frac{y_2 l_a}{2t_2}\right) \text{ henry}$$

2.11 MAGNETIC CIRCUITS CONTAINING PERMANENT MAGNETS

The second type of excitation source that is commonly used for supplying energy to magnetic circuits used in rotating machines and other types of electromechanical devices is the permanent magnet. The nature of permanent magnets has been briefly described in Section 2.2. There is obviously a great deal of difference in physical appearance between an electrical exciting coil and a permanent magnet source of excitation, so we should expect some differences in methods of analyses used in the two types of magnetic circuits. Actually, these differences are relatively minor and are related to the use of the permanent magnet itself rather than to the other portions of the magnetic circuit.

An electrical excitation coil energized from a constant voltage or constant current source is relatively unaffected by the magnetic circuit that it excites, except during transient conditions when changes are occurring in the magnetic circuit or in the external electrical circuit. Under steady state conditions with a constant voltage source, the current in the coil is determined solely by the magnitude of the voltage source and the dc resistance of the coil.

In a circuit excited by a permanent magnet, the operating conditions of the permanent magnet are largely determined by the external magnetic circuit. Also, the operating point and subsequent performance of the permanent magnet are a function of how the magnet is physically installed in the circuit and whether it is magnetized before or after installation. In many applications, the magnet must go through a stabilizing routine before use. These considerations are, of course, meaningless for electrical excitation sources. The details necessary to find the excitation required to establish a flux density in an air gap of known dimensions, are demonstrated for electrical exciting coils in Examples 2.1 and 2.2. For permanent magnet excitation, the object is to determine the size (length and cross section) of the permanent magnet. The first step in this process is to choose a specific type of permanent magnet, for each type of magnet has a unique characteristic that will partially determine the size of the magnet required. In a practical design, this choice will be based upon cost factors, availability, mechanical design (hardness and strength requirements), available space in the magnetic circuit, as well as the magnetic and electrical performance specifications of the circuit. Most permanent magnets are nonmachinable and usually must be used in the circuit as obtained from the manufacturer. Table 2.3 summarizes some of the pertinent characteristics of common permanent magnets.

Permanent magnet excitation is chosen for a specified air gap flux density with the aid of the second quadrant *B-H* curve, often called the *demagnetization curve*, for a specific type of magnet. This curve has been introduced in Section 2.2 as Figure 2.3. The *B-H* characteristics of a number of alnico permanent magnets are shown in Figure 2.13, and Figure 2.14 shows the characteristics of several ferrite magnets. Also shown on these figures are curves of *energy product*, the product of *B* in gauss (G) and *H* in oersteds (Oe), and *permeance ratio*, the ratio of *B* to *H*. The energy product is a measure of the magnetic energy that the permanent magnet is capable of supplying to an external circuit as a function of its flux density and field intensity. In general, a permanent magnet is used most efficiently when operated at conditions of *B* and *H* that result in the maximum energy product. Permeance coefficients are useful in the design of the external magnetic circuit. It should be noted that this parameter is acutally "relative permeability" as defined previously, since μ_0 is 1.0 in the CGS system of units. The symbols, B_d for flux density, and H_d for field intensity, are used to designate the coordinates of the demagnetization curve.

Once the permanent magnet type has been chosen, the design of the magnet's size follows the general approach taken in Section 2.4. From Ampere's law,

$$H_d l_m = H_g l_g + V_{mi} \tag{2.38}$$

Table 2.3 **Characteristics of Permanent Magnets**

Type	Residual Flux Density—B_r (gauss)	Coercive Force H_c (oersted)	Max-Energy Product (gauss-oersted $\times 10^6$)	Recoil Permeability (average)
1% Carbon steel	9,000	50	0.18	
$3\frac{1}{2}$% Chrome steel	9,500	65	0.29	35
36% Cobalt steel	9,300	230	0.94	10
Alnico I	7,000	440	1.4	6.8
Alnico IV	5,500	720	1.3	4.1
Alnico V	12,500	640	5.25	3.8
Alnico VI	10,500	790	3.8	4.9
Alnico VIII	7,800	1,650	5.0	—
Cunife	5,600	570	1.75	1.4
Cunico	3,400	710	0.85	3.0
Vicalloy 2	9,050	415	2.3	—
Platinum-cobalt	6,200	4,100	8.2	1.1
Barium ferrite-isotropic	2,200	1,825	1.0	1.15
Oriented type A	3,850	2,000	3.5	1.05
Oriented type B	3,300	3,000	2.6	1.06
Strontium ferrite				
Oriented type A	4,000	2,200	3.7	1.05
Oriented type B	3,550	3,150	3.0	1.05
Rare earth-cobalt	8,600	8,000	18.0	1.05

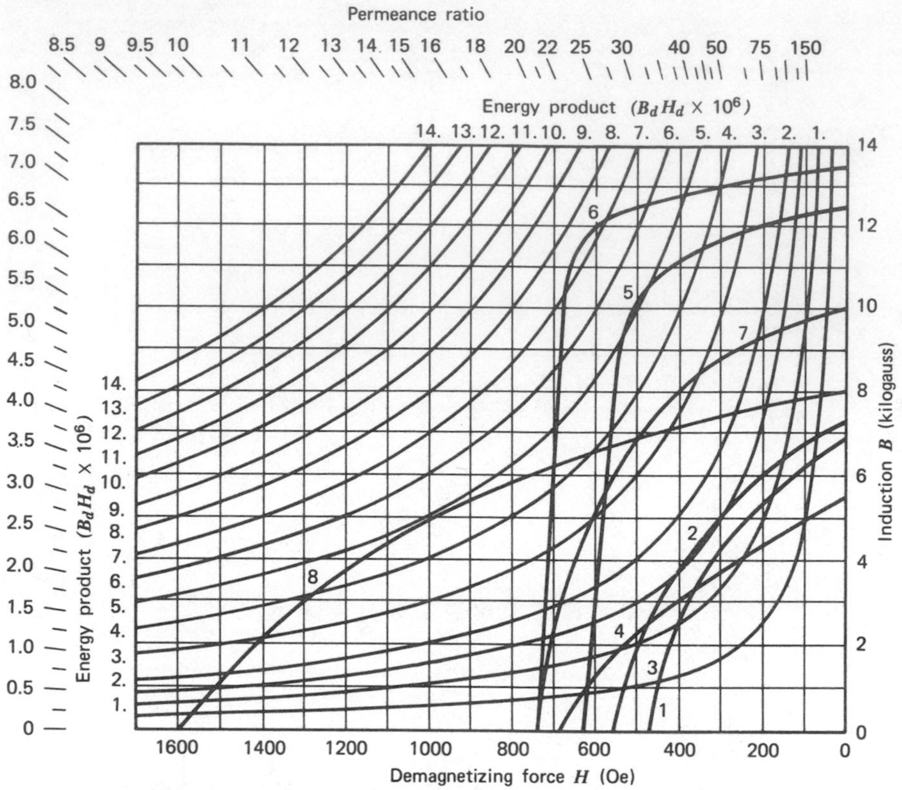

Figure 2.13. Demagnetization and energy product curves for Alnicos 1 to 8. *Key:* 1, Alnico I; 2, Alnico II; 3, Alnico III; 4, Alnico IV; 5, Alnico V; 6, Alnico V-7; Alnico VI; and 8, Alnico VIII.

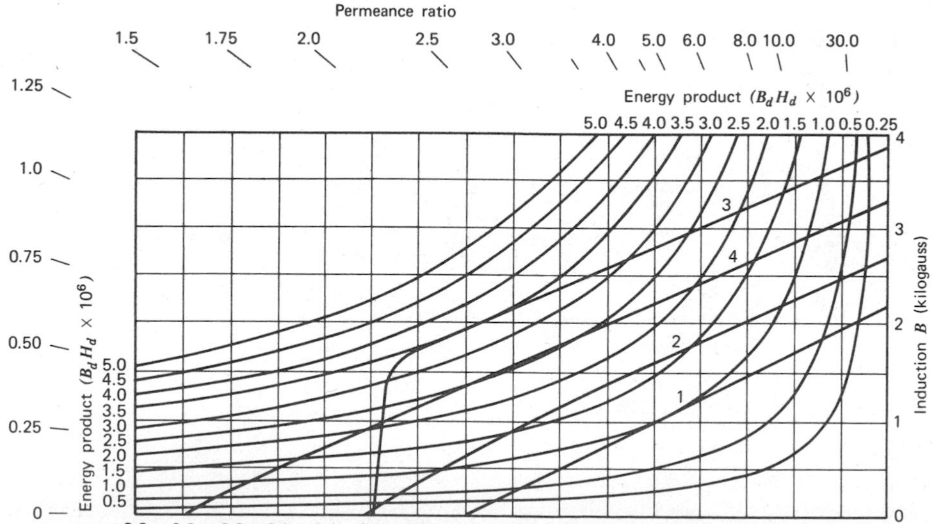

Figure 2.14. Demagnetization and energy products curves for Indox ceramic magnets. *Key:* 1, Indox 1; 2, Indox II; 3, Indox V; and 4, Indox VI-A.

where

H_d = magnetic field intensity of the magnet, oersted

l_m = length of magnet, cm

H_g = field intensity in the gap = flux density in gap (in CGS units)

l_g = length of gap, cm

V_{mi} = reluctance drop in other ferromagnetic portions of the circuit, gilbert

The cross-sectional area of the magnet is calculated from the flux required in the air gap as follows:

$$B_d A_m = B_g A_g K_1 \tag{2.39}$$

where

B_d = flux density in the magnet, gauss

A_m = cross-sectional area of magnet, cm^2

B_g = flux density in the gap, gauss

A_g = cross-sectional area of gap, cm^2

K_1 = leakage factor

The leakage factor, K_1, is the ratio of flux leaving the magnet to flux in the air gap. The difference between these two fluxes is the leakage flux in the regions of space between the magnet and the air gap. The leakage factor can be determined by the methods described in previous sections of this chapter or by other standard and more accurate methods. A number of formulas have been developed through the years for the common configurations used in permanent magnet magnetic circuits and can be found in References 1, 2, and 3. Two of these formulas will be illustrated in the following examples.

Example 2.8

Determine the length and cross-sectional area of the magnets in Figure 2.15 to produce a flux density in the air gap of 2500 gauss. The permanent magnet to be used is Alnico V; the dimensions of Figure 2.15 are as follows: $l_g = 0.4$ cm, $W = 6.0$ cm; and the gap area = 4.0 cm^2(2 cm×2 cm). We assume that the reluctance in the soft iron portions of the circuit is negligible, giving a reluctance drop, V_{mi}, of zero; we estimate that the leakage factor is 4.0 and that the magnet is to be operated at its maximum energy product condition (knee of the demagnetization curve in Figure 2.13.)

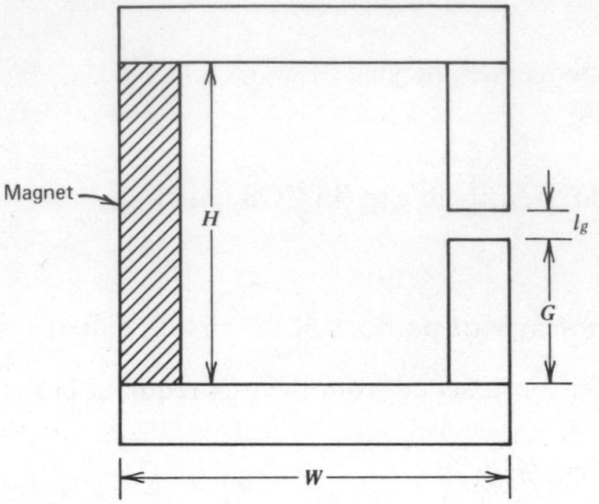

Figure 2.15. A magnetic circuit with a permanent magnet.

From (2.39),

$$A_m = \frac{B_g A_g K_1}{B_d} = \frac{(2500)(4.0)(4.0)}{10{,}500} = 3.8\ \text{cm}^2$$

From (2.38), noting the $H_g = B_g$ in CGS units,

$$l_m = \frac{(2500)(0.4)}{450} = 2.22\ \text{cm} = H\ \text{(in Figure 2.15)}$$

The dimension, l_m, determines the dimensions, H and G in Figure 2.15. We must now check the assumption of leakage factor. From Reference 2, a leakage factor is given for this configuration, based upon experimental measurements and calculations similar to those described earlier in this chapter, as

$$K_1 = 1 + \frac{l_g}{A_g}\left[1.7C_G\frac{G}{G+l_g} + 1.4W\sqrt{\frac{C_W}{H}} + 0.67C_H\right] \tag{2.40}$$

where

C_H = the perimeter of the cross-section of the circuit of length, H

C_W = the perimeter of the cross-section of the circuit of length, W

C_G = the perimeter of the cross-section of the circuit of length, G

The factor, 0.67, in the third term within the brackets in (2.40) arises from the fact that permanent magnets have a "neutral zone" that does not contribute to leakage. From the length and area calculations above we can determine the length parameters in (2.40) and estimate the perimeters of the various cross sections: $H=2.22$ cm, $G=0.91$ cm, $W=6.0$ cm, $C_G=8$ cm, $C_W=8.0$ cm, and $C_H=7.8$ cm. Substituting these values into (2.40) gives

$$K_1=4.062$$

This value could now be put back into (2.39), giving a slight change in magnet area, A_m. In turn this calculation may require a few changes in other dimensions used in (2.40), resulting in a new value of leakage factor. A few iterations of these formulas are usually necessary to obtain a consistent set of dimensions for the total magnetic circuit.

The high value for leakage factor obtained for the configuration of Figure 2.15 indicates that this is not a very efficient magnetic circuit. Stated in another manner, the permanent magnet is in the wrong location within the magnetic circuit. The high leakage in this configuration can be readily explained by means of Ampere's law and the simple magnetic circuit theory presented earlier in this chapter. This explanation is left as a problem at the end of this chapter. A much more efficient use of the permanent magnet in this configuration is to locate it adjacent to the air gap, as shown in Figure 2.16. From Reference 2, the leakage factor for Figure 2.16 is

$$K_1=1+\frac{l_g}{A_g}0.67C_G\left(1.7\frac{0.67G}{0.67G+l_g}+\frac{l_g}{2G}\right) \tag{2.41}$$

Using the same dimensions for all sections of the circuit in Figure 2.16 as were used for Figure 2.15 (even though this might result in an over-sized permanent

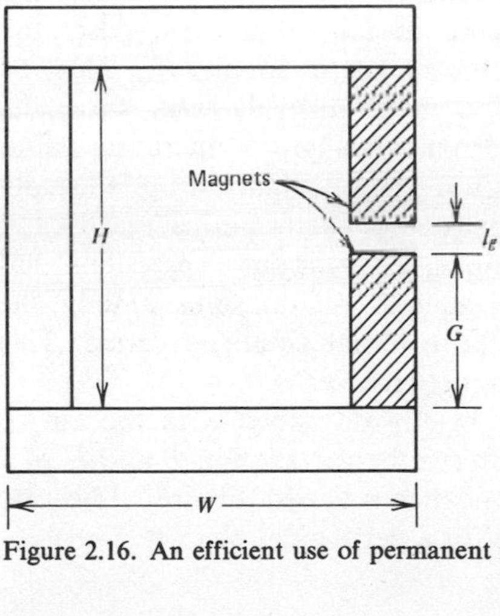

Figure 2.16. An efficient use of permanent magnets in a magnetic circuit.

magnet), and substituting these into (2.41) gives

$$K_1 = 1.624$$

which is less than half the leakage factor found for the configuration of Figure 2.15.

It is interesting to observe the volume of permanent magnet material required to establish a given flux in an air gap. Solving for A_m in (2.39) and for l_m in (2.38) (neglecting V_{mi}), and noting that in the CGS system, $H_g = B_g$, we obtain

$$\text{vol} = A_m l_m = \frac{B_g^2 A_g l_g K_1}{B_d H_d} \tag{2.42}$$

It is seen that magnet volume is a function of the square of the air gap flux density. The importance of the leakage factor in minimizing the required magnet size is also apparent from this equation. The denominator of (2.42) is the energy product that is a function of the permanent magnet material and the operating point on the demagnetization curve of the magnet.

The parameter, permeance ratio, shown on Figures 2.13 and 2.14, is the ratio of the equivalent permeance of the external circuit, $A_g K_1 / l_g$, to the permeance of the space occupied by the permanent magnet, A_m / l_m, in the CGS system of units. This can be seen by solving for B_d from (2.39) and for H_d from (2.38) (neglecting V_{mi}), and taking the ratio,

$$\frac{B_d}{H_d} = \frac{A_g l_m K_1}{A_m l_g} = \frac{P_{ge}}{P_m} = \tan\alpha \tag{2.43}$$

Equation (2.43) is deceptively simple in appearance, for the task of obtaining analytical expressions for K_1 is very difficult, as has been seen. Also, the reluctance drop in the soft iron portions of the magnetic circuit, V_{mi}, must be included somehow in (2.43). This is even a more difficult task, since the reluctance drop is a function of both the permanent magnet's operating point, $B_d H_d$, and the effects of leakage flux in the iron. The reluctance drop is usually introduced by means of a factor similar to the leakage factor and is based upon measurements in practical circuit configurations. The various expressions that make up (2.43) are of value in observing general relationships among the magnetic parameters as the permeance of the external circuit is varied. This leads us to the second type of permanent magnet circuit.

Circuits with a *varying air gap* will be briefly described with the aid of Figure 2.17. Keeping (2.43) in mind, let us observe the variation of B and H of a permanent magnet as the external circuit permeance is varied. Figure 2.17 shows a typical second-quadrant B-H characteristic for a permanent magnet. Theoreti-

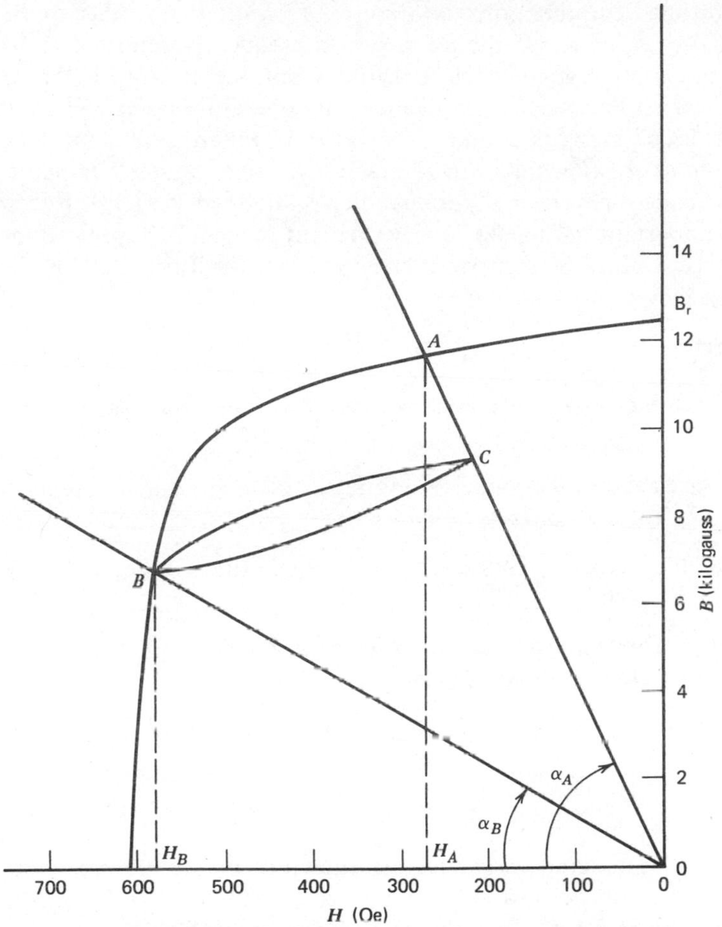

Figure 2.17. Second quadrant *BH*-characteristic of a permanent magnet.

cally, it is possible to have infinite permeance in the external magnetic circuit that would correspond to $\alpha = 90°$ in (2.43), and the magnet operating point would be at $B_d = B_r$ and $H_d = 0$ in Figure 2.17. This situation is approximated by a permanent magnet having an external circuit consisting of no air gap and a high permeability soft iron member, often called a "keeper." In practice, however, there is always a small equivalent air gap and a small reluctance drop in the keeper, and the operating point is to the left of B_r and α is less than 90°.

For a finite air gap, the operating point will be at some point, A, on the B-H curve, and the permanent magnet will develop the magnetic field intensity, H_A, to overcome the reluctance drop of the air gap and other portions of the external magnetic circuit. If the air gap is increased, P_{ge} decreases and the magnet must develop a larger magnetic field intensity, H_d. From (2.43), it is seen that α decreases and the operating point on Figure 2.17 will move further to the left to some point, say B, at α_B. If the air gap is subsequently returned to its original value, the operating point will not return to A but, rather, to C. If the air gap is successively varied between the two values, the operating point will trace a "minor hysteresis loop" between B and C as shown in Figure 2.17. The slope of this loop is known as *recoil permeability*; since it is a slope on the B-H plane, it is also sometimes called incremental permeability, as defined in (2.11). Recoil permeability is an important parameter of a permanent magnet for applications with varying air gaps; values of this parameter are given in Table 2.3 for the permanent magnets shown.

REFERENCES

1. H. C. Roters, *Electromagnetic Devices*, John Wiley & Sons, Inc., New York, 1941.

2. "Design and Application of Permanent Magnets," *Indiana General Corporation Manual* No. 7, Valparaiso, Ind. 1968.

3. "Permanent magnet design," Thomas and Skinner, Inc., Bulletin M303, Indianapolis, Indiana 1967.

4. L. V. Bewley, *Two-Dimensional Fields in Electrical Engineering*, Macmillan Publishing Co., Inc., New York, 1948.

5. P. Silvester, *Modern Electromagnetic Fields*, Prentice-Hall, Inc., Englewood Cliffs, N. J., 1968.

PROBLEMS

2.1. From the magnetic material characteristics shown in Figure 2.4 determine the relative amplitude permeability at a flux density of 1.0 T for the following materials: (a) Deltamax, (b) M-19, and (c) AISI 1020.

2.2. It has been stated (Section 2.3) that the area enclosed by the dc hysteresis loop of a magnetic material is equal to the hysteresis loss of that material. Referring to Figure 2.2, derive the SI units of this loss from the coordinates of the hysteresis loop.

2.3. From Figure 2.4, determine the relative differential permeability for the material, 48NI, at a flux density of 0.8 T.

2.4. The 3% silicon steel designated M-19 in Figure 2.4 is widely used in small power transformers, induction motors, and other electromagnetic devices used in electronic circuits. In this chapter it is stated that there is a portion of the B-H characteristic over which the relationship is approximately linear. This does not necessarily show up in Figure 2.4, since the curves are plotted on semilog paper in order to describe the characteristic over a wider range. Replot the M-19 curve on rectangular coordinate paper and determine the range of B over which the characteristic can be considered linear. What is the relative differential permeability in this range?

2.5. Derive an equation to describe the straight-line or linear portion of the B-H curve for M-19 obtained in Problem 2.4.

2.6. If you needed to use the B-H characteristics of M-19 in a digital computer design program for an electronics transformer, how would you model this characteristic, including saturation and initial characteristics?

2.7. Discuss the physical significance of the first Maxwell equation (2.1) upon the nature of lines of magnetic flux (magnetic streamlines).

2.8. A certain air gap has an area of 1 in.^{2} and a length of 1 mm. What are the reluctance and permeance of this gap in SI units?

2.9. A flux of 0.02 Wb exists in the air gap of Problem 2.8. Assuming no fringing effects, what is the magnetic potential (MMF) across the gap?

2.10 Assume a region of space described by conventional cylindrical coordinates, r, ϕ, and z. An *infinitely* long solenoid of radius, r_1, extends along and is centered about the z axis. The solenoid has N turns/m of length and carries a current, I ampere. Neglect the small z-component of current and assume that the current is purely "circular" at any finite value of z. At any point in space (say, at r_a, ϕ_a, z_a), with which dimensional coordinate does the magnetic field intensity, H, vary? The Biot-Savart law may be of value in answering this question. What component of H exists at any point in space? Answer these questions for points both within and outside of the radius, r_1.

2.11. For the long solenoid of Problem 2.10, show that $H_z = \text{NI}$ for $r < r_1$.

2.12. A toroid of mean radius, 0.1 m, is constructed of 0.14 mm strips of 48NI magnetic material. The flux density in the magnetic material is 0.6 T. Including the effects of "stacking factor," determine the current in a 100-turn coil required to maintain this flux density.

2.13. The toroid of Problem 2.12 is cut to have an air gap 1 mm in length. What current is required in the coil to maintain the same flux density as in Problem 2.12? Neglect leakage and fringing.

2.14. For the magnetic circuit shown in Figure 2.18, determine the MMF of the exciting coil required to produce a flux density of 1.6 T in the air gap. Neglect leakage and fringing. In Figure 2.18, the mean lengths of portions of the magnetic circuit of constant cross-sectional area are $l_{m1} = 45$ cm, $A_{m1} = 24$ cm^2; $l_{m2} = 8$ cm, $A_{m2} = 16$ cm^2; $l_g = 0.08$ cm, and $A_g = 16$ cm^2. Material: M-19

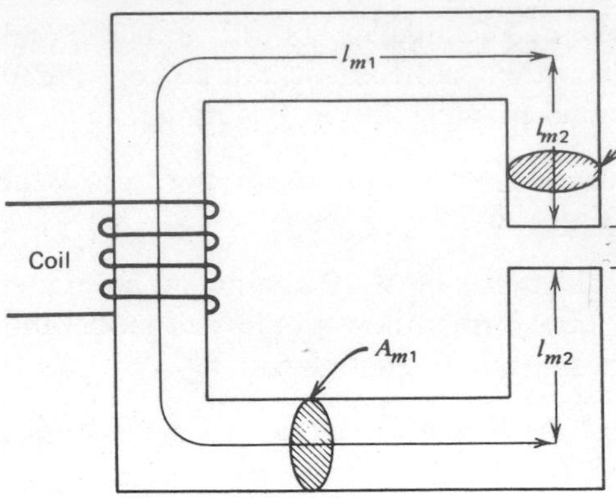

Figure 2.18. Figure Problem 2.14.

2.15. For the magnetic circuit shown in Figure 2.19 determine the current in a 100-turn coil to establish a flux in the air gap of 10^{-3} Wb. Assume a fringing factor of 1.09 and neglect the leakage flux.

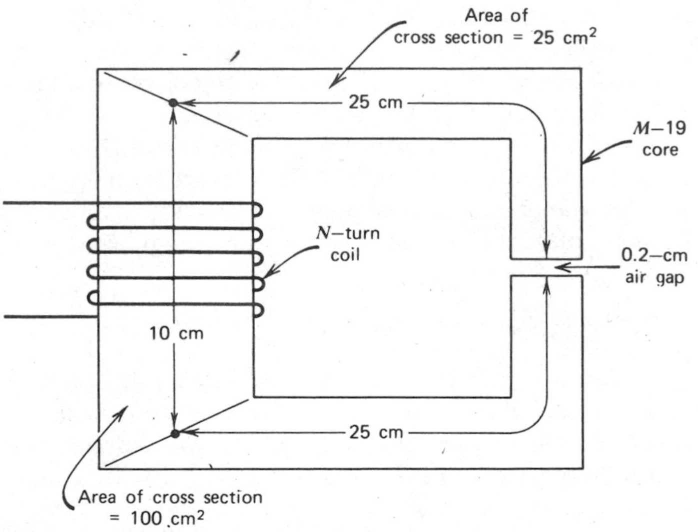

Figure 2.19. Figure Problem 2.15.

2.16. Figure 2.20 shows the cross section of a portion of the magnetic circuit found in certain types of reluctance stepper motors. The rectangular cross sections at the bottom of the magnetic circuit are cross sections of magnetic sectors that are located on rotating discs. There are three air gaps, all equal in length, separating these disc sectors from each other and from the U-shaped "return" magnetic circuit. The symbol, ϕ_1, represents the net equivalent of *all* leakage flux across the air spaces between the sides of the U magnet and can be assumed to be concentrated at the location in the magnetic circuit shown in Figure 2.20. Circuit dimensions are as follows:

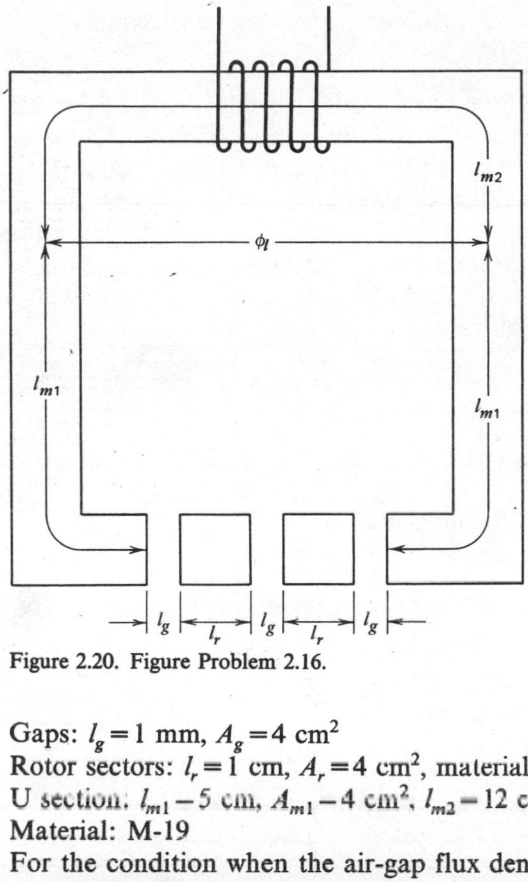

Figure 2.20. Figure Problem 2.16.

Gaps: $l_g = 1$ mm, $A_g = 4$ cm^2
Rotor sectors: $l_r = 1$ cm, $A_r = 4$ cm^2, material: 48NI
U section: $l_{m1} = 5$ cm, $A_{m1} = 4$ cm^2, $l_{m2} = 12$ cm, $A_{m2} = 6$ cm^2,
Material: M-19

For the condition when the air-gap flux density equals 1.2 T, it was found that $\phi_l = 0.0003$ Wb. For this condition,

(a) Find the magnetic potential across the equivalent leakage path.
(b) Determine the permeance in SI units of the equivalent leakage path.
(c) Determine the flux through the exciting coil.
(d) Determine the current required in a 10-turn coil to produce this flux.
(e) Calculate the "leakage inductance" of the coil.
(f) Draw the equivalent dc circuit (in the manner of Figure 2.9) for this magnetic circuit.

2.17. Referring to the two permanent magnet configurations shown in Figures 2.15 and 2.16, explain the large difference in leakage flux between these two configurations. As an aid in this analysis, assume a relative magnetic potential, V_m, across the air gap of, say 1.0, in each circuit, and relate this to the estimated potential between the top and bottom sections of the circuit (of width W).

2.18. As a result of the analysis made in Problem 2.17, where should the exciting coil in the circuit of Figure 2.8 (assuming only one coil required) be located to minimize leakage flux?

2.19. A *nonmagnetic* toroid has a square cross section of dimensions, 2 cm×2 cm. The toroid inner diameter is 8 cm and the outer diameter is 12 cm. A 1200-turn coil is wound uniformly and tightly about the toroid. Assuming zero leakage flux and a uniform flux distribution over the cross section of the toroid,

(a) Calculate the flux density at the mean diameter (center of the toroid's cross section).

(b) Calculate the inductance of the coil.

2.20. Repeat Problem 2.19 without the assumption of uniform flux distribution over the toroid's cross section.

2.21. A one-turn coil carries a current of 5 amperes. Determine the inductance of this coil if

(a) The total energy stored in the magnetic field is 0.01 joule.

(b) The total flux enclosed by the coil is 0.005 Wb.

2.22. Figure 2.21 shows the cross section of a typical solenoid used as a

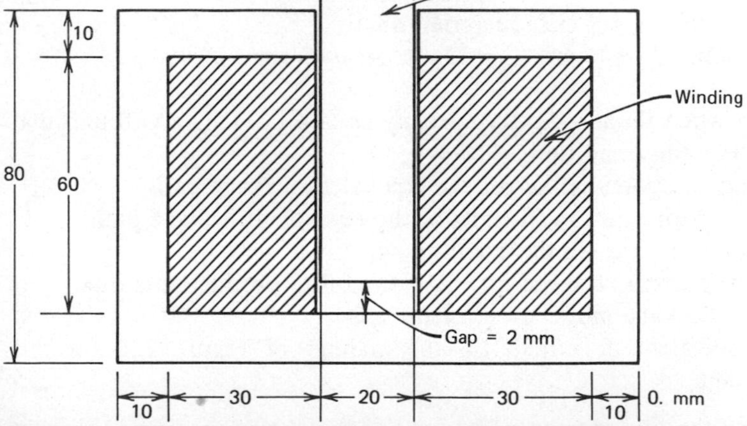

Figure 2.21. Figure Problem 2.22 Solenoid cross section. All dimensions in mm. Plunger and magnetic housing 40 mm perpendicular to the plane of paper.

mechanical actuator in many industrial applications, such as a hydraulic valve control. The top section of the plunger is composed of nonmagnetic material. Magnetic force for moving the plunger is created in the air-gap region at the bottom of the plunger. The nonmagnetic regions where the exciting coil is located is known as the "window" of the solenoid. The "fill factor" of a solenoid or transformer is the ratio of the total cross-sectional area of the copper (or aluminum) in the coil to the window area.

(a) For the dimensions shown, how many turns of No. 24 AWG copper magnet wire can be used in the exciting coil if the fill factor is 75%?

(b) The air gap shown in Figure 2.21 is the initial air gap for the solenoid. Determine the current in the coil chosen in (a) to establish a flux density of 1.0 T in the air gap beneath the plunger. Neglect leakage and fringing. The material is M-19.

(c) In the use of solenoids, the speed of response of the plunger is often important. The electrical circuit time delay is related to the time *constant*, τ, of the *R-L* circuit of the exciting coil. Assuming that the coil is energized from a constant voltage source, that a fixed "fill factor," air gap flux density and length are maintained, will changing the number of turns in the exciting coil affect the time constant?

(d) Assuming the conditions fixed as in (c), will changing the magnetic material from M-19 to Deltamax change the time constant?

2.23 A magnetic circuit of the configuration of Figure 2.16 is to be constructed of soft iron members and Indox V ferrite permanent magnets. Dimensions are $l_g = 1$ cm, $W = 5$ cm, gap area, $A_g = 3.5$ cm^2. If the flux density in the air gap is to be 0.1 tesla, determine the length and cross-sectional area of the Indox V magnets. Assume $K_1 = 1.5$ and neglect fringing.

2.24 Design a permanent magnet excitation for the magnetic circuit of Figure 2.18 (Problem 2-14), using any of the permanent magnet characteristics described in this chapter. The magnet will replace portions of the soft-iron members in Figure 2-18. Are all types of permanent magnets applicable to this problem?

Chapter 3
Transformers

In this chapter, we shall begin to apply the general principles and methods of analysis developed in the last chapter to a specific electromagnetic device, the transformer. Since the transformer has a relatively simple electromagnetic structure, it will be useful for illustrating these principles and for developing relationships that will be of value later in the analysis of more complex electromagnetic structures. The transformer is extremely important as a component in many different types of electrical circuits, from small-signal electronic circuits to high-voltage power transmission systems. A knowledge of the theory, design relationships, and performance capabilities of transformers is essential for understanding the operation of many electronic, control, and power systems. Therefore, both as a vehicle for understanding some basic electromagnetic principles and as an important component of electrical systems, the transformer deserves serious study.

The most common functions of transformers are (1) changing the voltage and current levels in an electrical system, (2) impedance matching, and (3) electrical isolation. The first of these functions is probably best known to the reader and is typified by the distribution transformer on the nearby electric pole which steps down the voltage on the distribution lines from, say, 2300 V, to the household voltage of 115/230 V. The second function is found in many communication circuits and is used, for example, to match a load to a line for improved power transfer and minimization of standing waves. The third feature is used to eliminate electromagnetic noise in many types of circuits, for blocking dc signals, and for user safety in electrical instruments and appliances.

Transformers are used in circuits of all voltage levels, from the microvolt level of some electronic circuits, to the highest voltage used in power systems—which today is approximately 750,000 V. In some pulse applications, even higher voltages may exist. Also, transformers are applied throughout the entire frequency spectrum found in electrical circuits, from near dc to hundreds of megahertz, with both continuous sinusoidal and pulse waveforms. The physical size and shape of transformers is also quite varied, and transformers come in sizes from not much bigger than a pea up to the size of a small house.

The transformer is basically a static device, although there are specialty transformers in which some motion occurs in portions of the electromagnetic structure. Examples of transformers with movable members include the constant-current transformer sometimes used in series streetlight systems, several types of voltage regulators used in power distribution systems, and the variable autotransformer in which only the tap between primary and secondary moves. In this chapter, we shall treat only static transformers. The transformer is not an energy converter in the conventional sense in that energy of both its input and output is electrical energy. In Chapter 1, a brief method for analyzing electromagnetic systems was presented. We shall now proceed with the first step of this method by describing the physical construction and electromagnetic structure of the two-winding transformer.

3.1 TRANSFORMER CONSTRUCTION AND ELECTROMAGNETIC STRUCTURE

The magnetic structure of a transformer consists of one or more electrical windings linked together magnetically by a magnetic circuit or core. The magnetic circuit of most transformers is constructed of a magnetic material, but nonmagnetic materials—often called an "air core"—are found in some applications. Also, in some special applications, the magnetic circuit may consist of a magnetic material in series with an air gap. When there are more than two windings on a transformer, two of the windings are usually performing the identical functions. Therefore, in terms of understanding the theory and performance of a multiwinding transformer, only the relationships in two of the windings need to be considered. The autotransformer, with only one winding, will be treated in Section 3.7.3. The two basic windings are often called primary and secondary. The meaning usually attached to this nomenclature is that the input or source energy is applied to the primary winding and the output energy is taken from the secondary winding. However, since a transformer is a bilateral device and often operated bilaterally, this meaning is not very significant and these words are used more as a way to distinguish the two windings. It is more common to designate the windings by numbered subscripts, and we shall use this practice in this chapter. A simple two-winding transformer model is shown in Figure 3.1.

The construction of transformers varies greatly depending upon their applications, winding voltage and current ratings, and operating frequencies. Many

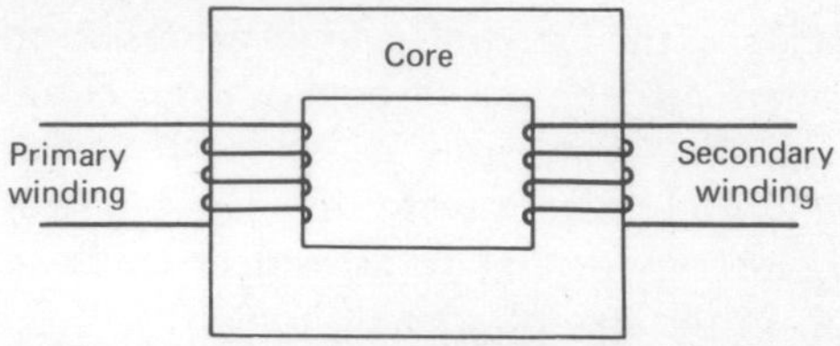

Figure 3.1. Elementary model of a transformer.

electronics transformers consist of little more than the electromagnetic structure itself with a suitable means of mounting to a frame. In general, the electromagnetic structure is contained within a housing or case for safety and protection purposes. In several types of transformers, the space surrounding the electromagnetic structure is filled with an electrically insulating material to prevent damage to the windings or core and to prevent their movement or to facilitate heat transfer between the electromagnetic structure and the case. In electronics transformers, a viscous insulating material called "potting compound" is used and, in many power transformers, a nonflammable insulating oil, called transformer oil, is used.

Transformer oil serves an added function of improving the insulation characteristics of the transformer, since it has a higher electric strength than air. In most oil-filled transformers, the oil is permitted to circulate through cooling fins or tubes on the outside of the case to further improve the heat transfer characteristics. The fins or tubes are often cooled by forced air. Figures 3.2*a* to *c* and 3.3 illustrate practical transformers of various types and applications. In larger transformers operating at high voltage and current levels, there are other important structural components, some of which can be seen in Figure 3.3. Such components include porcelain bushings through which the winding leads are brought for external connection, oil pressure and temperature gauges, and internal structural supports to prevent movement of the leads or windings due to electromagnetic forces resulting from high current levels.

The magnetic core of a transformer must be constructed in a manner to minimize the magnetic losses. These core constructions are described briefly in Chapter 2. Power transformer cores are generally constructed from soft magnetic materials in the form of punched laminations or wound tapes. Lamination or tape thickness is a function of the transformer frequency. Pulse transformer and high-frequency electronics transformer cores are often constructed of soft ferrites. Laminations used in electronics transformers are often termed "alphabet" laminations, since they are in the shape of several letters of the alphabet and designated as *E, C, I, U* cores and so forth. Ferrite cores are designated by the names cup, pot, sleeve, rod, slug, and so forth, which also describe the general shape of these cores. The most common lamination materials are silicon-iron, nickel-iron, and cobalt-iron alloys. Powdered permalloy is used for many communication transformer cores.

Figure 3.2*a*. Elements of an electronics power transformer showing magnetic core with wound strip of laminations and typical windings. (Courtesy GTE Lenkurt.)

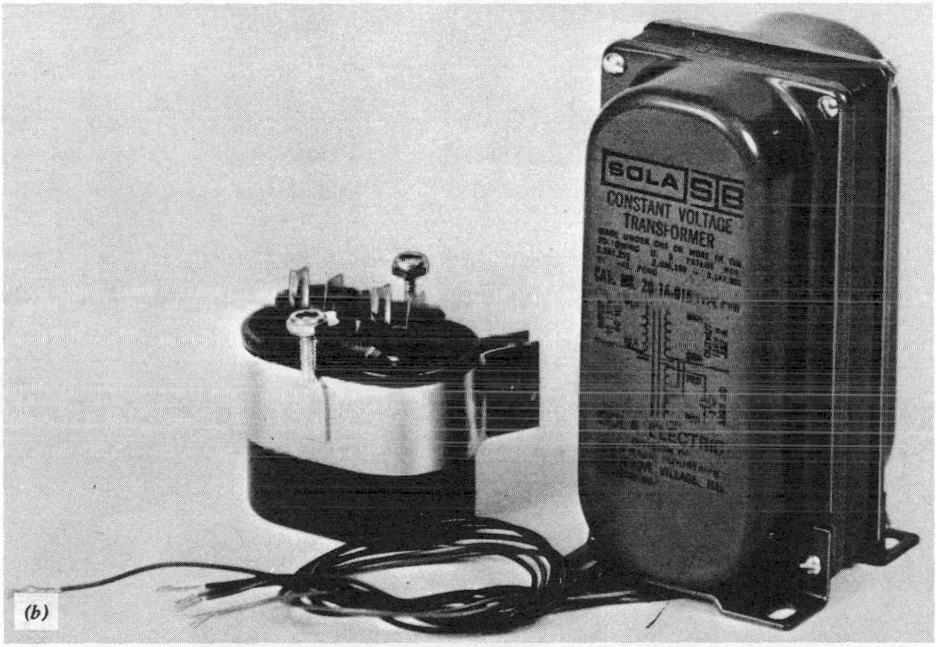

Figure 3.2*b*. Constant-voltage ferroresonant transformer. Capacitor shown on left. (Courtesy Sola Electric Division of Sola Basic Industries.)

Figure 3.2*c*. Pulse transformer, frequently used in thyristor grate trigger circuits. (Courtesy GTE Lenkurt.)

Figure 3.3. A power transformer (Courtesy General Electric Company)

Transformer windings are constructed of solid or stranded copper or aluminum conductors. The conductor used in electronics transformers—as well as in many small and medium-size motors and generators—is known as "magnet wire." It consists of copper wire coated with a thin enamel layer for electrical insulation. Magnet wire is classified by an insulation class symbol, *A, B, C, F,* and *H*, which designate the safe operating temperature in which the wire can be used. The lower letters represent the lower operating temperature (105° *C* for Class *A*) and Class *H* has the highest operating temperature of the standard classifications at 180° *C*. Magnet wire can also be obtained with cloth coverings (or "servings") outside of the enamel to provide more protection from abrasion or cutting. Conductor sizes are designated by a system known as the American Wire Gauge, abbreviated AWG, with increasing gauge number designating decreasing conductor cross section. A portion of the AWG wire tables is given as Appendix II.

In transformers having split cores of the *E, C, U,* or toroidal configurations, the windings are usually wound on an insulating spool or reel known as a bobbin. The wound bobbin can then be conveniently slipped over one leg of the core section; the remaining sections of core can be assembled; and the completed core assembly containing the winding is generally clamped together by means of a metal tape. The purpose of the bobbin is to provide a structural support for the winding, to electrically insulate the winding from the core, and to prevent abrasion or cutting of the winding at the core edges. Transformer bobbins are constructed of nylon, teflon, and various paper and fiber products. To improve the magnetic coupling between the two windings, a bifiler winding technique is frequently used. This technique consists of laying the conductors to be used in the windings side-by-side and winding them on the core or bobbin simultaneously. Figure 3.4 illustrates the various stages of assembling a winding on an electronics transformer core.

Figure 3.4. Various stages of the assembly of an electronics transformer (Courtesy GTE Lenkurt.)

The windings of large power transformers generally use conductors with heavier insulation than magnet wire insulation. The windings are assembled with much greater mechanical support, and winding layers are generally insulated from each other. Larger, high-power windings are often preformed, and the transformer is assembled by stacking the laminations within the preformed coils. For further details of the structural features of electronics and power transformers, References 1 to 4 may be consulted.

3.1.1 TRANSFORMER CLASSIFICATION

Because of the great diversity in size, shape, and application of transformers, there has been some attempt to designate types or classes of transformers. However, the terms used to classify transformers are very loosely defined and there is much overlap among the meanings of these terms. The terms have developed more from the types of circuits in which the transformers are used rather than from a delineation of transformer characteristics. Since these descriptive terms are widely used in manufacturers' catalogues of transformers and since it will aid in the subsequent description of transformer characteristics and structural features, some of the terms which categorize basic transformer types are listed below:

A. General Application Classification
   1. Power systems transformers: transformers used in power distribution and transmission systems. This class has the highest power or volt-ampere ratings and the highest continuous voltage ratings.
   2. Electronics transformers: transformers of many different types and applications used in electronic circuits. Sometimes electronics transformers are considered as those transformers with ratings of 300 volt-amperes and below. A large class of electronics transformers are called "power transformers" and are used in supplying power to other electronic systems—which confuses this nomenclature considerably.
   3. Instrument transformers: transformers used to sense voltage or current in both electronic circuits and power systems—often called potential transformers and current transformers. The latter are series-connected devices and are operated in a configuration which is, in many respects, the inverse of the conventional voltage or potential transformer. The current transformer will be discussed in Section 3.7.4.
   4. Specialty transformers: this designation covers a gamut of styles and operating features and includes such devices as saturating, constant-voltage, constant-current, ferroresonant, and variable-tap transformers.

B. Classification by Frequency Range

This is probably the most significant method of classifying transformers for describing electromagnetic design features and includes the following:

1. Power: these are generally constant frequency transformers that operate at the power frequencies (50, 60, 400 Hz, etc.), although other frequency components, including dc, may be present in electronics power transformers and power semiconductor transformers.
2. Audio: used in many communication circuits to operate at audio frequencies.
3. Ultra high frequency (UHF).
4. Wide-band: electronics transformers operating over a wide range of frequencies.
5. Narrow-band: electronics transformers designed for a specific frequency range.
6. Pulse: transformers designed for use with pulsed or chopped excitation, both in electronics and power systems applications.

C. Classification by Number of Windings

This has already been alluded to and includes one-winding (autotransformers); two-winding (or conventional); and multiwinding transformers, where a winding is defined as a two-terminal electrical circuit with no electrical connection to other electrical circuits.

D. Classification by Polyphase Connection

This classification applies mainly to transformers in power systems and refers to the method of connecting individual windings in polyphase applications. In some transformers, the entire polyphase set of transformers is contained within a single housing or case with some resulting reduction of the weight of the magnetic core. The most common connections are the wye (or star) and the delta connections of three-phase systems. These and other connections are briefly discussed in Section 3.7.2.

3.2 TRANSFORMER THEORY

We have so far introduced various terms and descriptions of transformers. Let us now look at the principle of operation of the simple two-winding model of Figure 3.1. Despite our reservations about the use of the words, "primary" and "secondary," as stated in the introduction, we shall find it convenient to use these terms and shall associate the subscript "1" with the primary and "2" with the secondary. However, keep in mind that these are arbitrary terms and are in no way inherent properties of a transformer. Because a number of these arbitrary designations are required in the analysis of the transformer—and are subsequently transferred to the analysis of many other electromagnetic devices

—we shall adhere to the conventions used by international standards organization (IEC) and the Institute of Electronics and Electrical Engineers[5]. Also, all formulas are given in terms of the Standard International (SI) or MKS units. The simple transformer model of Figure 3.1 has been repeated as Figure 3.5 with the addition of pertinent parameters for this analysis.

A fundamental parameter of the transformer is turns ratio, defined as

$$a = \frac{N_1}{N_2} \tag{3.1}$$

The value of the turns ratio is generally known only to the transformer manufacturer or the person winding a laboratory transformer. It can be measured in the laboratory by measuring the induced voltages in the two windings but always with a certain degree of inaccuracy, which depends upon the coefficient of coupling between the two windings. This should be kept in mind in the analysis of a transformer. The turns ratio is frequently given as part of the nameplate data by the manufacturer in both electronics and power systems transformers.

In this section, we shall consider a transformer excited from a single-frequency, sinusoidal voltage source, represented as v_1 in Figure 3.5. We shall deal primarily with the *steady state* voltages, currents, and magnetic fluxes. *Transient* transformer characteristics will be dealt with in later sections of this chapter. We begin with the electromagnetic field relationships in the transformer as an aid in the understanding of the physical phenomena that result in the external transformer characteristics.

Magnetic field theory need not be used to describe external transformer characteristics, which can be developed entirely by means of electrical circuit theory. The circuit theory approach is adequate for a great many problems in electrical engineering, such as problems in which the transformer is to be represented as a mathematical model or as a two-terminal pair network or "black box." This circuit approach will be discussed in some detail later in this chapter following the development of the transformer equivalent circuit.

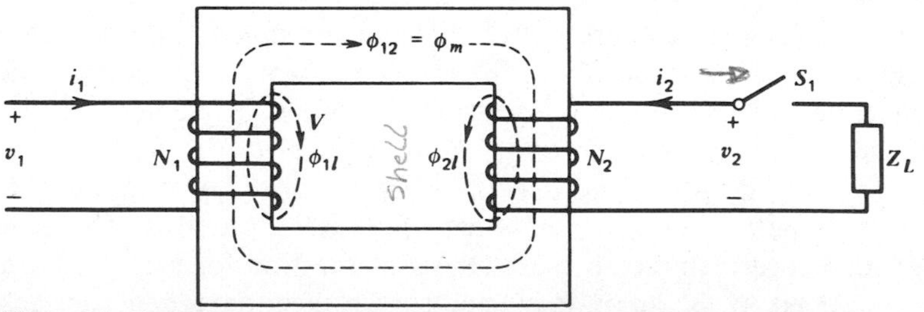

Figure 3.5. Transformer model showing instantaneous voltages, currents, and fluxes.

3.2.1 EXCITATION CHARACTERISTICS

In Figure 3.5, assume that the switch in the load circuit, S, is open and $i_2=0$. This condition is termed the transformer *no-load* condition and will facilitate the development of the no-load or *exciting* transformer voltage and current relationships. Assume that the applied voltage is

$$v_1 = V_{1m} \sin \omega t \tag{3.2}$$

where

V_{1m} = maximum value of applied voltage

$\omega = 2\pi f$ = angular frequency of applied voltage

f = frequency, Hz

For the moment, assume that the resistance of winding, N_1, is zero. Therefore,

$$v_1 + e_1 = 0$$

or

$$v_1 = -e_1 = -E_{1m} \sin \omega t \tag{3.3}$$

where

e_1 = electromagnetically induced voltage in winding, N_1

From (2.27) of Chapter 2, this voltage can be expressed as

$$e_1 = -\frac{d\lambda_1}{\partial t} = -\frac{d(N_1\phi_1)}{dt} = -N_1\frac{d\phi_1}{dt} \tag{3.4}$$

where ϕ_1 = the total magnetic flux linking the turns, N_1, Wb.

From Figure 3.5, it is seen that total magnetic flux is composed of two components

$$\phi_1 = \phi_{1l} + \phi_m \tag{3.5}$$

where

ϕ_{1l} = leakage flux linking only winding N_1, Wb

ϕ_m = "mutual" flux, that is, flux linking both windings, Wb

The concept of leakage and mutual fluxes in a magnetic circuit is discussed in Sections 2.4 to 2.7. Another representation of the flux relationships shown by (3.5) is in common use and leads to the concept of self-and mutual inductances as developed in Chapter 2:

$$\phi_{12} = k\phi_1 \tag{3.5a}$$

where k = the coefficient of coupling between windings 1 and 2. The meaning of the subscripts of the mutual flux term, ϕ_{12}, is the "flux linking winding 2 due to a current in winding 1." If the transformer magnetic circuit is isotropic and homogeneous, there is a symmetrical relationship among mutual flux components, as noted by (2.33),

$$\phi_{21} = \phi_{12} \tag{3.5b}$$

The relationships between flux and voltage can be obtained by taking the indefinite integral of (3.4) as

$$\int e_1 dt = -N_1 \int d\phi \tag{3.4a}$$

Solving for flux, and using (3.3), gives

$$\phi_1 = -\frac{E_{1m}}{N_1 \omega} \cos \omega t = -\phi_{1m} \cos \omega t \tag{3.6}$$

It is seen that, with a sinusoidal applied voltage, the resultant magnetic flux is cosinusoidal. This relationship is represented by means of a phasor diagram in Figure 3.6. It is seen that the flux phasor lags the applied voltage by $\pi/2$ radians or 90°.

The current component flowing in winding 1 is known as the *magnetizing current*. An expression for this current can be obtained by evaluating the

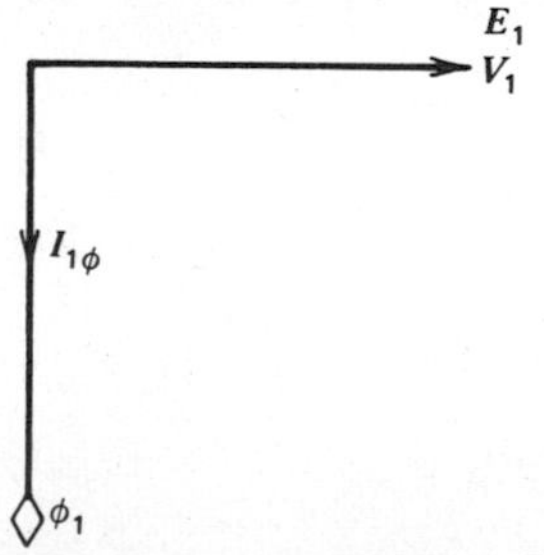

Figure 3.6. Phase relationships among excitation components.

magnetic potential as defined in (2.17):

$$V_m = i_{1\phi} N_1 = R\phi_1 = R\phi_{1m} \cos \omega t \tag{3.7}$$

where

$i_{1\phi}$ = magnetizing current

R = reluctance of the transformer core

Replacing ϕ_{1m} from (3.6) and solving for $i_{1\phi}$ gives

$$i_{1\phi} = \frac{E_{1m} R}{N_1^2 \omega} \cos \omega t \tag{3.8}$$

This equation is not too useful in an analytical sense, since we have not yet introduced the role that saturation of the magnetic circuit may play. However, it does illustrate three fundamental relationships of the transformer and all other inductive circuits, such as inductors and induction motors: (1) the magnitude of the magnetizing current is a function of the *volts-to-frequency ratio*, E_{1m}/ω; (2) the magnetizing current varies inversely with the square of the number of turns in the exciting winding, N_1; and (3) the reluctance of the magnetic circuit of the transformer may profoundly influence the magnitude of the magnetizing current. The first of these observations from (3.8) is a fundamental principle in the use of inductive devices and must be considered in all variable-frequency or variable-voltage applications. It will be seen that saturation in the magnetic circuit tends to greatly accentuate the current increase that will result when this ratio is increased over the design or nominal ratio.

From the two forms of (3.6) another useful relationship between voltage and flux in a sinusoidal circuit is available:

$$E_{1m} = N_1 \omega \Phi_{1m} \tag{3.9}$$

This will be expressed in more general nomenclature as the relationship between the maximum values of induced voltage and magnetic flux in a winding of N turns excited at frequency, f:

$$E_m = 2\pi f N \Phi_m \tag{3.10}$$

The rms induced voltage ($E_m/\sqrt{2}$) is given by

$$E = 4.44 f N \Phi_m \tag{3.11}$$

Let us now examine the role of saturation in the magnetic circuit of a transformer. The basic concepts of saturation and nonlinearities of magnetic materials have been developed in Chapter 2. The effect of saturation in a transformer will be illustrated by an example.

Example 3.1
The flux versus ampere-turn relationship of the magnetic circuit of a certain transformer is given in Figure 3.7. This transformer has a primary winding of 120 turns, which is excited from a sinusoidal voltage source of 110 V (rms) and 60 Hz. Determine the maximum value of magnetic flux and the general waveform of the magnetizing current. Neglect the winding resistance.

The flux is found from (3.11) and is

$$\Phi_m = \frac{110}{4.44 \cdot 60 \cdot 120} = 0.00343 \text{ Wb}$$

The waveform of the flux is sinusoidal; phase relationships are not required in this problem. The resulting magnetizing current is found by the graphical construction illustrated in Figure 3.7. Values of flux from the sine wave of flux with maximum value of 0.00343 Wb are transferred to the magnetic saturation curve and the resultant magnetic potential or MMF for each value of flux is read from the abscissa of this curve. The magnetizing current is found by dividing the MMF by the turns, that is, by 120.

It is seen from the above example that the magnetizing current in a transformer with saturation will be nonsinusoidal in waveform. There are other circuit parameters that will also affect this waveshape in an actual practical transformer: Hysteresis of the magnetic characteristic will alter the waveform somewhat; and there are usually other circuit parameters in series with the winding, such as winding resistance, which will influence the waveform. In some cases, the nature of the excitation source may change the waveform, especially if the source is a voltage- or current-regulated electronic power supply. The relative magnitude of the magnetic core losses may also influence the current waveform. In many transformers, the magnetic circuit is designed so that the maximum flux required at rated applied voltage results in operation at approximately the knee of the saturation curve. This will minimize the amount of distortion from a sine wave of the magnetizing current. If the transformer is operated at a volt/frequency ratio in excess of its design value, however, the distortion of the magnetizing current may be greatly increased. For example, if the transformer in Example 3.1 were to be excited from a 110 V, 50 Hz source, the maximum flux would be, from (3.11), 0.0041 Wb. Inspection of the graphical construction of Figure 3.7 shows that this operation would greatly increase the spike at the center of the current half-wave.

The magnetizing current is required to set up the mutual flux that makes transformer action possible and it is, therefore, a necessary parameter in every transformer. However, it is desirable to minimize this current component in order to minimize the copper loss due to this current, to improve the transformer

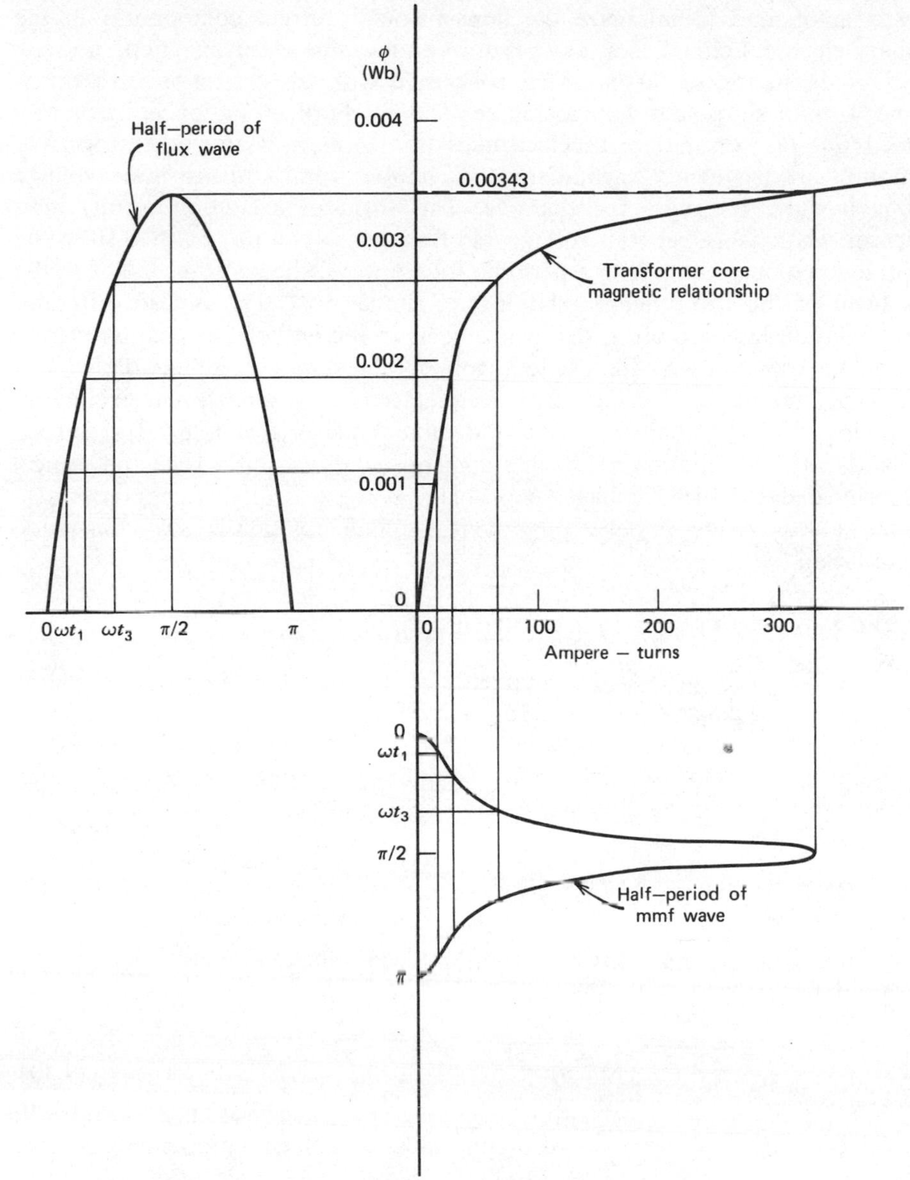

Figure 3.7. Construction for obtaining mmf or exciting current waveform.

power factor, and to minimize the nonsinusoidal current components in the primary electrical circuit that may produce undesirable electromagnetic noise.

Our discussion so far has been concerned with the excitation of a transformer from a sinusoidal voltage source. This mode of excitation is frequently termed *sine-flux* excitation. In electronics transformers, many other forms of excitation are frequently encountered. Excitation from a *square wave* voltage source is quite common, for example. This form of excitation results in a different relationship between voltage and flux, which can be obtained from the integral equation, (3.4a). With a periodic square wave, the constant factor in the rms form of the flux voltage relationship, similar to (3.11), is also different. These relationships are presented as problems at the end of this chapter. Many electronics transformers are excited from a constant current source. When sinusoidal, this mode of excitation is frequently termed *sine-current* excitation. By analogy to the graphical procedure demonstrated in Example 3.1, it is seen that sine-current excitation results in a nonsinusoidal flux in the core and, hence, a nonsinusoidal induced voltage. A problem at the end of this chapter is also given to aid the reader in developing the flux-voltage relationships for this mode of excitation.

3.2.2 NO-LOAD VOLTAGE RELATIONSHIPS

Equation (3.5) introduces two components of flux that link the primary coil of Figure 3.5: a leakage component, ϕ_{1l}, that links only coil 1, and a mutual component that links both coil 1 and coil 2, ϕ_{12}. It is convenient for the purpose of analyzing a transformer to split the induced voltages into two related components. This can be obtained by substituting (3.5) into (3.4) giving

$$e_1 = -N_1 \frac{d\phi_{1l}}{dt} - N_1 \frac{d\phi_{12}}{dt} = e_{11} + e_{12} \tag{3.12}$$

The mutual flux component results in an induced voltage in coil 2 which, from (2.27) of Chapter 2 can be expressed as

$$e_{21} = -N_2 \frac{d\phi_{12}}{dt} \tag{3.13}$$

The relationships between the two mutual induced voltage components is seen to be

$$\frac{e_{12}}{e_{21}} = \frac{N_1}{N_2} = a \tag{3.14}$$

This relationship is the basis for the voltage transformation characteristics of the transformer.

We have gone into considerable detail to develop an exact relationship between transformer induced voltages and the turns ratio of the two coils or windings. For a transformer with close magnetic coupling between the two coils, the leakage component of induced voltage, e_{1l}, is negligibly small and is usually ignored. When a high coupling coefficient can be assumed, the double subscripts are dropped from (3.14) and it becomes

$$\frac{e_1}{e_2} = \frac{N_1}{N_2} = a \tag{3.14a}$$

As noted above, the primary induced voltage component, e_1 or e_{12}, is not measurable; only the primary terminal voltage, v_1, can be measured. The difference between v_1 and e_1 is due to the leakage voltage, e_{1l}, and the voltage drop across the resistance of coil 1 (to be discussed in the next section). Since both of these voltages are small, in many types of transformers under the open secondary or no-load condition, (3.14a) is frequently written as

$$\frac{v_1}{e_2} = a \tag{3.14b}$$

The set (3.14) is also written in terms of rms components when the transformer is excited from a sinusoidal voltage source. Equation (3.14b) is the most common relationship for obtaining the turns ratio from measured data. However, it must be kept in mind that this relationship is always an approximation that may be quite inaccurate for some transformers in which the resistance of coil 1 is relatively high or in which the coefficient of coupling is relatively low.

3.2.3 NO-LOAD LOSS COMPONENTS

So far we have ignored the lossy components of the transformer. In any practical transformer, these components must be considered. The principal component of no-load loss is the core loss, sometimes called magnetization loss. The nature of this loss is discussed in Section 2.3, which should be reviewed for the present discussion. Core loss is classified as a no-load loss, since it is not a function of the load current in the transformer and can be considered as constant for all conditions of transformer loading. Core loss is a function of the flux density in the magnetic core, the frequency of the excitation signal, and the type of material and construction of the core. In an air-core transformer, this loss is zero.

There are several formulas for approximate calculations of the two components of core loss—eddy current loss and hysteresis loss—but it is generally necessary to obtain the magnitude of the core loss from tests. Since the core loss is a function of frequency and flux density, it is also implicitly a function of the induced voltage as can be seen from (3.4) or (3.11). Therefore, as in the case of the magnetizing current, it is dependent upon the volts to frequency ratio of the excitation signal and may increase disproportionally if this ratio is increased significantly beyond the rated value. The core loss is introduced into transformer analysis by means of an equivalent resistance. This resistance is a nonlinear function of the applied voltage and frequency in the primary winding but is generally invariant as a function of the transformer load current. This representation will be further discussed in connection with the transformer equivalent circuit.

The ohmic resistance of the primary winding is obviously another element that results in a power loss. Under no-load conditions, the loss associated with the primary winding resistance is often negligibly small and can be ignored. The voltage drop across this resistance is generally included in a phasor diagram representation.

In high-voltage transformers, a third component of no-load loss often exists. This is known as corona loss and is due to the ionization of the air surrounding the high-voltage windings caused by the high electric field strength in these regions. Corona loss can occur in electronics transformers used in television and radar applications and the large power transformers connected to high-voltage transmission lines. We shall not consider this loss in the transformer models and analytical methods to be developed in later sections of this chapter, but further descriptions and test methods for evaluating corona loss may be found in References 6 and 7.

3.2.4 NO-LOAD PHASOR DIAGRAM

The phasor diagram of Figure 3.8 will aid in summarizing the transformer no-load components that have been described in the previous discussion. In Figure 3.8, the commonly used convention of separating the primary and secondary voltages is applied. This is a convention of simple convenience to

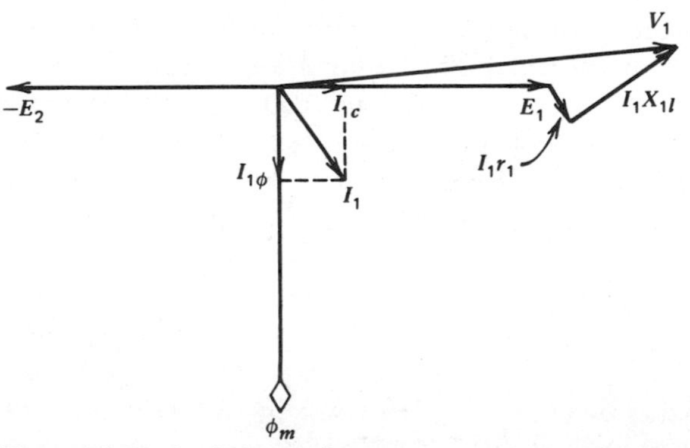

Figure 3.8. No-load phasor diagram.

make the diagram less congested and does not necessarily give the directional characteristics of the voltages. The directional relationships between flux and voltage can be found from the directional properties of Faraday's law as discussed in Section 2.8. Directional properties *with respect to the transformer terminals* are treated by the concept of *transformer polarity*; this is presented in Section 3.4.1. In Figure 3.8, sine-flux (or constant-voltage) excitation is assumed, and the voltage, flux, and current parameters are rms values.

3.2.5 TRANSFORMER LOAD CHARACTERISTICS

Referring once again to Figure 3.5, assume now that the switch, S, is closed, and a closed electrical circuit is connected across the secondary (coil 2) terminals. The induced voltage in coil 2, e_2, will cause a current, i_2, to flow in the external or load circuit connected across the secondary terminals. The *direction* of this current in the schematic diagram of Figure 3.5 follows from the basic laws of electromagnetics: It is in such a direction that its magnetic flux *opposes* the flux that produced the secondary induced voltage, e_2, that is, the mutual flux component, ϕ_{12}. These directional properties can be found by means of the simple "hand rules" given in Chapter 2. For the winding layout shown in Figure 3.5, the directions of the secondary current flow is as shown. The reader should verify the directional characteristics shown in Figure 3.5 and should also work out secondary current directions for different winding layouts. These directional properties are the basis for the polarity conventions of transformers to be described later.

The *magnitude* of the secondary current follows from another fundamental electromagnetic relationship, Ampere's law, which is discussed in Section 2.5. From (2.22), assuming zero air gap, it is seen that

$$N_1 i_1 - N_2 i_2 = \phi_{12} R \tag{3.15}$$

This magnetic circuit approach shows that the *difference* between the primary and secondary winding MMF's is equal to the reluctance drop in the mutual magnetic circuit, that is, the transformer core. This MMF difference is the exciting MMF, expressed in magnetic circuit nomenclature, that has been discussed in the preceding paragraphs. There are two magnetic paths in parallel with the mutual path—the primary and secondary leakage paths, which are largely through nonmagnetic regions and therefore of high reluctance. Two additional equations could be written relating the two MMF's on the left-hand side of (3.15) to their respective leakage flux reluctance drops. However, the leakage path reluctances are difficult to describe analytically and these equations are not of much significance. It is much simpler to handle the leakage magnetic relationships in terms of electrical circuit parameters based upon electrical measurements described later in this chapter.

The difference in the MMF's on the left-hand side of (3.15) has already been defined in (3.7). The magnetizing current, $i_{1\phi}$, is a component of the total current in the primary (winding 1 in the nomenclature of Figure 3.5). The difference between this total current, i_1, and the magnetizing current, $i_{1\phi}$, is often termed the *load component* of primary current, i_{1L}. Rewriting (3.15) to include the two primary current components give

$$N_1(i_{1L}+i_{1\phi})-N_2i_2=\phi_{12}R \tag{3.16}$$

This equation can be separated into two equations which lead to an important relationship that further describes transformer action:

$$Ni_{1\phi}=\phi_{12}R \tag{3.17}$$

$$N_1i_{1L}=N_2i_2 \tag{3.18}$$

or

$$\frac{i_{1L}}{i_2}=\frac{N_2}{N_1}=\frac{1}{a} \tag{3.19}$$

Equation (3.19) gives the relationships between the load components of primary and secondary currents and the turns ratio. It is seen to be the inverse of the voltage relationships. Stated in another manner, if a transformer is a step-up voltage transformer, it is a step-down current transformer, and vice versa. In many power transformers, the exciting component of primary current is very small compared to the load component; therefore,

$$\frac{i_1}{i_2}\approx\frac{1}{a} \tag{3.20}$$

A useful parameter for sizing and rating a transformer is the product of load voltage and load current, termed volt-amperes or apparent power. This can be found from (3.14a) and (3.19):

$$\text{volt-amperes}=e_1i_{1L}=e_2i_2 \tag{3.21}$$

It is seen that the required load volt-ampere capabilities of the primary and secondary windings must be identical. In addition, the primary must be capable of supporting the required exciting volt-amperes.

Example 3.2
The sinusoidal flux in the core of a transformer is given by the expression, $\phi_1=0.0012\sin 377t$ Wb; this flux links a primary winding of 150 turns. Determine the rms induced voltage in the primary winding.

With $\omega=377$, the frequency is 60 Hz; from (3.11),

$$E_1=(4.44)(60)(150)(0.0012)=48 \text{ V}$$

Example 3.3
The flux in the core of a transformer is expressed as $\phi=0.002\sin 377t+0.00067\sin 1131t$. Determine the induced voltage in an 80-turn coil linking this flux.

From (3.4),

$$e=-N\frac{d\phi}{dt}=-60.5\cos 377t-60.5\cos 1131t$$

Example 3.4
The voltage, $v=100\sin 377t-20\sin 1885t$, is applied to a 200-turn transformer winding. Derive the equation for the flux in the core, neglecting leakage flux and winding resistance. Determine the rms values of the voltage and the flux.

From (3.4a),

$$\phi=-\frac{1}{200}\left[-\frac{100}{377}\cos 377t+\frac{20}{1885}\cos 1885t\right]$$

$$=0.00133\cos 377t-5.3\times 10^{-5}\cos 1885t$$

The rms values are

$$\Phi=\sqrt{\left[\frac{(.00133)^2}{2}+\frac{(5.3\times 10^{-5})^2}{2}\right]}=0.00094\text{ Wb}$$

$$V=\sqrt{\left[\frac{100^2}{2}+\frac{20^2}{2}\right]}=72.1\text{ V}$$

Example 3.5
In a certain transformer tested with open secondary (no-load), a core loss of 150 W is measured when the no-load current is 2.5 A and the induced voltage is 400 V (voltage and current are rms values). Neglect the winding resistance and leakage flux. Determine (1) the no-load power factor, (2) rms magnetizing current, and, (3) the rms core loss component of current.

1. Power factor, $PF=(150)/(2.5)(400)=0.15$.
2. Magnetizing current, $I_{1\phi}=2.5\sin(\cos^{-1}0.15)$
   $=2.5(\sin 81.4°)=2.42\text{ A}$
3. Core-loss current, $I_{1c}=150/(400)=2.5(\cos 81.4°)$
   $=0.375\text{ A}$

Example 3.6
A transformer is rated 100 kVA (kilovolt-amperes), 400-V primary and 200-V secondary voltages. Neglecting the magnetizing and core-loss volt-amperes, determine the turns ratio and the winding current ratings.

The turns ratio is approximately equal to 400/200 or $a=2$; assuming that both windings are rated for the transformer volt-ampere rating,

$$I_1 = \frac{100{,}000}{400} = 250\,\text{A};\ I_2 = \frac{100{,}000}{200} = 500\,\text{A}$$

Note that transformer ratings are always given in terms of rms parameters.

Example 3.7
The transformer of Example 3.6 has 1000 turns in its primary winding; the conductor used in the primary winding is copper AWG No. 1/0. If the secondary winding is designed to have exactly the same "copper loss" (ohmic resistance loss), determine the size of the wire to be used in the secondary winding.

For the copper losses of the secondary to equal those of the primary, the secondary resistance must be one-fourth that of the primary, since the secondary current is twice the primary current. However, since the number of turns on the secondary is half the number of primary turns, it can be assumed that the length of the secondary winding is half that of the primary. Therefore, in order to reduce the secondary resistance to one-fourth that of the primary, the cross-sectional area of the secondary conductors must be twice that of the primary. The primary 1/0 (AWG) conductor has an area of 105,600 CM (circular mils), from Appendix II; the nearest standard conductor size approximately twice this value is 4/0 (211,600 CM). In practice, flat or square conductor, rather than round, would be used.

3.3 TRANSFORMER EQUIVALENT CIRCUITS

Many concepts have been developed through the years to aid in the understanding of the theory and operation of electromagnetic devices and to facilitate their analysis. One of the most convenient of such concepts is an equivalent electrical network that "models" the voltage, current, and magnetic relationships within the actual device. Transformer equivalent circuits were conceived and widely used long before the "age of computers" in which we now live, but these equivalent circuit representations are particularly suited for computer simulation of electrical systems of all types.

Among the earliest computer analysis of complex electrical systems was the modeling of interconnected power transmission lines, transformers, and synchronous generators by means of a type of analog computer known as the AC Network Calculator. This analysis was the basis for the development of the large interconnected power systems that are so necessary for an industrial economy.

Transformer equivalent circuits were used in the four-terminal network theory representations for telephone systems and other higher-frequency communication systems long before the advent of the modern digital computer. Today, there are many software programs available on most digital computers, some of which are listed in Appendix III, for which most forms of the transformer equivalent circuit are well suited. Equivalent circuits are also very helpful in understanding the physical operation of electromagnetic devices and in establishing systematic techniques for device design.

In the last chapter, a steady-state or dc equivalent circuit was developed to represent magnetic circuits of the type found in transformers. This type of a circuit is useful in sizing transformer cores, in studying the effects of core geometry upon leakage flux paths and reluctances, and in calculating leakage and mutual inductances. Such an equivalent circuit contains only one type of circuit element, resistance, which represents magnetic reluctance. The magnetic equivalent circuit does not represent the lossy components of the magnetic core, although it can be extended to do so. Transformer windings appear only as they affect the steady state magnetic potentials or MMFs.

The transformer equivalent circuit, which we now wish to discuss, may contain all three types of electrical circuit elements and can be used to describe almost every aspect of transformer electrical performance at all excitation frequencies and with many variations of current and voltage waveforms. We shall begin with the low-frequency equivalent circuit in which transformer capacitance can be neglected.

3.3.1 THE IDEAL TRANSFORMER: IMPEDANCE RELATIONSHIPS

A useful strategem in the development of a transformer equivalent circuit is the concept of the *ideal transformer*. The ideal transformer is defined as a transformer having zero winding resistance, zero leakage reactance, zero magnetizing current [which, from (3.8) represents zero magnetic reluctance of the core, or a core with infinite permeability], and zero core loss. The ideal transformer contains the ideal magnetic circuit discussed in Chapter 2. The ideal transformer is simply a turns ratio.

Let us look at the current and voltage relationships in the ideal transformer. In Figure 3.5, assume that the switch is closed and the secondary current, i_2, is flowing in the load impedance, Z_L. From (3.19), this current is balanced by a primary current equal to ai_1 (the double subscript notation is not required, since the magnetizing and core loss components are assumed zero). The voltage across the load impedance is i_2Z_L. This voltage is equal to the secondary induced voltage in magnitude, since the secondary resistance and leakage inductance are assumed zero. From (3.14a), the primary induced voltage is related to the secondary induced voltage as $e_1 = ae_2$. The primary terminal voltage equals this induced voltage, for the ideal transformer, and is expressed as

$$v_1 = e_1 = ae_2 = a(i_2 Z_L) = a(ai_1 Z_L) = a^2 i_1 Z_L \tag{3.22}$$

The ratio of primary voltage to primary current, v_1/i_1, gives an equivalent impedance which is the load impedance referred to the primary winding—sometimes called the load impedance "as seen from the primary." From (3.22),

$$Z_L' = \frac{v_1}{i_1} = a^2 Z_L \tag{3.23}$$

By an analysis similar to the above, it can be shown that a primary impedance referred to the secondary winding or "as seen from the secondary" has the inverse relationship, or

$$Z_L'' = \frac{Z_L}{a^2} \tag{3.24}$$

Equations (3.23) and (3.24) can be generalized to state that *any impedance* in one winding is referred to the other winding as a function of the square of the transformer turns ratio, as shown in these equations. The superscripts in these two equations are used to designate the winding to which the impedance is being referred. This is a most useful characteristic of transformers and is the basis for the transformer application termed "impedance matching."

Example 3.8
A transformer is rated at 3 kVA and 100/400 V. A 50-ohm load resistance is connected across a 400-V winding. With this load, determine the current in the two windings and the equivalent load resistance referred to the 100-V winding; assume that the transformer is ideal.

The current in the 400-V winding is $400/50 = 8$ A; call this winding 2. Therefore, $a = 1/4$, and from (3.19), $i_1 = 8/(1/4) = 32$ A. The impedance referred to winding 1 is $a^2(50) = 50/16 \doteq 3.125$ ohms.

Example 3.9
A transmission line is to be terminated with a resistance of 100 ohms to minimize standing waves. The actual load resistance is 500 ohms. Determine the required transformer ratio to match the actual load to the transmission line.

Using the nomenclature of (3.23), let $Z_L = 500$, $Z_L'' = 100$. Therefore, $a^2 = 100/500$ and $a = 1/\sqrt{5} = 0.446$.

3.3.2 EQUIVALENT CIRCUIT CONFIGURATION

The basic configuration of the transformer equivalent circuit referred to winding 1 is shown in Figure 3.9. This representation is adequate for most power and

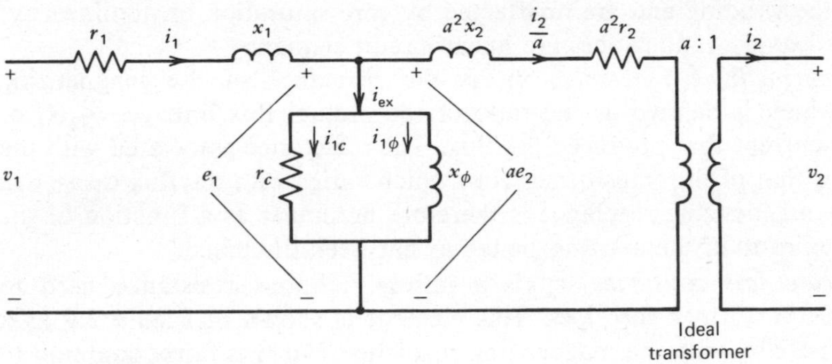

Figure 3.9. Transformer equivalent circuit in terms of winding 1.

audio frequency transformers. In transformers operating at higher frequencies, the interwinding capacitances often are significant and must be included in the equivalent circuit. The circuit modifications to include this parameter will be discussed in Section 3.5.

The background theory for the circuit elements shown in Figure 3.9 has been presented in previous sections of this chapter and in Chapter 2. The relationships of these parameters to transformer theory and the significance of each parameter will now be summarized. First note that this circuit is a passive lumped "T" network of the type used to represent many other electrical systems and devices, such as transmission lines, filters, and other electromagnetic devices.

The circuit elements in Figure 3.9 represent equivalent circuit elements in the actual transformer that are usually obtained from test measurements or can be calculated if the dimensions and materials of the actual transformer are well defined. The circuit shown in Figure 3.9 is based upon the assumption of sinusoidal excitation, and the impedances are in the form of conventional impedances of sinusoidal circuit theory. The circuit of Figure 3.9 is the most common form of the circuit, but it can easily be adapted for use with other types of excitation and for transient analysis by replacing the inductive reactances with inductances. The series elements of this model follow:

1. The *effective resistances* of the two windings, r_1 and r_2, are sometimes called "ac resistance." In a great many transformers, the use of ohmic (or dc) resistance is adequate to represent r_1 and r_2. However, in power transformers with windings that have a large conductor cross section or in some electronics transformers operating at high frequencies or with pulsed excitation, skin effect may be significant and the effective resistance should be used.

2. The *leakage reactances* of the two windings, x_1 and x_2 are defined in the conventional manner as $2\pi f L_1$ and $2\pi f L_2$, respectively. The leakage inductances, L_1 and L_2, are defined as the ratio of leakage flux linkage of one winding to the current in that winding that produced the flux, as discussed in Chapter 2. The reluctance of the leakage paths are primarily through the nonmagnetic regions

surrounding the winding and are unaffected by core saturation or nonlinearity; therefore, the leakage inductances are linear circuit elements.

3. The *magnetizing reactance*, x_ϕ, is the reactance of the magnetizing inductance, which is defined as the ratio of the mutual flux linkage, $\phi_{12}N_1$ or $\phi_{21}N_2$, to the current that produced the flux. The reluctance associated with the mutual path is that of the transformer core which varies with the flux density in the core. The magnetizing reactance is therefore nonlinear as a function of the volt/frequency ratio of the exciting signal, as has been discussed.

4. The *core loss resistance*, r_c, is a purely fictitious resistance used to represent the transformer core loss. This element is shown in Figure 3.9 as a resistance in parallel with the magnetizing reactance, but it is fairly common to use a series representation of these two elements—with different magnitudes, of course. The parallel representation lends itself to phasor diagram analysis somewhat easier than a series circuit, and it is slightly easier to calculate values for the parallel representation from test data. Either representation is acceptable. The core loss resistance, like the magnetizing reactance, is nonlinear as a function of the volt/frequency ratio, and, also, of the frequency.

5. The *ideal transformer* is included in the equivalent circuit configuration to add completeness to this representation by making the voltage and current of winding 2 to have the correct magnitude. Its specific function in the equivalent circuit is simply to multiply or divide the winding 2 voltage or current by the turns ratio, a. If the equivalent circuit were to be expressed in reference to winding 2, the ideal transformer would be located at the extreme left of the circuit, adjacent to the winding 1 terminals. Or, in some cases, it may be desirable to place it near the center of the T network, on one side or the other of the magnetizing impedance branch. However, wherever it is located, it supplies no knowledge about the voltage drops, exciting current, phase angle, power factor, losses, or efficiency of the transformer, which are the characteristics for which the equivalent circuit is generally used. Therefore, it is often dropped from the representation when analyzing a single transformer, since it tends to "clutter up" a nice, symmetrical T network anyway. Since the purpose of the ideal transformer is to refer a voltage or current from one winding to the other, this cataloging can be done in simpler ways. We are including discussion of this topic here because in systems studies, where a number of electrical systems are interconnected by means of transformers, the ideal transformer must be included in the representation. In computer models, this is done by means of simple multiplication or division by a constant.

Another important means of representing systems interconnected by transformers is the *per unit notation*, in which voltages, currents, and impedances in each circuit are expressed as a decimal portion of a set of base values of voltage, current, and impedance. For further description of per unit notation, see Reference 7. In models using per unit notation, the need for the ideal transformer in the transformer representation is eliminated.

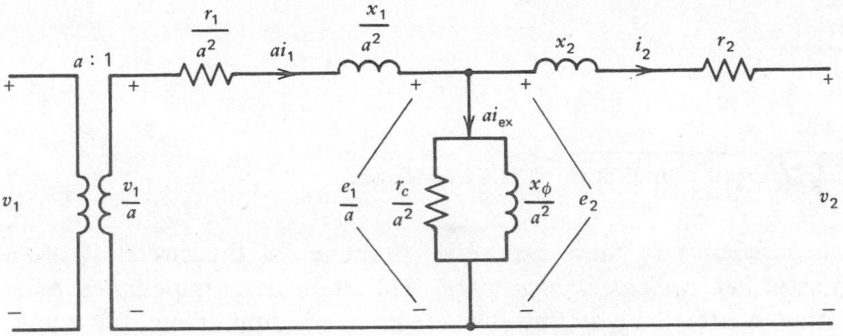

Figure 3.10. Transformer equivalent circuit in terms of winding 2.

There are many modifications and simplifications of this basic complete equivalent circuit. The use of these variations depends upon the requirements of the particular problem being solved and the degree of accuracy required. Figure 3.10 illustrates the same complete equivalent circuit referred to winding 2. Note that in all presentations of the transformer in this text the turns ratio, a, is defined as in (3.1).

It has already been noted that the ideal transformer is often neglected in the equivalent circuit. If the winding 1 representation is used as in Figure 3.9, this requires that the load impedance, Z_L, be referred to winding 1. This simplification is shown in Figure 3.11. If this same simplification is made in Figure 3.10, the winding 1 input voltage, v_1, must be expressed in winding 2 terms.

Another common simplification of the equivalent circuit is to ignore the exciting branch containing x_ϕ and r_c. When the magnetizing and core-loss current components are very small compared to the load components, as is true in many power transformers, this approximation is often justified. Also, in many problems it is the series impedances that are of interest for calculating voltage regulation in a power transformer. In this approximation, the equivalent circuit is reduced to the two "arms" of the T network, as shown in Figure 3.12.

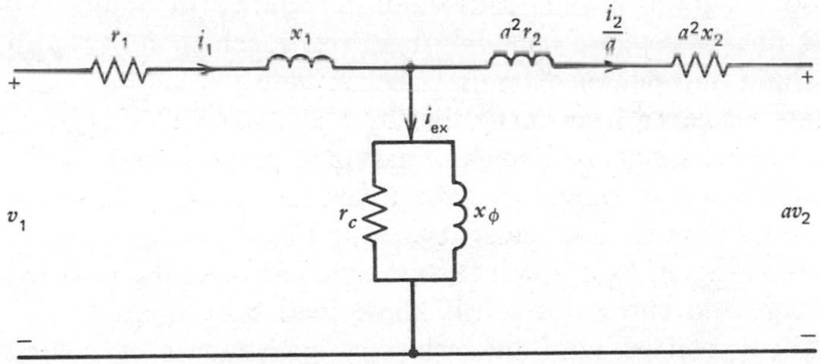

Figure 3.11. A commonly used equivalent circuit of transformers.

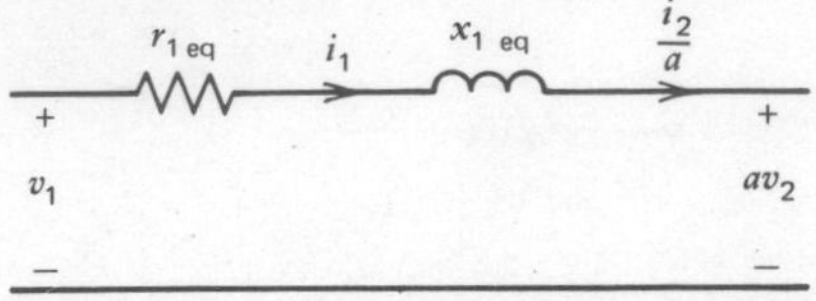

Figure 3.12. An approximate equivalent circuit of a transformer.

The total impedance of these two series branches of the circuit is often called the transformer *equivalent impedance*. The equivalent impedance must always be further qualified by stating the reference winding. Thus, the equivalent impedance in terms of winding 1 is, from Figure 3.12:

$$z_{1eq} = r_1 + a^2 r_2 + j(x_1 + a^2 x_2) \tag{3.25}$$

Similarly, the equivalent impedance referred to winding 2 is

$$z_{2eq} = \frac{r_1}{a^2} + r_2 + j\left(\frac{x_1}{a^2} + x_2\right) \tag{3.26}$$

3.3.3. TRANSFORMER PHASOR DIAGRAM

The phasor diagram is representation of the phase relationships of the steady state current and voltage components of the equivalent circuit, and is a convenient guide to use when solving for voltages and currents by means of the equivalent circuit. It is also a good visual representation of the theoretical relationships that have been expressed as equations and should help in the reader's understanding of transformer theory. Before the advent of computers and calculators, or even before the widespread use of slide rules, the phasor diagram was often drawn to scale so that the *magnitudes*—as well as phases—of the components were accurately represented and graphical methods for solving for various components were used. This approach is seldom used today, and phasor magnitudes are generally not drawn to scale.

The no-load phasor diagrams have already been introduced, in Figures 3.6 and 3.8. The complete phasor diagram including load relationships is shown in Figure 3.13. The sketching of a phasor diagram is begun with the load phasors, aV_2 and I_2/a, which are displaced from each other by the power factor angle of the load, Θ_L. These parameters must be known or assumed in order to draw the diagram. In certain problems it is general practice to assume rated transformer load current—(or some fraction of rated current) at rated load voltage—and to assume several different values of load power factor; then calculate the required transformer input voltage and current to satisfy these load conditions. In this chapter, we have already started with the arbitrary convention of calling winding 2 the load or output or secondary winding and winding 1 the input or

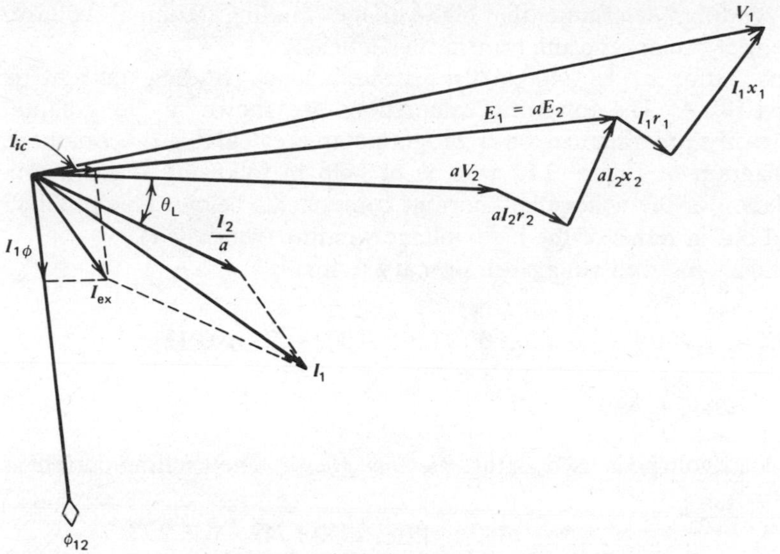

Figure 3.13. Phasor diagram of a transformer on load.

primary winding. This convention is in no way necessary, and the reader should use whatever convention in notation that feels most comfortable.

Once the load phasors have been established, the winding 2 or secondary impedance drops are added phasorially to the load voltage giving the secondary induced voltage, aE_2. The magnetizing current components are also added to the secondary current giving the primary or input current, I_1. The primary voltage drops are then calculated giving the input voltage, V_1. Note that the phasor diagram is always drawn in terms of one winding or the other (winding 1 was used in Figure 3.13); therefore, the magnitude changes in voltage and current attributable to the ideal transformer do not appear and in no way illuminate what the phasor diagram is used to illustrate.

Some of the uses of the equivalent circuit and phasor diagram will now be illustrated.

Example 3.10

A transformer is rated 500 kVA, 2400:480 V, 60 Hz. The impedances (in ohms) of the equivalent circuit for this transformer are as follows, where the subscript 1 is used for the high-voltage winding and 2 for the low-voltage winding:

$$r_1 = 0.058 \qquad r_2 = 0.002 \qquad x_\phi = 400$$
$$x_1 = 0.29 \qquad x_2 = 0.012 \qquad r_c = 2000$$

The load is connected across the low-voltage winding and draws rated current at 0.866 pf lagging with rated voltage at the terminals of this winding. For this condition of loading, determine the high-voltage winding terminal voltage, current, and power factor, and the transformer efficiency.

The turns ratio, a, is $2400/480=5$; rated load winding current is $500{,}000/480=1040$ A. The following calculations are shown in "long-hand" form but are easily programmed on a programmable calculator or computer. The phasor diagram of Figure 3.13 will be of help in following the complex algebra calculation of the voltage and current components below. This problem will be worked out in terms of the high-voltage winding (winding 1).

The secondary induced voltage (in primary terms) is

$$E_2' = 5\times 480 + \frac{1040}{5}\angle -30^\circ \times (25\times 0.002 + j25\times 0.012)$$

$$= 2440 + j49 \text{ V}$$

Note that the load voltage is used as the *reference phasor*. The exciting current is

$$I_{ex} = I_c + jI_\phi = \frac{2440+j49}{2000} + \frac{2440+j49}{j400}$$

$$= 1.3425 - j6.0755 \text{ A}$$

The primary (winding 1) current is

$$I_1 = I_{ex} + \frac{I_2}{a} = 1.3425 - j6.0755 + 208\angle -30^\circ = 182.2425$$

$$-j110.0755 = 213\angle -31.2^\circ \text{ A}$$

The primary voltage is

$$V_1 = 2440 + j49 + (213\angle -31.2)\times(0.058 + j0.29)$$

$$= 2482.5 + j95.5 = 2483\angle 2.2^\circ \text{ V}$$

The power factor at the terminals of winding 1 is the cosine of the angle between V_1 and I_1, or $\cos 33.4^\circ = 0.832$. The efficiency of the transformer is

$$= \frac{1040\times 480\times 0.866}{(1040)\times 480\times 0.866 + (1040)^2\times 0.002 + (213)^2\times 0.058 + \dfrac{2440^2}{2000}}$$

$$= \frac{433{,}000}{440{,}760}$$

$$= 0.98$$

Example 3.11
Determine the transformer series equivalent impedance (1) in high-voltage terms (winding 1), and (2) in low-voltage terms, for the transformer of Example 3.10.

1. The series equivalent impedance in winding 1 terms is

$$z_{1\text{eq}} = 0.058 + 25 \times 0.002 + j(0.29 + 25 \times 0.012)$$

$$= 0.108 + j0.59 = 0.6 \angle 79.6°.$$

2. In winding 2 terms,

$$z_{2\text{eq}} = \frac{0.058}{25} + 0.002 + j\left(\frac{0.29}{25} + 0.012\right)$$

$$= 0.00432 + j0.0236 = 0.024 \angle 79.6°.$$

Example 3.12
For the transformer of Example 3.10, determine the Thevenin equivalent impedance of the transformer load circuit as seen from the primary terminals.

For this calculation, we shall need the impedance of the load in primary (winding 1) terms. This is found as follows:

$$z_L = \frac{480}{1040 \angle -30°} = 0.461 \angle 30° = 0.392 + j0.226 \text{ ohm}$$

and in primary terms,

$$z_{1L} = 25 \times z_L = 11.3 \angle 30° = 9.8 + j5.65$$

The total impedance of the secondary circuit, including load and winding impedance, in primary terms is,

$$z'_{2t} = 25(0.002 + j0.012) + 9.8 + j5.65 = 9.85 + j5.95$$

$$= 11.5 \angle 31.2°$$

This impedance is in parallel with the two exciting impedances (x_ϕ and r_c):

$$z_a = \frac{1}{1/2000 + 1/j400 + 1/11.5 \angle 31.2°}$$

$$= 11.22 \angle 32.3°$$

$$= 9.5 + j6$$

This equivalent impedance is in series with the primary (winding 1) impedance:

$$z_{TH} = 0.058 + j0.29 + 9.5 + j6 = 9.558 + j6.29 = 11.45 \angle 33.4°$$

Note that the angle of this equivalent input impedance is equal to the angle between the input voltage and current phasors found in Example 3.10.

Tests for obtaining the components of the equivalent circuit and several other transformer parameters will now be described.

3.4 TRANSFORMER TESTS

3.4.1 TRANSFORMER POLARITY

Transformer polarity is the transformer characteristic describing the relative direction of induced voltage and load-current components in the two transformer windings. The relative direction of these parameters in the two windings depends, first of all, upon the inherent property described by the Lenz law relationship given in Section 2.8. The directional characteristics of this relationship are defined by the vector properties discussed following (2.2) and the "hand rules" associated with the vector properties. A second factor also enters into the directional characteristics of the two windings, and that is the relative direction around the core in which the two windings are wound. This can be seen by observing the schematic winding layouts in Figures 3.1 and 3.5. Either winding can be wound in a clockwise or counterclockwise direction around the core. This choice is made on the basis of practical considerations during the transformer assembly process. In the transformer assembly as purchased, there is usually no visual evidence of the relative winding directions, and, therefore, no visual means of determining the relative directions of induced voltage and load current.

In most transformers, there is some form of marking on the transformer, supplied by the manufacturer, to indicate these directional properties. Such markings are known as *polarity markings*. The simplest of these polarity markings is a dot on one terminal of each winding, as illustrated in Figure 3.14. The

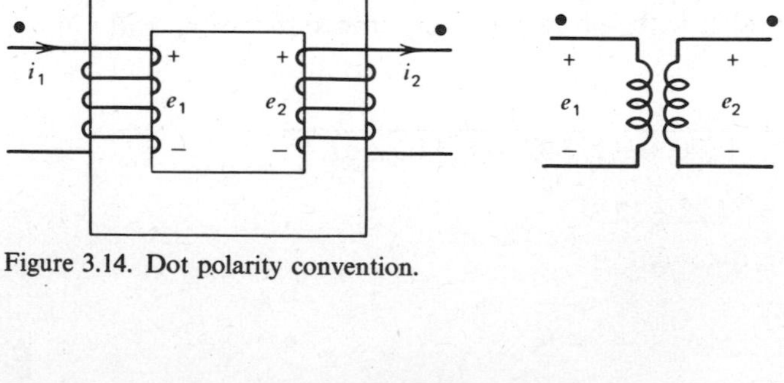

Figure 3.14. Dot polarity convention.

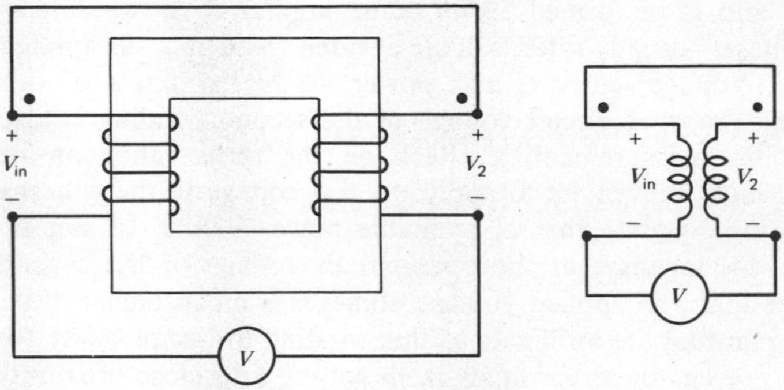

Figure 3.15. Polarity test. As shown, the voltmeter will read ($V_{in} - V_2$) and the polarity will be subtractive.

meaning of the dot-polarity convention is as follows: At the instant of time when the induced voltage in one winding is positive from the dotted terminal to the other terminal, it is positive from the undotted terminal to the dotted terminal of the other winding. Or, in terms of the steady state load current components, with load current flowing into the dotted terminal of one winding, it will be flowing out of the dotted terminal of the other winding. Dot-polarity conventions should be verified by the reader by means of the "hand-rules" or vector notation of (2.2) described in Chapter 2 using the simple transformer schematic diagrams of Figures 3.1 or 3.5.

In large power transformers used in power systems, another type of polarity convention is used. This convention identifies all four terminals of the transformer: high-voltage terminals are identified as H_1 and H_2; low-voltage terminals as X_1 and X_2. The terminals with the subscript, 1, in this convention are equivalent to the dotted terminals in the dot-polarity convention.

When there is a doubt about the transformer polarity, it can be checked by a simple test, requiring only voltage measurements with the transformer unloaded. In this polarity test, rated voltage is applied to *one winding*, usually the winding for which a voltage source is most conveniently available. An electrical connection is made between one terminal from one winding and one from the other. Usually, the terminals in closest physical proximity of each winding are connected. The voltage across the two remaining terminals, one from each winding, is then measured. If this measured voltage is *larger* than the input test voltage, the polarity is said to be *additive*; if smaller, the polarity is *subtractive*. This test can easily be related to the polarity convention, as shown in Figure 3.15.

3.4.2 OPEN-CIRCUIT TEST (NO-LOAD TEST)

The purpose of this test is to obtain the *exciting-branch* impedances in the equivalent circuit, the no-load loss, and the no-load current and power factor.

This is a simple test and is performed as its name implies: One winding is open-circuited and voltage—usually rated voltage at rated frequency—is applied to the other winding. Voltage, current, and power at the terminals of this winding are measured. The open-circuit voltage of the second winding is also measured and from this measurement, a check on the turns ratio can be obtained. It is usually most convenient to apply the test voltage to the winding that has a voltage rating equal to that of available power source. In step-up voltage transformers, this means that the open-circuit voltage of the second winding will be higher than the applied voltage, sometimes much higher. Care must be exercised in guarding the terminals of this winding to insure safety for test personnel and to prevent these terminals from getting into close proximity with other electrical circuits, instrumentation, grounds, and so forth.

In presenting the no-load parameters obtainable from test data, it will be assumed that winding 1 is the winding to which voltage is applied; winding 2 is open-circuited. The no-load power loss is equal to the wattmeter reading in this test; core loss is found by subtracting the ohmic loss in winding 1, which is usually small and may be neglected in some cases:

$$P_c = W_{\mathrm{NL}} - I_{1\mathrm{NL}}^2 r_1 \tag{3.27}$$

The no-load power factor is equal to $W_{\mathrm{NL}}/V_1 I_{1\mathrm{NL}}$; the winding 1 induced voltage is

$$E_{1\mathrm{NL}} = V_1 - I_{1\mathrm{NL}} \angle \Theta_{\mathrm{NL}} (r_1 + jx_1) \tag{3.28}$$

where Θ_{NL} is the no-load power factor angle. The equivalent core-loss resistance, r_c, and magnetizing inductive reactance are found as follows:

$$r_c = \frac{E_{1\mathrm{NL}}^2}{P_c} \tag{3.29}$$

$$I_c = \frac{P_c}{E_{1\mathrm{NL}}} \tag{3.30}$$

$$I_\phi = \sqrt{I_{1\mathrm{NL}}^2 - I_c^2} \tag{3.31}$$

$$x_\phi = \frac{E_{1\mathrm{NL}}}{I_\phi} \tag{3.32}$$

$$a \cong \frac{V_1}{E_{2\mathrm{NL}}} \tag{3.33}$$

Many electronics transformers are excited from nonsinusoidal waveform sources, such as pulses or other irregular waveforms. If possible, it is desirable to perform the no-load test with the input excitation equal to the actual excitation under which the transformer will be operated. This is particularly important in measuring the no-load power loss which is highly dependent upon excitation waveform. Care must be taken to use proper instrumentation when such irregular waveforms are used.

Thermocouple instruments are the most applicable to these measurements since they read in terms of rms values irrespective of waveform. However, the frequency response of the instrumentation must also be noted, since the pulse-type excitation may contain very high frequency components even though the fundamental frequency is relatively low. Power measurements are the most difficult to obtain accurately not only because of the nonsinusoidal waveforms, but also because of the very low power factor associated with the no-load test in most transformers. Power factor may often be in the range of 0.05 to 0.2. Dynamometer wattmeters are generally not applicable in such situations. Electronic multiplier wattmeters are probably the most general means of making no-load transformer power measurements under highly nonsinusoidal, low-power factor conditions,[8] and can often be constructed from readily available laboratory components. For further discussion of such tests, see Reference 5.

3.4.3 SHORT-CIRCUIT TEST

The short-circuit test is well described by its name. One winding is short-circuited across its terminals, and a *reduced* voltage is applied to the other winding. This reduced voltage is of such a magnitude as to cause a specific value of current—usually rated current or a fraction thereof—to flow in the short-circuited winding. Again, the choice of the winding to be short-circuited is somewhat arbitrary and is usually determined by the measuring equipment to be used in the test. However, care must be taken to note which winding is short-circuited, for this determines the reference winding for expressing the impedance components obtained by this test. These components can easily be referred to the other winding by the a^2-relationships derived in previous sections of this chapter. To describe this test, we shall assume that winding 2 is short-circuited and the reduced voltage is applied to winding 1.

The principal purpose of the short-circuit test is to obtain the series impedances. A secondary purpose is to obtain the short-circuit power loss. This is useful in evaluating skin effect in the windings, temperature rise, and per unit equivalent resistance. The short-circuit test is usually performed at rated transformer frequency. In pulse transformers, it may be desirable to perform this test over a range of frequencies related to the frequency components in the pulses at which the transformer is to be operated. The discussion of instrumentation given in the previous section applies also to short-circuit tests.

In the short-circuit test with a very low voltage applied to the unshorted winding, the core-loss and magnetizing current components become very small.

This implies that, under this condition, the core-loss and magnetizing impedances become relatively large. Since these impedances are electrically in parallel with the series impedance of the shorted winding, the magnetizing impedances can usually be neglected in the short-circuit test. The reader should verify this conclusion from the equations and theory developed in earlier sections of this chapter. With this assumption, the short-circuit impedances are found as follows:

$$Z_{1SC}=\frac{V_{1SC}}{I_{1SC}} \tag{3.34}$$

$$r_{1SC}=\frac{P_{1SC}}{I_{1SC}^2} \tag{3.35}$$

$$x_{1SC}=\sqrt{Z_{1SC}^2-r_{1SC}^2} \tag{3.36}$$

These short-circuit impedances are approximately equal to the equivalent impedance given in (3.25) for winding 1 terms. The approximation is due to effects of the exciting branch of the transformer that may or may not be significant. It is usually assumed, for engineering analysis, that the short-circuit impedances are equal to the equivalent impedances. It is also common to assume that the two winding impedances, referred to the same winding, are equal. This permits obtaining the individual winding impedances from the short-circuit test and is valid for many engineering problems. It will be used, unless otherwise stated, for the problems at the end of this chapter. On the basis of these two assumptions,

$$r_1=a^2r_2=\frac{r_{1SC}}{2} \tag{3.37}$$

$$x_1=a^2x_2=\frac{x_{1SC}}{2} \tag{3.38}$$

$$z_1=a^2z_2=\frac{z_{1SC}}{2} \tag{3.39}$$

A more accurate separation of winding 1 and 2 resistances can be obtained by measuring the ohmic resistance of each winding and assuming that the effective resistances divide according to the same ratio as that of the measured ohmic resistances.

3.4.4 OTHER TRANSFORMER TESTS

There are several other tests of importance in electronics transformers described in Reference 5 that may be of interest to the reader.

1. Return loss test is a terminated impedance measurement used to limit reflections on transmission lines to minimum values. This requires a specialized bridge circuit known as a return-loss bridge.

2. The frequency response is of value in applying transformers in electronic circuits operating at variable frequencies. Usually, the frequency response is determined for a system, of which the transformer is one component.

3. The pulse response is a test to determine the dynamic characteristics of a transformer with pulsed excitation. Conventional pulse measuring techniques are applicable.

Example 3.13
No-load measurements on a certain transformer give the following readings: measured on winding 1, 115 V, 60 Hz, 80 W, 3 A; the winding 1 series impedances are known to be $r_1=0.02$, $x_1=0.1$ ohm. Determine r_c and x_ϕ.

The no-load power factor angle is

$$\Theta_{NL}=\cos^{-1}\left(\frac{80}{115\times3}\right)=76.6°$$

The no-load winding 1 induced voltage is

$$E_{1NL}=115-(3\angle-76.6°)(0.02+j0.1)=114.7\angle0°$$

$$I_c=\frac{80}{114.7}=0.7\,\text{A}$$

$$I_\phi=\sqrt{3^2-0.7^2}=2.93\,\text{A}$$

$$r_c=\frac{114.7}{0.7}=164\text{ ohms}$$

$$x_\phi=\frac{114.7}{2.93}=39.2\text{ ohms}$$

It is seen that ignoring the winding 1 voltage drops would make an insignificant change in the values of r_c and x_ϕ.

Example 3.14
A transformer rated at 10 kVA, 7200: 120 V, 60 Hz is short-circuit tested on the high-voltage winding with the following measured values: 220 V, 1.39 A, 200 W, 60 Hz. Determine the series equivalent impedances.

Since the applied voltage for this test is about 3% of rated, the exciting components can be neglected.

$$z_{SC} = \frac{220}{1.39} = 156 \text{ ohm}$$

$$r_{SC} = \frac{200}{1.39^2} = 104 \text{ ohm}$$

$$x_{SC} = \sqrt{156^2 - 104^2} = 116 \text{ ohm}$$

By assuming that the impedances divide equally,

$$r_1 = a^2 r_2 = \frac{104}{2} = 52 \text{ ohm}$$

$$x_1 = a^2 x_2 = \frac{116}{2} = 58 \text{ ohm}$$

$$r_2 = \frac{52}{3600} = 0.144 \text{ ohm}$$

$$x_2 = \frac{58}{3600} = 0.161 \text{ ohm}$$

3.5 TRANSFORMER CAPACITANCE

Capacitance exists among many physical parts of all transformers. The principal elements of capacitance are between turns of the windings, between layers of the windings, between the core and the turns, between turns and the housing—or case if metallic, and between the terminals and external leads. These capacitive elements are distributed throughout the transformer volume in general, but the combined effect of the distributed capacitance can often be treated in terms of lumped capacitances and can be combined with the equivalent circuit previously discussed. The effects of capacitance are significant only in higher-frequency and pulse applications, although the voltage gradients between turns attributable to interturn capacitance must be considered in designing the insulation system for high-voltage transformers at power frequencies.

A lumped circuit containing three capacitances is commonly used to describe the distributed capacitance of a transformer.[3] This circuit is shown in Figure 3.16*a*. The ideal transformer can be eliminated from the equivalent circuit in a manner similar to that discussed with the other equivalent circuit, except for the mutual capacitance term, C_{12}. This variation is shown in Figure 3.16*b*.

These equivalent circuits containing capacitances are suitable for modeling transformers at frequencies where the distributed capacitance is significant.

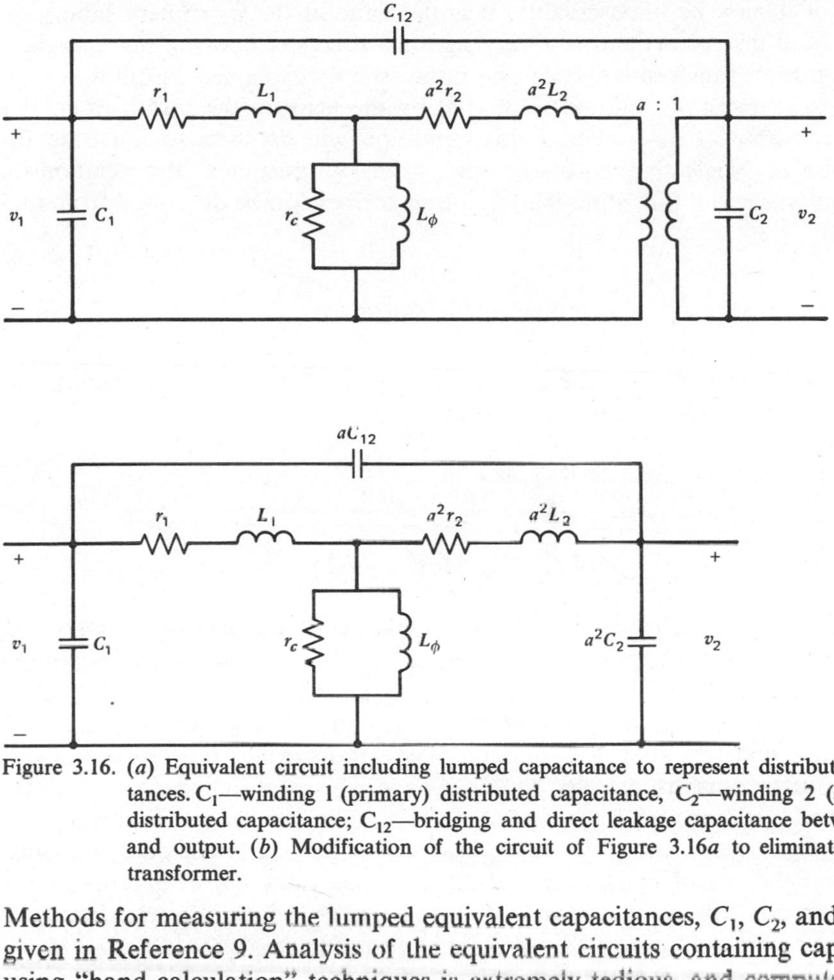

Figure 3.16. (*a*) Equivalent circuit including lumped capacitance to represent distributed capacitances. C_1—winding 1 (primary) distributed capacitance, C_2—winding 2 (secondary) distributed capacitance; C_{12}—bridging and direct leakage capacitance between input and output. (*b*) Modification of the circuit of Figure 3.16*a* to eliminate the ideal transformer.

Methods for measuring the lumped equivalent capacitances, C_1, C_2, and C_{12}, are given in Reference 9. Analysis of the equivalent circuits containing capacitance using "hand-calculation" techniques is extremely tedious, and computer circuit analysis programs are almost a necessity.

3.6 INRUSH CURRENT

Much of this chapter has been devoted to the steady state theory and analysis of transformers. However, there is one aspect of transformer transient characteristics that deserves special mention, and this is the potentially high current that may appear briefly upon initial energizing of a transformer from a sinusoidal voltage source. This initial current peak, termed *inrush current*, is somewhat like the fabled "will-o'-the wisp," since it is a function of the instant at which the transformer is connected to the voltage source which, for practical purposes, is uncontrollable. It may result in a current peak many times the rated transformer current or it may be unobservable. It is the bane of the machinery laboratory student (and instructor) and is a very significant concern during the energizing of large power transformers. It can be rather simply explained as follows.

Inrush current is modified somewhat by the load on the transformer. It is most observable at no-load and this condition will be used to illustrate this phenomenon. Neglecting core loss and primary resistance, the relationship between voltage and flux at no-load in a transformer can be described from (3.3) and (3.4):

$$v_1 = N\frac{d\phi}{dt} = V_{1m}\sin(\omega t + \alpha) \tag{3.40}$$

where α = the argument of the sine term at $t=0$. The solution to this linear, first-order differential equation can be found to be

$$\begin{aligned} \phi &= \frac{V_{1m}}{\omega N}\left[\cos\alpha - \cos(\omega t + \alpha)\right] \\ &= \Phi_m\left[\cos\alpha - \cos(\omega t + \alpha)\right] \end{aligned} \tag{3.41}$$

The flux is seen to consist of two components, the transient and steady state. Since we have neglected core loss and primary resistance, the transient term, $\Phi_m \cos\alpha$, contains no decrement factor; in any practical transformer, of course, there is such a term and the transient component decays exponentially with a time constant that is a function of the transformer resistance and core loss. In small transformers, this time constant is usually in the order of a few milliseconds; in large power transformers, it may be one second or more.

Keeping in mind the fact that the transient component does decay, we look at the implications of (3.41). First of all, it is seen that the magnitude of the transient component is a function of the instant at which the transformer is connected to the voltage source, which is given by the angle, α, in (3.41). If α is $\pi/2$ (which means that the source voltage at $t=0$ is at its positive maximum), the transient term is zero and there is no inrush current. If α is zero, the transient term is maximum and a severe current peak can occur during the first few cycles after the transformer is energized. This is the worst case and is illustrated in Figure 3.17. It is seen that the two flux components are additive after one-fourth of a period and, at $\omega t = \pi$, the flux in the transformer core would be $2\Phi_m$, neglecting decay in the transient term. Actually, this first peak of flux may be more than twice the normal flux maximum, since there is often residual magnetism in the core when initially energized. If the residual flux is in the proper direction, it will add to the flux calculated from (3.41).

The current peak associated with this flux peak is proportionately much larger due to the effect of core saturation. In most transformers, the core and windings are designed so that, under rated voltage, the maximum steady state

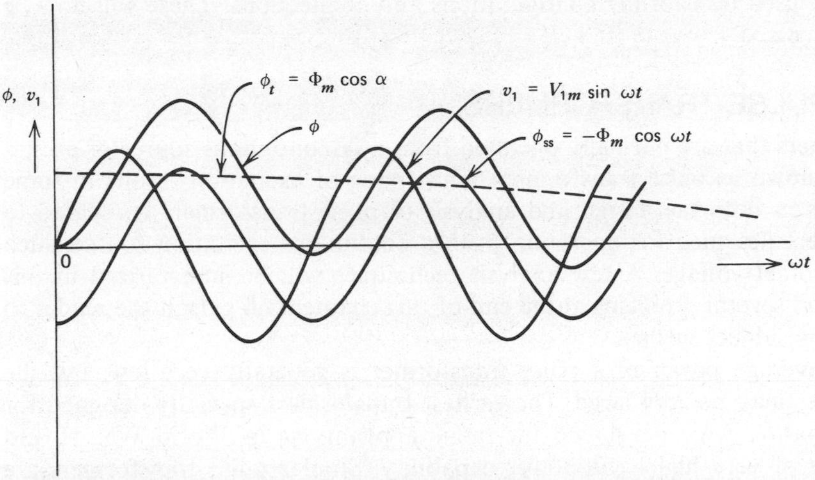

Figure 3.17. Flux relationships following initial energizing of a transformer switched on at $\alpha=0$, Equation 3.41.

flux is around the knee of the saturation curve. Therefore, the condition described above during the initial energizing of a transformer, which requires a flux of twice or more times the maximum steady state flux, may demand an exciting current many hundreds of times the normal excitation current because of core saturation. This possibility must be considered in choosing the source for the circuit and for protection in any circuit involving a transformer. Reference 2 gives a convenient approximation for relating the first inrush current peak (in amperes) to transformer design parameters:

$$I_{pK}=\frac{bA_c(B_r+2B_m-1.95)}{\mu_0 A_s N} \tag{3.42}$$

where b = coil length along coil axis, m
A_s = total area of the space enclosed by the mean turn of the excited coil, m^2
A_c = net cross-sectional area of the magnetic core, m^2
N = number of turns of the excited coil winding
B_r = residual flux density at $t=0, T$
B_m = design maximum, steady state flux density, T

3.7 SOME PARTICULAR TRANSFORMER TYPES AND CONNECTIONS

The basic transformer theory and testing as presented in the previous section generally applies to all types of transformers. However, there are many additional concepts of theory and practice that are essential in the analysis of many commonly used transformer configurations and connections. These will now be briefly discussed.

3.7.1 PULSE TRANSFORMERS

Transformers that are normally operated from a discontinuous source of excitation are known as pulse transformers. This mode of excitation results in some differences in both the design and analysis of pulse transformers compared to transformers designed for operation from a continuous excitation source, such as a sinusoidal voltage. A few analysis techniques will be summarized in this section, and several problems at the end of this chapter will permit the reader to explore this subject further.

The average power of a pulse transformer is generally very low, but the peak power may be very large. The earliest transformers specially designed for pulse excitation were developed for radar applications in World War II and these were of very high peak-power capability. Similar pulse transformers are used in linear accelerators and similar equipment. Most recently, medium and low peak-power pulse transformers have been developed for many electronic and control applications, such as in the gate firing circuits of power thyristors. Pulse excitation typically consists of a rectangular voltage pulse or a short burst of high-frequency sine pulses, but many other pulse shapes are commonly used. Both the magnitude and the repetition rate of the pulse determine the average transformer power. A few pulse waveforms are described in connection with electronic motor control in Chapter 8.

When the excitation pulses are repetitive with a steady pulse rate, it is possible to resolve this mode of excitation into continuous exponential functions by means of Fourier series methods. It might appear that the theory presented earlier in this chapter, which is based upon sinusoidal excitation, could then be used in the analysis of the pulse transformer. This is generally not feasible due to the transformer nonlinearities resulting from saturation of the magnetic core. Also, the use of Fourier series methods gives less insight into the physical operation of the transformer.

Rather than use Fourier methods, pulse transformer analysis is generally based upon solving the differential equations in a piecewise manner using the equivalent circuits of Figure 3.16 and modifications of these circuits. Pulse transformer parameters are defined in terms of a standard pulse waveform, shown in Figure 3.18. This waveform can represent either a voltage or current waveform resulting from a rectangular pulse excitation, and therefore is described by a generalized symbol, A. The principal pulse parameters related to this wave are also shown in Figure 3.18. It is seen that these parameter definitions follow those used in circuit analysis and control theory. The actual shape and magnitude of the output voltage or current pulses resulting from excitation from a rectangular pulse in a specific transformer can be found by analysis of the complete equivalent circuit of Figure 3.16.

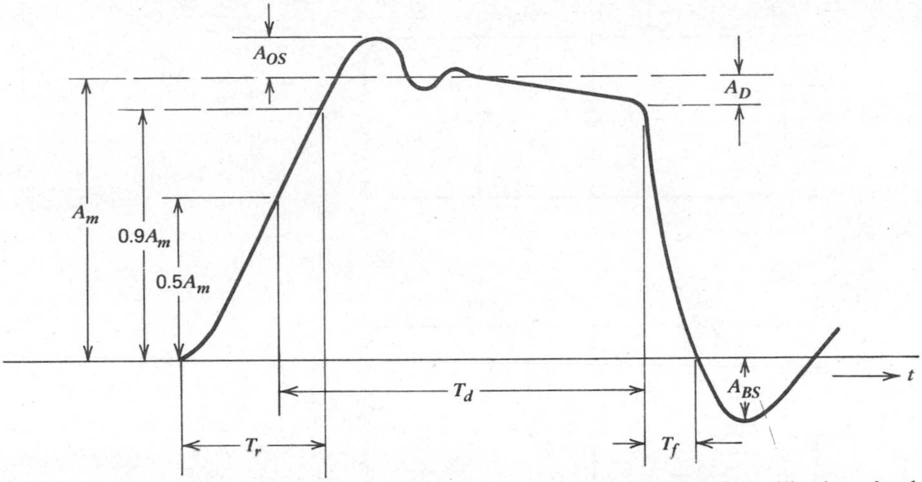

Figure 3.18. Standard voltage or current pulse shape used in the analysis and specification of pulse transformers. A_m = pulse amplitude, A_{os} = overshoot, A_D = droop or tilt, A_{BS} = back-swing, T_d = pulse duration, T_r = pulse rise time, and T_f = pulse fall time.

As has been noted, such analysis requires computer simulation techniques and is tedious and time-consuming. It has been observed, however, that considerable simplification of the complete equivalent circuit is possible during various portions of the pulse. These simplifications are shown in Figures 3.19*a*, *b*, and *c*, with all parameters assumed to be expressed in secondary (output winding) terms.[9, 10] All of these parameters have been defined previously except for C_D, which is the distributed capacitance of the secondary winding, and C_L, which is the load and terminating cable capacitance.

Both the average and peak volt-ampere and power ratings must be considered in the design and use of pulse transformers, and these may be several orders of magnitude apart. Another rating of significance in pulse transformers is the *voltage-time product* rating. This rating is defined[5] as the maximum voltage-time integral of a rectangular voltage pulse that can be applied to a winding before core saturation effects cause the resultant exciting current pulse waveform to deviate from a linear ramp by a given percentage. The definition of the permeability of the magnetic core is also modified for pulse application to be "the value of amplitude permeability when the rate of change flux density is held substantially constant over a period of time during each cycle,"[11] or

$$\mu_p = \frac{1}{\mu_0}\frac{\Delta B}{\Delta H} \tag{3.43}$$

where ΔB = change in flux density during the stated time interval
ΔH = associated change in magnetic field strength

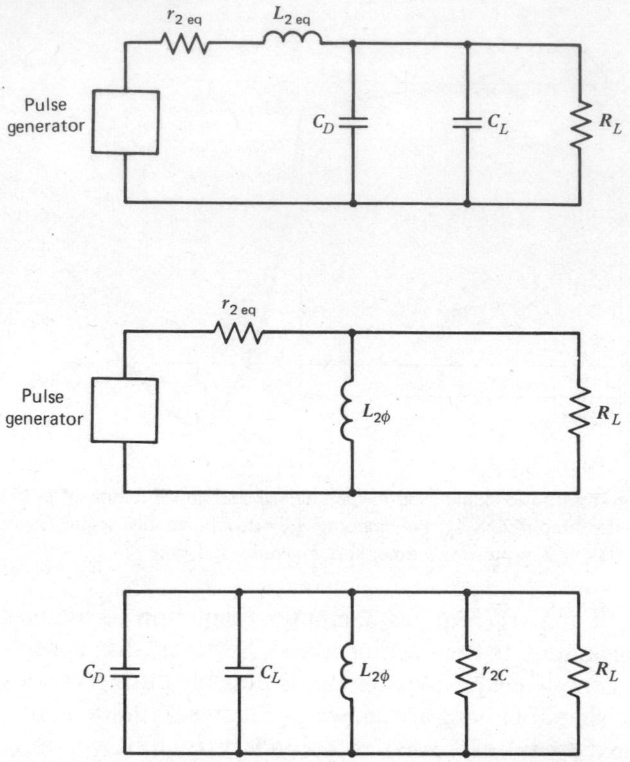

Figure 3.19. (*a*) Simplified equivalent circuit for evaluating the leading edge of a transformer pulse (rise time) with a resistive load. (*b*) Simplified equivalent circuit for evaluating the top of a transformer pulse. (*c*) Simplified equivalent circuit for evaluating the trailing edge (fall time and backswing) of a transformer pulses.

3.7.2 POLYPHASE CONNECTIONS

In many power, rectifier, and motor control applications, polyphase circuits are used to reduce the losses in conductors and machines. The most common polyphase system is the three-phase system, but 2-phase, 6-phase, and even 12-phase systems are used in various applications. Most power generation, transmission and distribution systems, and many ac motor control systems are three-phase systems. In balanced three-phase systems, the voltage and current phasors in the three phases are displaced from each other by an electrical angle of 120°.

When voltage or current transformations are required in polyphase systems, polyphase transformer connections must be used. Three-phase transformers may consist of three individual transformers connected appropriately or a single core on which three sets of windings are placed in the proper manner. The latter scheme, known as a three-phase transformer, results in some reduction in the size and weight of the core and case as compared to three individual transformers of the same total rating. In oil-cooled transformers, there is also some reduction in the amount of oil required. The analysis of the two types of transformer systems, however, is approximately the same except for the magnetic circuit analysis of the cores and thermal analysis of the overall systems.

The analysis of polyphase transformers is essentially identical with analysis of single-phase transformers, which has been described in the previous sections of this chapter. Some of the principal concerns in the application of polyphase transformers are the harmonic components of exciting current, the effect of unbalanced loading of the phases, and the effects of dc components in the transformers. Since transformers used in power transmission and distribution applications are often of very large volt-ampere rating, their exciting current magnitudes can be relatively large. The higher-frequency harmonics of these currents can result in electromagnetic interference with adjacent telephone lines, undue heating in certain portions of the polyphase circuits, and unbalanced voltages in some transformer connections.

Another concern in the application of three-phase transformer connection is to have a reasonable grasp of the power, voltage, and current relationships in three-phase systems, which has become almost a lost art to the modern-day electronics engineer. Therefore, as an aid to those readers who are caught in the glamor of electronics, computers, and integrated circuits but who still have to occasionally explain why the fuses are constantly blowing on the plant's air conditioner or why an induction motor will not start, a brief summary of three-phase transformer voltage, current, and power relationships will be given.

The two most common three-phase transformer connections are known as *delta* and *wye*, so named because of the phasor diagrams representing these connections. These connections are shown schematically in Figure 3.20*a*. Note that the transformer windings are oriented to show the phasor relationships. In general, the primary or secondary windings of a three-phase transformer can be connected in either delta or wye. The choice depends upon the requirements of the particular application, the need for a neutral wire (as in power distribution

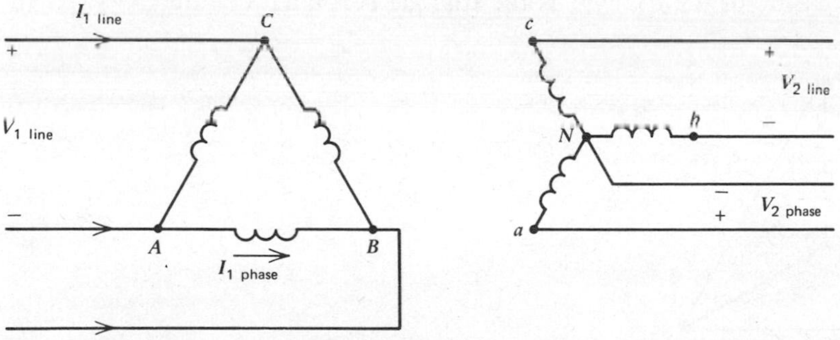

Figure 3.20a. Delta-wye transformer connection.

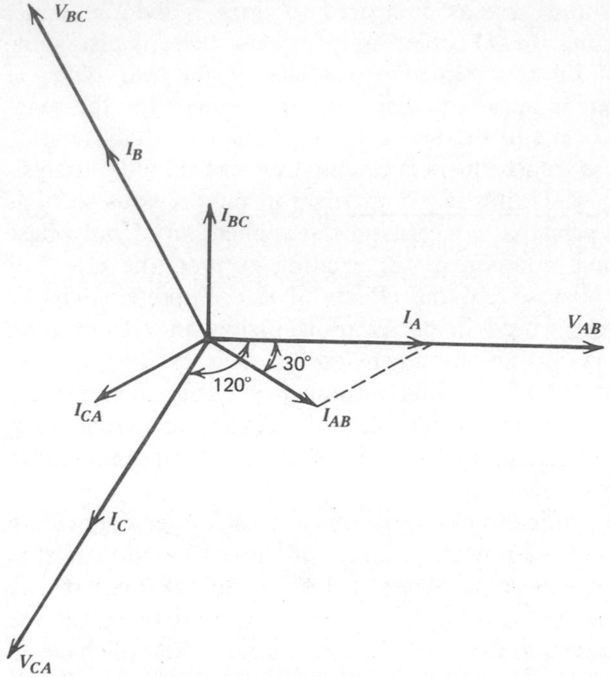

Figure 3.20b. Phasor Diagram for Delta Winding of Figure 3.20a (See Example 3.15)

systems), the need for harmonic suppression, and grounding considerations. In power systems, delta/wye connections are the most common.

The phasor relationships in any balanced three-phase transformer can *all* be described by the symmetrical diagram shown in Figure 3.21. The outer sides of this diagram are the sides of an equilateral triangle, and N represents the geometric center of this triangle. The physical connections of a delta/wye transformer are shown in Figure 3.22. The lettering of the triangle in Figure 3.21 and the connection diagram of Figure 3.22 is purely arbitrary except for the matter of *phase sequence*. Phase sequence is the order in which the letters pass by an observer outside of the triangle as the triangle is rotated counterclockwise in

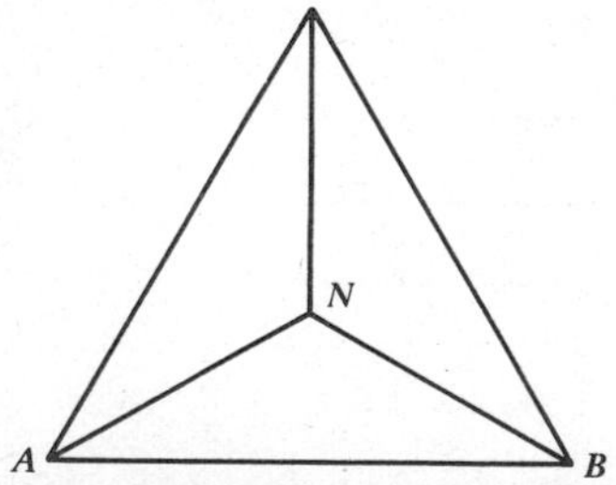

Figure 3.21. Diagram for analyzing three-phase transformers.

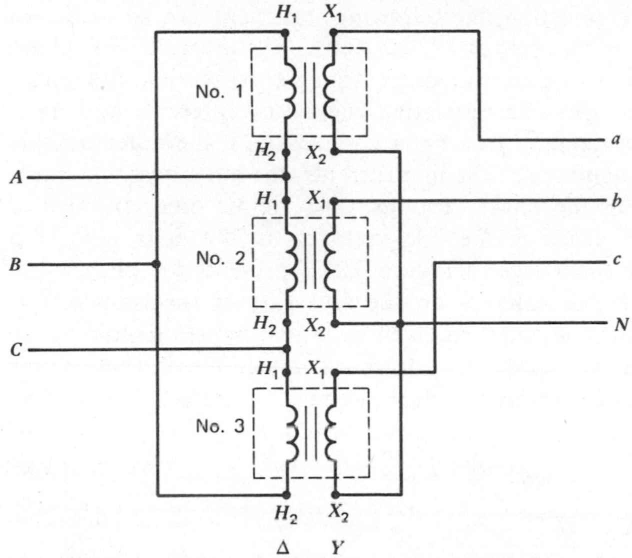

Figure 3.22. Connections for three identical transformers connected in delta-wye.

Figure 3.21. The phase sequence in Figure 3.21 is A-C-B. Phase sequence is of no importance in the analysis of balanced three-phase transformers, but it does determine the direction of rotation of polyphase motors that may be connected to the secondary of the transformer. Since we are studying the transformer in this chapter, we shall proceed by ignoring the phase sequence.

In polyphase systems, the voltages between the lines coming into or out of the transformer are called the *line voltages*; the currents in these lines are called *line currents*. The voltages across the transformer windings are called *phase voltages* and the currents in these windings are called *phase currents*. In delta connection, it is seen that the phase and line voltages are one and the same. In a wye connection, it is seen that the phase and line currents are the same. In a wye connection, the phase currents and voltages are often also called *line-to-neutral* currents and voltages.

Returning now to Figure 3.21, the line segments in this diagram can be used to represent the phase relationships of both voltages and currents in either delta or wye connections. To make sure what is meant by phase relationships, in Figure 3.21, line A-to-B represents a phasor at 0°; line B-to-C represents a phasor at 120°; line N-to-A represents a phasor at 210°; line C-to-N represents a phasor at $-90°$; and so forth.

The final basic concept needed for analysis of three-phase connections requires the Kirchoff law relationships which state that whether line or phase,

$$\Sigma\,(\text{voltages}) = 0 \tag{3.44}$$

$$\Sigma\,(\text{currents}) = 0 \tag{3.45}$$

In a wye transformer connection, the following statement can be made: In Figure 3.21, if the wye in the center of the triangle represents the phase relationships of the phase voltages, the outer triangle represents the phase relationships of the line voltage. The respective lines also represent the *magnitudes* of the phase and line voltages. In a delta connection, if the outer triangle represents the phase and magnitude of the line currents, the inner wye represents the phase and magnitude of the phase currents. Also, if the outer triangle of Figure 3.21 represents the phase of the line voltages on the delta side of a delta/wye transformer, the inner wye of Figure 3.21 represents the phase—but not the magnitude—of the line voltages on the wye side of the transformer. There are several more games that can be played with this simple diagram which are useful in the analysis of three-phase transformer connections. But let us now look at magnitude relations in three-phase transformers. In a delta connection,

$$I_{\text{line}} = \sqrt{3}\ I_{\text{phase}} \tag{3.46}$$

In a wye connection,

$$V_{\text{line}} = \sqrt{3}\ V_{\text{phase}} \tag{3.47}$$

The power in three-phase connections is

$$\begin{aligned} P &= 3(V_{\text{phase}})(I_{\text{phase}})(\cos\Theta_{\text{phase}}) \\ &= \sqrt{3}\ (V_{\text{line}})(I_{\text{line}})(\cos\Theta_{\text{phase}}) \end{aligned} \tag{3.48}$$

Note that the only meaningful power factor angle in polyphase systems is that between *phase components*.

Example 3.15

The high-voltage windings of three 100 kVA, 19,000 V transformers are connected in delta. Determine the magnitude and phase of the phase and line voltages and currents of this connection when the phase windings are carrying rated current at a power factor of 0.866 lagging.

The winding connection can be assumed to be that of the delta in Figure 3.20*a*. The rated winding current is 100,000:19,000 = 5.26 A. The rated line current is $\sqrt{3}\ \times 5.26 = 9.1$ A. To develop the phasor diagram, we shall arbitrarily assume that the three-phase winding voltages are the consistent set taken from Figure 3.21, $V_{AB}\angle 0°$, $V_{BC}\angle 120°$, and $V_{CA}\angle 240°$. The phase currents are located with respect to these phase voltages by the power factor angle, which in this example is $\cos^{-1} 0.8666 = 30°$. To obtain the phase of the line currents, we have to assume a direction of current flow in the diagram of Figure 3.20*a*. A consistent set of line currents is either *into or out of the delta*. Any other assumption on current flow (e.g., one current into and the other two out of the delta terminals) will not give the three-phase phasors. We shall arbitrarily assume that current is into the delta terminals on all three lines. The line currents are obtained by applying Kirchoff's current law at the corners of the delta. For example, $I_A = I_{AB} - I_{CA} = 5.3\angle -30° - 5.3\angle 210° = 9.1\angle 0°$. The phasor diagrams are shown in Figure 3.20*b*.

Note that all of the above relationships apply only to *balanced* three-phase systems, that is, systems in which all phase and line voltages and currents are exactly equal in magnitude and separated from each other in phase by exactly 120°. In such systems, the neutral wire plays a very minor role, serving as the ground connection and as a good roosting place for birds. Unbalance of three-phase systems is caused by unbalanced loading of the individual phases and by the effects of the harmonics of transformer exciting currents. The two connections react differently to these unbalances.

In the delta connection, unbalance shows up as circulating current around the delta. This causes additional ohmic losses in the transformer windings and may cause excessive heating of the transformers. However, this is generally considered one of the advantages of the delta winding, in that it "traps" harmonic currents and tends to maintain equal phase and line voltages on the wye side of the connection. Both types of unbalances (unbalanced loads and harmonics) tend to cause neutral wire current over on the wye side of the connection. This current is minimized if the other side of the connection is a delta connection. If there is no neutral wire connection, the voltage of the neutral tends to stray from the geometric center of the triangle in Figure 3.21. This results in abnormally high voltage across at least one of the transformers, or sections of a three-phase transformer, and is an undesirable condition. Ungrounded wye connections are rarely used today for this reason.

The above discussion demands an explanation of the phase relationships of harmonics in three-phase connections. It can be shown[7, 12] that the phase relationships of harmonics in polyphase transformers vary as a function of the phase order. For example, in a three-phase transformer, the third harmonic of the phase displaced from the reference phase by 120° is displaced by $3 \times 120° = 360°$. Also, the third harmonic in the phase displaced from the reference phase by 240° is displaced by $3 \times 240° = 720°$. Therefore, in three-phase transformers, the third harmonics in the three transformers are *in phase*. This also applies to all odd multiples of the third harmonic.

In a delta connection, third (and odd-multiple) harmonics of current add to each other and form circulating currents around the delta, which is why this connection is called a "trap" for harmonics. Similarly in an ungrounded wye, the third harmonics of voltage add to each other and tend to push the neutral in Figure 3.21 far off of geometric neutral. This is a simplistic explanation of why the delta connection is useful in the suppression of harmonics and why the ungrounded wye is undesirable. It probably should have been preceded by a statement that the exciting current of polyphase transformer can contain *only odd harmonics*, which goes back to the fundamentals of Fourier series theory.[12, 13]

This conclusion is left for the reader to ponder with the help of References 2, 7, 12, and 13.

A final word on three-phase connections concerns their symmetry. First of all, it should be noted from the above discussions that the delta and wye are *dual* connections, in the sense that voltage and current relationships in the two systems are the inverse of each other. Discussion of symmetry of three-phase connections introduces some of the more esthetic characteristics of many engineering systems. The geometric symmetry of three-phase systems has been described by many as "beautiful," and this insight should not be lost by the student or practicing engineer as he struggles to grasp the mathematical and physical meaning of these systems. The three-phase power systems are the basis for the industrial economy of most of the world and, as energy conservation becomes more requisite, the optimization of these systems will become even more significant. Therefore, although this phase of transformer and power engineering matured many years ago, its significance and possibilities for future improvement should not be ignored.

Several other polyphase connections are of some importance. The *open-delta* connection, shown in Figure 3.23, as the name implies, is a delta connection with one transformer missing. Since, in a three-phase system, any one phase voltage is the phasor sum of the other two (3.44), the voltage across the open side of the delta is the third-phase voltage. This permits a three-phase transformation with only two transformers. The open-delta connection is an unbalanced connection, and the two transformer windings will operate at different power factors even when connected to a balanced source and balanced load impedances. Therefore, the permissible rating of the open-delta connection is only $\sqrt{3}$ times the volt-ampere rating of one transformer.

The *Scott* connection, shown in Figure 3.24, is a means of converting three-phase to two-phase, or vice versa. It requires an 0.866 tap on one transformer and a midwinding tap on the other. Like the open delta, it is an unbalanced connection. The *six-phase star* connection is shown in Figure 3.25. It is a means of converting three-phase to six-phase and is used in many rectifier

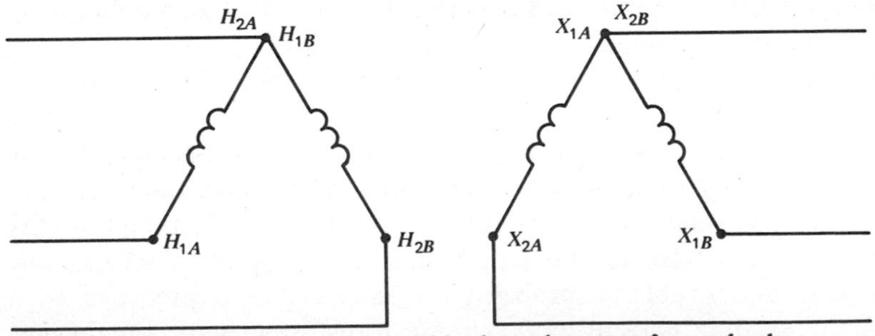

Figure 3.23. Open-delta connection used in three-phase transformer banks.

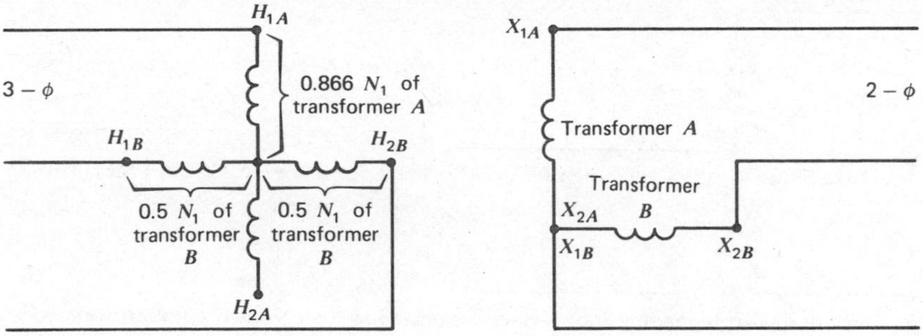

Figure 3.24. Scott connection for three-phase to two-phase transformations.

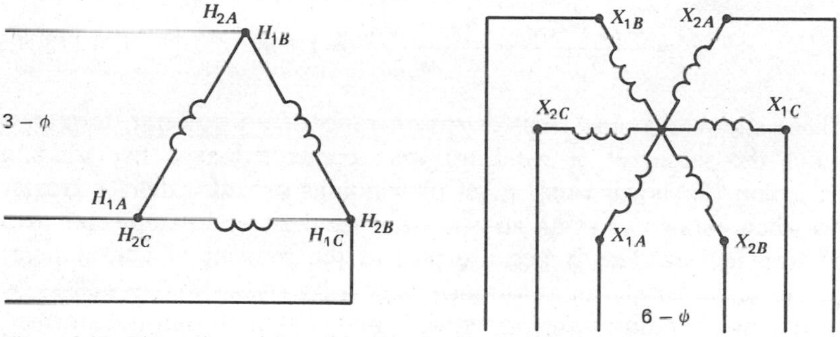

Figure 3.25. Three-phase delta to six-phase star connection.

and thyristor circuits where a path for dc current flow is required. The characteristic angle of a six-phase system is 60°.

3.7.3 AUTOTRANSFORMERS

The single-winding or autotransformer is a very useful device for some applications because of its simplicity and relatively low cost compared to multiwinding transformers. However, it does not provide electrical isolation and therefore cannot be used where this feature is required. The autotransformer circuit is shown in Figure 3.26. The autotransformer circuit can be developed from a two-winding transformer by connecting the two windings electrically in series so that the polarities are additive. Assume that this has been done in the circuit of Figure 3.26, where the primary of the two-winding transformer is winding *A-B* and the secondary of the two-winding transformer, is winding *B-C*. The primary of the autotransformer is now the sum of these two windings, namely *A-C*. The autotransformer secondary is winding *B-C*. The *two-winding* voltage and turns ratio is

$$a = \frac{N_{AB}}{N_{BC}} = \frac{E_{AB}}{E_{BC}} \tag{3.49}$$

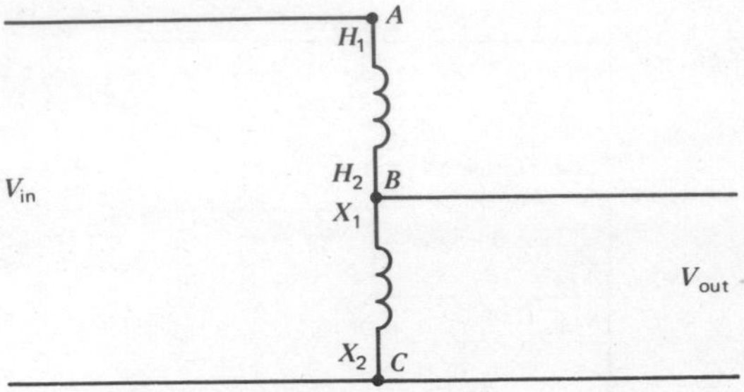

Figure 3.26. Conventional two-winding transformer connected as a step-down autotransformer.

The autotransformer voltage and turns ratio is

$$a' = \frac{E_{AB} + E_{BC}}{E_{BC}} = \frac{N_{AB} + N_{BC}}{N_{BC}} = 1 + a \tag{3.50}$$

Equation (3.50) shows that the transformation ratio of an autotransformer is greater than if the same set of windings were connected as a two-winding transformer. It can be shown[12] that a set of windings can also deliver greater volt-amperes when connected as an autotransformer than when connected as a two-winding transformer. This is because part of the transfer of volt-amperes from primary to secondary in an autotransformer is by *conduction* as well as by *induction* as in a two-winding transformer. A common type of autotransformer, found in most laboratories, is the variable-ratio autotransformer in which the point, B, in Figure 3.26 is movable.

3.7.4 CURRENT TRANSFORMERS

A current transformer is a two-winding transformer used in instrumentation applications to measure or sense current. The primary is connected in series with a source or load, which can be considered as a constant-current mode of excitation. The current transformer is essentially excited from a constant-current source. The current transformer is operated essentially in a dual or inverse manner to the voltage transformer: The "no-load" condition of a current transformer occurs when the secondary is shorted on itself. An open secondary can result in dangerously high voltages.

The impedance or load on the secondary is known as the transformer "burden," since it is an undesirable, although necessary element, and is always minimized. The purpose in the design and application of current transformers is accuracy of both phase and magnitude of the current ratio, rather than any load-carrying capabilities. It is seen from the phasor diagrams of Figure 3.13 that the source of current ratio inaccuracy is the excitation current and, therefore, this component should be minimized. Current transformer ratios are almost invariably current step-down ratios, and often the primary is not an integral part of the transformer itself but is part of the wire whose current is being measured.

REFERENCES

1. D. G. Fink and J. M. Carroll, *Standard Handbook for Electrical Engineers*, 10th ed., McGraw-Hill Book Company, New York, 1968.
2. L. F. Blume, et al., *Transformer Engineering*, John Wiley & Sons, Inc., New York, 1951.
3. N. R. Grossner, *Transformers for Electronic Circuits*, McGraw-Hill Book Company, New York, 1967.
4. O. Kiltie, "Design Shortcuts and Procedures for Electronics Power Transformers and Inductors," Harris Publishing Company, Cleveland, Ohio, 1975.
5. "Tests for Electronics Transformers and Inductors," *IEEE Standard* No. 389-1978, New York, 1978.
6. *Electrical Transmission and Distribution Reference Book*, 4th. ed., Westinghouse Electric Corporation, East Pittsburgh, Pa., 1950.
7. E. Clarke, *Circuit Analysis of A-C Power Systems*, Vol. 1, John Wiley & Sons, Inc., New York, 1943.
8. D. H. Hamburg and L. E. Unnewehr, "An Electronic Wattmeter for Nonsinusoidal Low Power Factor Power Measurements," *IEEE Transactions on Magnetics*, Vol. Mag-7, New York, September 1971, pp. 438-442.
9. "Low-Power Pulse Transformers," *IEEE Standard*, No. 390-1975, New York, 1975.
10. H. W. Lord, "Pulse Transformers," *IEEE Transactions on Magnetics*, Vol. Mag-7, No. 1, New York, March 1971 pp. 17-28.
11. "Test Procedures for Magnetic Cores," *IEEE Standard*, No. 393-1977. 1977.
12. G. V. Mueller, *A.C. Machinery*, McGraw-Hill Book Company, New York, 1948.
13. P. Franklin, *Fourier Methods*, McGraw-Hill Book Company, New York, 1949.

PROBLEMS

3.1. A symmetrical square wave of voltage is applied to the primary winding of a transformer with the secondary winding open (no load). Determine and sketch the waveform of the magnetic flux in the core. Neglect primary resistance and core losses.

3.2. Find the relationship among the rms voltage applied to the primary, the maximum flux in the core, and the square-wave frequency for the transformer of Problem 3.1. This equation will be analogous to the relationship for sinusoidal excitation (3.11).

3.3. Repeat Problems 3.1 and 3.2 for an applied voltage whose waveform is a symmetrical triangular wave.

3.4. A transformer is to be assembled using a magnetic core constructed of M-19 transformer steel laminations (see Figure 2.4). The *net* cross-sectional area of the core is 10 cm^2. The transformer will be excited from a sinusoidal voltage source of 100 V (rms) at 400 Hz. Determine the number of primary winding turns required so that the maximum core flux density at rated applied voltage is at the knee of the saturation curve for M-19 steel (1.2 T). Neglect winding resistance and core loss.

3.5. The mean magnetic length of the core of Problem 3.4 is 40 cm. Determine the rms magnetizing current with rated voltage applied, using the turns calculated in Problem 3.4.

3.6. The alternating flux in the core of a transformer increases as the square of time from zero at the beginning of the flux period to 2.4×10^{-3} Wb at $t = 10$ ms. Then it decreases to zero at $t = 20$ ms along a curve that is symmetrical to the rising flux curve about the one-fourth period time. The negative half-period is symmetrical to the positive half-period about the time axis.

(a) Sketch the curve showing the voltage applied to a 500-turn winding that will result in this flux characteristic.

(b) Determine the maximum, rms, half-period average, and form factor of this applied voltage wave.

3.7. The transformer of Example 3.1 is excited from a current source (sine-current excitation). The rms value of the exciting current is 2.0 A. Using a graphical procedure similar to that used in Example 3.1, determine the waveform over a half-period of the flux in the core. Sketch the approximate half-period waveform of the primary induced voltage.

3.8. For the transformer of Problem 3.4, estimate the peak inrush current from (3.42). Use the turns calculated in Problem 3.4 and assume a coil length of 10 cm, a mean turn area (A_s) of 12 cm^2, and a residual density (B_r) of 0.3 T.

3.9. The nonsinusoidal voltage, $v = 120\sin 377t - 60\sin 1885t$, is applied to the 200-turn winding of a transformer. What is the equation of the time variation of the flux in the core? Determine the maximum value of the flux. Sketch the voltage and flux waveforms.

3.10. A transformer is rated 100 kVA, 11,000:2200 V, and 60 Hz. The no-load test on the low-voltage winding gives 2200 V, 2 A, 100 W, and 60 Hz. If this test were performed on the high-voltage winding at rated voltage, determine the current and power that would be measured.

3.11. The voltage, $v = 100 \sin 377t$, is applied to a transformer winding in a no-load test. The resulting current is found to be $i = 5\sin(377t - 60°) + 2\sin(1131t - 120°)$ A. Determine the core loss and the rms value of the exciting current.

3.12. A no-load test on a certain transformer gives the following data: 120 V, 2.3 A, 75 W, and 60 Hz. Neglecting the winding resistance and leakage reactance, determine the value of the magnetizing reactance, x_ϕ, and the core-loss equivalent resistance, r_c, and the no-load power factor.

3.13. If the winding resistance and reactance of the transformer of Problems 3.12 are 0.4 ohm and 1.5 ohm, respectively, recalculate x_ϕ and r_c including the effects of these winding impedances.

3.14. A transformer rated 220:440 V, 25 kVA, and 60 Hz is tested as follows: A no-load test, performed on the 220-V winding, gives 220 V, 10 A, 700 W, and 60 Hz; the short-circuit test, performed on the 440-V winding, gives 37 V, rated current, 1000 W, and 60 Hz. Determine the impedances of the complete equivalent circuit (Figure 3.9) in terms of the 440-V winding. State what assumptions were made in obtaining these impedances.

3.15. Express the complete equivalent circuit of Problem 3.14 in terms of the 220-V winding.

3.16. Voltage regulation of a transformer is defined as the secondary (or load voltage) at no-load, minus the secondary voltage at a given load condition, divided by the load voltage. In percent this becomes

$$\text{voltage regulation} = 100\frac{V_2\,(\text{no-load}) - V_2\,(\text{load})}{V_2\,(\text{load})}$$

This calculation is usually made neglecting the exciting components, that is, using the approximate equivalent circuit of Figure 3.12. For the transformer of Problem 3.14, determine the voltage regulation for the following:

(a) Rated load on the 440-V winding at a power factor of 0.85 lagging.
(b) Rated load on the 440-V winding at a power factor of 0.85 leading.
(c) 1/2 rated load on the 440-V winding at a power factor of 0.85 lagging.

3.17. A certain transformer is rated 1000 kVA, 11,000:2200 V, and 60 Hz. The short-circuit test on the 11,000-V winding gives 1000 V, rated current, and 9 kW. Determine the equivalent series resistance, reactance, and impedance in terms of both windings.

3.18. A transformer rated 500 kVA, 2400 : 120 V, and 60 Hz, has a no-load loss (at rated voltage) of 1600 W, and a short-circuit loss (at rated current) of 7500 W. Determine the efficiency of this transformer under the following conditions of load:
(a) Rated current at 0.8 power factor lagging.
(b) 300 kW at 0.8 power factor leading.
(c) 100 kW at 0.8 power factor lagging.

3.19. Three identical transformers are each rated 200 kVA, 13,200 : 2300 V, and 60 Hz. The high-voltage windings are connected in delta and the low-voltage windings in wye. Determine the rated voltages and currents of the lines and phase windings on both sides of this polyphase connection.

3.20. A balanced three-phase load of 300 kVA at 460 V is to be supplied from a 2300-V, three-phase system by a delta bank of transformers. Specify the current and voltage and kVA rating of the windings of each transformer.

3.21. Two transformers, of the type described in Problem 3.19, are connected in open-delta on both primary and secondary.
(a) Determine the load kVA that can be supplied from this transformer connection.
(b) A delta-connected three-phase load of 300 kVA, 0.866 pf, 2300 V is connected to low voltage terminals of this open-delta transformer. Determine the transformer currents on the 13,200-V side of this connection.

3.22. The transformer of Problem 3.14 is connected as an auto-transformer to transform 660 V to 220 V.
(a) Determine the autotransformer ratio, a'.
(b) Determine the volt-ampere rating of the autotransformer.
(c) With a load of 25 kVA, 0.866 pf lagging connected to the 220 V terminals, determine the currents in the load and the two transformer windings.

3.23. A load of 12 kV at 0.7 pf lagging is to be supplied at 110 V from a 120 V supply by an autotransformer. Specify the voltage and current ratings of each section of the autotransformer.

3.24. A transformer with additive polarity is rated 15 kVA, 2300 : 115 V, 60 Hz. Under rated conditions, the transformer has an excitation loss of 75 W and a short-circuit loss of 250 W. The transformer is to be connected as an autotransformer to transform 2300 V to 2415 V. With a load pf of 0.8 lagging, what volt-ampere load can be supplied without exceeding the current rating of any winding? Determine the efficiency at this load.

Chapter 4
Electromechanical Systems

In our discussions up to this point we have not included any mechanical motion. In earlier chapters we pointed out that the process of electromechanical energy conversion necessarily involves motion—both mechanical and electrical. By mechanical motion we mean variations of position and velocity of one circuit —electric, magnetic, or electromagnetic—with respect to the other. Electrical motion implies changes in current, voltage, and flux linkage in the circuit. Often in practice more than one circuit is likely to be involved, and the terminal voltages and currents are not allowed to vary simultaneously.

The mechanical motion in a device may take place because of externally applied mechanical forces, as in electric generators, or the motion may be caused by forces of electromagnetic origin, as in electric motors. It must be realized, however, that electromagnetic forces are present in generators also but tend to oppose the external mechanical forces. Also, internal induced voltages arise in motors because of mechanical motion. An understanding of force production from electromagnetic effects aids in the development of electric machines. Before we evaluate these forces quantitatively, we shall consider them qualitatively in terms of working principles demonstrated in a few examples.

In this chapter, we shall restrict our discussions to lumped-parameter electromechanical systems, where the fields are quasi-static and where the electrical and mechanical effects can be expressed in terms of a finite number of variables. Often, we shall assume incremental motion to simplify the analysis. Typical examples of incremental-motion electromechanical systems include various transducers such as loudspeakers, microphones, electromagnetic relays, and so forth.

4.1 MECHANICAL FORCES DUE TO MAGNETIC FIELDS

The two basic magnetic field effects resulting in the production of mechanical forces are (a) *alignment* of flux lines, and (b) *interaction* between magnetic fields and current-carrying conductors. Examples of "alignment " are shown in Figure 4.1*a* to *c*. In Figure 4.1*a* the force on the ferromagnetic pieces causes them to align with the flux lines, thus shortening the magnetic flux path and reducing the reluctance. Figure 4.1*b* shows a simplified form of a reluctance motor, in which electrical force tends to align the rotor axis with that of the stator. Figure 4.1*c* shows the alignment of two current-carrying coils. A few examples of "interac-

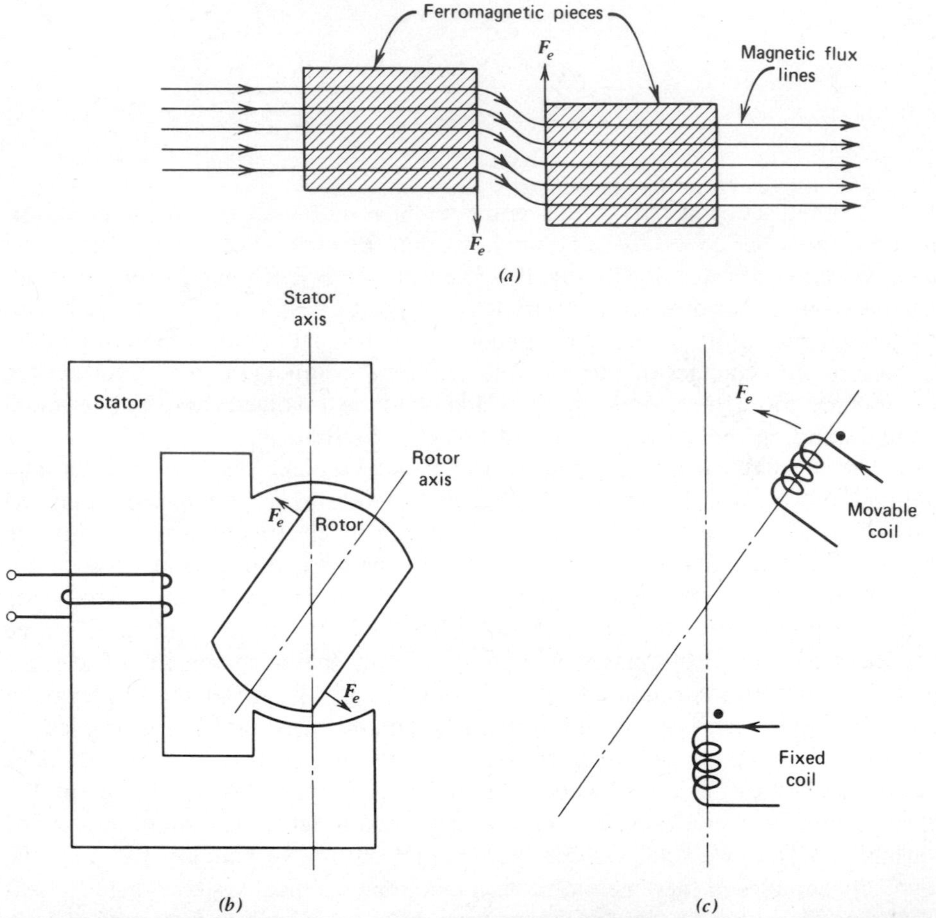

Figure 4.1. Magnetic field effects producing electrical force, F_e. (*a*) Alignment of ferromagnetic pieces in a magnetic field. (*b*) A reluctance motor. (*c*) Alignment of two current-carrying coils.

tion" are shown in Figures 4.2*a* to *c* in which current-carrying conductors experience mechanical force when placed in magnetic fields. For instance, in Figure 4.2*b* a force is produced by the interaction between the flux lines and coil current, resulting in a torque on the moving coil. This mechanism forms the basis of a variety of electrical measuring instruments. Almost all industrial dc motors work on the "interaction" principle.

Quantitative evaluation of the mechanical force of electromagnetic origin will be considered later. Here, we have simply considered some of the principles of force production. We wish to point out that the force is always in such a direction that the net magnetic reluctance is reduced or the energy stored in the magnetic field is minimized. Forces of relatively small magnitude may also be produced by the deformation of a ferromagnetic material by *magnetostriction*, which is not included in this book.

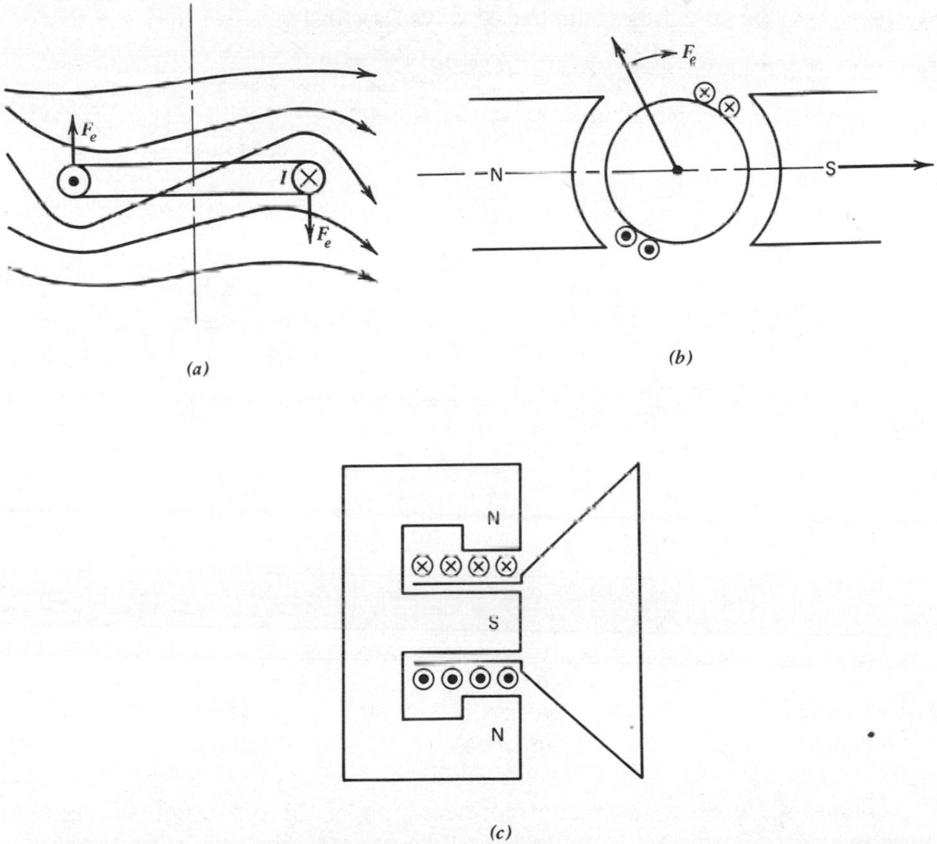

Figure 4.2. Electrical force produced by interaction of current-carrying conductors and magnetic fields. (*a*) A one-turn coil in a magnetic field. (*b*) A permanent magnet moving coil ammeter. (*c*) A moving coil loudspeaker.

4.2 ENERGY CONSERVATION AND ENERGY CONVERSION

We have given a few examples showing how mechanical forces are produced by magnetic fields. Clearly, for energy conversion, that is, for doing work, mechanical motion is as important as mechanical force. Thus, during mechanical motion the energy stored in the coupling magnetic field is disturbed. In Figure 4.2*b* for instance, most of the magnetic field energy is stored in the air-gap separating the rotor from the stator. The air-gap field may be termed the *coupling field*. Electromechanical energy conversion occurs when coupling fields are disturbed in such a way that the energy stored in the fields changes with mechanical motion. A justification of this statement is possible from energy conservation principles, which will enable us to determine the magnitudes of mechanical forces arising from magnetic field effects.

The energy conservation principle in connection with electromechanical systems, may be stated in a number of ways. For instance, we may say that

$$\begin{array}{c}\text{input}\\ \text{electrical}\\ \text{energy}\end{array} + \begin{array}{c}\text{input}\\ \text{mechanical}\\ \text{energy}\end{array} = \begin{array}{c}\text{increase}\\ \text{in stored}\\ \text{energy}\end{array} + \begin{array}{c}\text{energy}\\ \text{dissipated}\\ \text{as heat}\end{array} \tag{4.1}$$

or

$$\begin{array}{c}\text{input}\\ \text{electrical}\\ \text{energy}\end{array} = \begin{array}{c}\text{mechanical}\\ \text{work}\\ \text{done}\end{array} + \begin{array}{c}\text{increase}\\ \text{in stored}\\ \text{energy}\end{array} + \begin{array}{c}\text{energy}\\ \text{dissipated}\\ \text{as heat}\end{array} \tag{4.2}$$

Or, if only the conservative (or lossless) portion of the system is considered, we have

$$\begin{array}{c}\text{sum of}\\ \text{input energy}\end{array} = \begin{array}{c}\text{change in}\\ \text{stored energy}\end{array} \tag{4.3}$$

or

$$\begin{array}{c}\text{input}\\ \text{electrical}\\ \text{energy}\end{array} = \begin{array}{c}\text{mechanical}\\ \text{work}\\ \text{done}\end{array} + \begin{array}{c}\text{increase}\\ \text{in stored}\\ \text{energy}\end{array} \tag{4.4}$$

Figure 4.3 gives a schematic representation of the separation of the conservative part of a system from its dissipative portion. The total stored energy is the sum of the energy stored in the electric and magnetic fields. For practical purposes, however, the energy stored in the electric field is almost negligible.

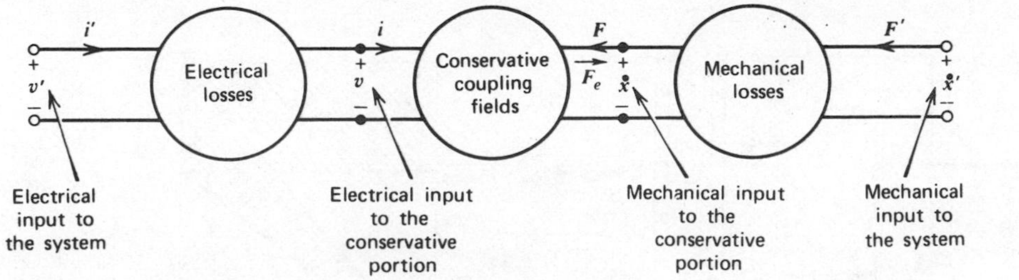

Figure 4.3. A representation of an electromechanical system.

4.3 THE FORCE EQUATION

From the preceding remarks we see that energy conversion is possible because of the interchange between electrical and mechanical energy via the coupling fields. This fact in turn leads us to a method of determining mechanical forces that arise from changes in stored energy. Thus, referring to Figure 4.4, (4.3) implies that

$$F\,dx + vi\,dt = dW \tag{4.5}$$

where

$$F\,dx = \text{mechanical energy input}$$

$$vi\,dt = \text{electrical energy input}$$

$$dW = \text{increase in stored energy.}$$

Now, if F_e is the force of electrical origin and acts against F (Figure 4.3), that is $F\text{dx} = -F_e\text{dx}$, and if dW_m is the energy stored in the magnetic field,—energy stored in the electric field being negligible—,(4.5) may be rewritten as

$$F_e\,dx = -dW_m + vi\,dt \tag{4.6}$$

From Faraday's law voltage v may be expressed in terms of the flux linkage λ as

$$v = \frac{d\lambda}{dt} \tag{4.7}$$

so that (4.6) becomes

$$F_e\,dx = -dW_m + i\,d\lambda \tag{4.8}$$

Before going any further, we should point out that (4.8) is a restatement of (4.4).

In an electromechanical system, either (i,x) or (λ,x) may be considered as independent variables. If we consider (i,x) as independent, the flux linkage λ is

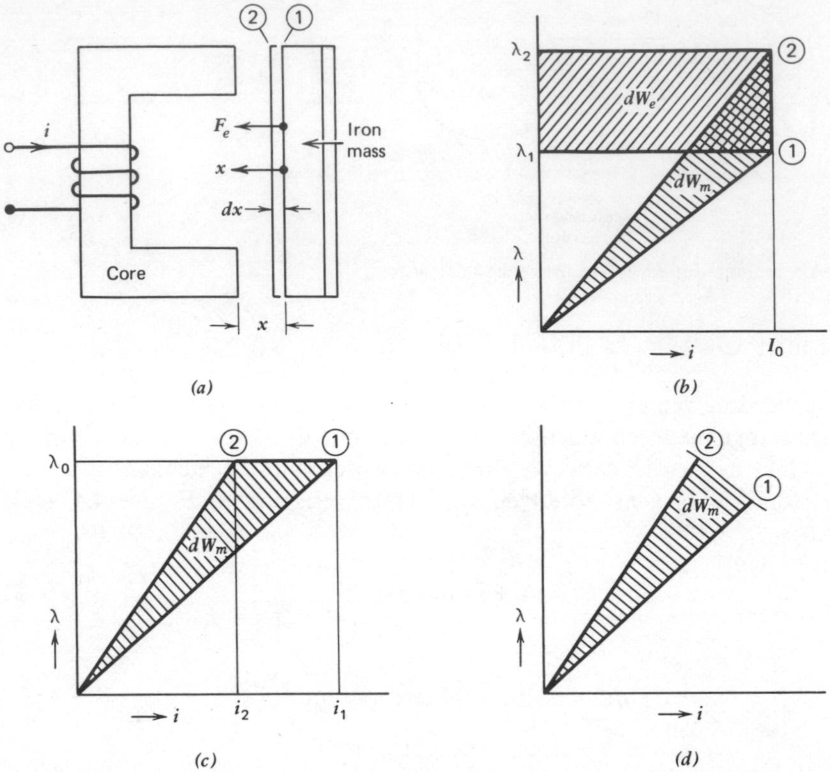

Figure 4.4. Energy balance in an electromechanical system. (*a*) A simple system. (*b*) Constant-current operation. (*c*) Constant-voltage (or flux linkage) operation. (*d*) A general case.

given by $\lambda = \lambda(i, x)$, which can be expressed in terms of small changes as

$$d\lambda = \frac{\partial \lambda}{\partial i} di + \frac{\partial \lambda}{\partial x} dx \tag{4.9a}$$

Also, we have $W_m = W_m(i, x)$ so that

$$dW_m = \frac{\partial W_m}{\partial i} di + \frac{\partial W_m}{\partial x} dx \tag{4.9b}$$

Thus (4.9*a*) and (4.9*b*) when substituted in (4.8) yield

$$F_e dx = -\frac{\partial W_m}{\partial x} dx - \frac{\partial W_m}{\partial i} di + i\frac{\partial \lambda}{\partial x} dx + i\frac{\partial \lambda}{\partial i} di$$

or

$$F_e dx = \left(-\frac{\partial W_m}{\partial x} + i\frac{\partial \lambda}{\partial x} \right) dx + \left(-\frac{\partial W_m}{\partial i} + i\frac{\partial \lambda}{\partial i} \right) di \tag{4.10}$$

Because the incremental changes di and dx are arbitrary, F_e must be independent of these changes. Thus, for F_e to be independent of di, its coefficient in (4.10) must be zero. Consequently, (4.10) becomes

$$F_e = -\frac{\partial W_m}{\partial x}(i,x) + i\frac{\partial \lambda}{\partial x}(i,x) \tag{4.11}$$

which is the force equation. This equation holds true if i is the independent variable. If, on the other hand, λ is taken as the independent variable, that is, if $i = i(\lambda, x)$ and $W_m = W_m(\lambda, x)$, then

$$dW_m = \frac{\partial W_m}{\partial \lambda} d\lambda + \frac{\partial W_m}{\partial x} dx$$

which when substituted in (4.8) gives

$$F_e\, dx = -\frac{\partial W_m}{\partial x} dx - \frac{\partial W_m}{\partial \lambda} d\lambda + i\, d\lambda \tag{4.12}$$

But $\int i\, d\lambda = W_m$ so that (4.12) finally becomes (since coefficients of $d\lambda = 0$)

$$F_e = -\frac{\partial W_m}{\partial x}(\lambda, x) \tag{4.13}$$

We notice that the preceding derivation assumes either i or λ as an independent variable. The case in which i is the independent variable corresponds to a current-excited system. For an electromechanical system such as that shown in Figure 4.4*a*, the current in the coil is held constant at I_o during the period that the armature undergoes motion from position (1) to position (2). During this motion, the flux linkage changes from λ_1 to λ_2 and the electrical energy input becomes $dW_e = I_o(\lambda_2 - \lambda_1)$. The electrical energy comes from the current source, as shown in Figure 4.4*b*. Also during the process, the increase in field energy is $dW_m - \frac{1}{2} I_o(\lambda_2 - \lambda_1)$.

Thus, from (4.5) we have

$$dW_e + F\,dx = dW_m$$

or

$$I_o(\lambda_2 - \lambda_1) + F\,dx = \frac{1}{2} I_o(\lambda_2 - \lambda_1)$$

or

$$F\,dx = -\frac{1}{2} I_o(\lambda_2 - \lambda_1) \tag{4.14}$$

where F may be considered as an externally applied mechanical force, and the negative sign is associated with dx, indicating that motion is against the positive x-direction. Clearly, the right-hand side of (4.14) is negative, $\lambda_2 > \lambda_1$, and $F\,dx = -F_e\,dx$. Thus, (4.14) becomes

$$-F_e\,dx = -\frac{1}{2}I_o(\lambda_2 - \lambda_1) \tag{4.15}$$

indicating that for a current-excited system the electrical energy input divides equally between increasing stored energy and doing mechanical work.

Next, we consider the case in which the flux linkage is kept constant at λ_o and the current is allowed to vary from i_1 to i_2 ($i_2 < i_1$) during motion, as shown in Figure 4.4c. In this case there is no electrical energy input from the source, as may be seen by comparing Figures 4.4b and c. The change in stored energy is

$$dW_m = \frac{1}{2}\lambda_o(i_2 - i_1)$$

which is negative. Thus, from (4.5) we get

$$F\,dx = dW_m$$

or

$$F\,dx = \frac{1}{2}\lambda_o(i_2 - i_1)$$

or

$$-F_e\,dx = \frac{1}{2}\lambda_o(i_2 - i_1) \tag{4.16}$$

indicating that the mechanical work done equals the reduction in stored energy.

In the preceding theoretical discussions we have considered that either i or λ remains constant. In reality, however, neither condition holds and the change from position (1) to (2) follows a path such as that in Figure 4.4d. Nevertheless, the principles of energy conservation are still useful in determining the electrical force.

Reconsidering (4.11) and (4.12), neglecting saturation, we have

$$W_m = \frac{1}{2}\lambda i = \frac{1}{2}Li^2 \tag{4.17}$$

so that

$$F_e = \frac{1}{2}i^2\frac{\partial L}{\partial x} = -\frac{1}{2}\lambda\frac{\partial i}{\partial x} \tag{4.18}$$

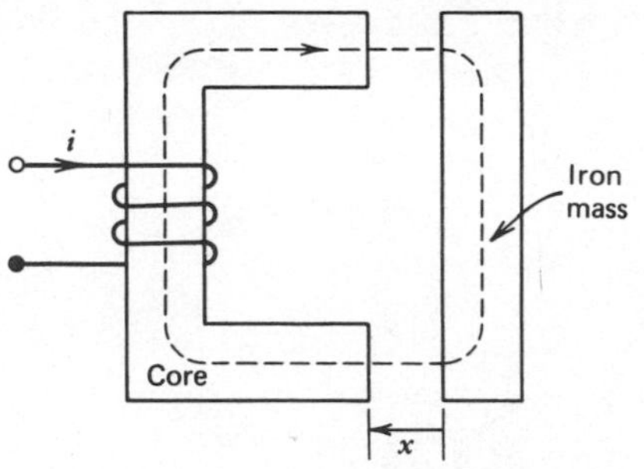

Figure 4.5. An electromechanical system.

We shall illustrate the application of the force equation by the following examples.

Example 4.1
In an electromechanical system (Figure 4.5) the current, flux linkage, and position are related by

$$i=\lambda^2+2\lambda(x-1)^2, \qquad x<1$$

Find the force on the iron mass at $x=0.5$.

Recall that $W_m=\int i\,d\lambda$, and from the given expression for i, we get

$$W_m=\frac{1}{3}\lambda^3+\lambda^2(x-1)^2$$

Consequently, the electrical force calculated from (4.13) becomes

$$F_e=-2\lambda^2(x-1)$$

The same result can also be achieved by evaluating $\partial i/\partial x$ as

$$\frac{\partial i}{\partial x}=4\lambda(x-1)$$

and substituting this in (4.18) to obtain

$$F_e=-2\lambda^2(x-1)$$

At $x=0.5$, $F_e=\lambda^2$, which indicates that the force is proportional to the square of the voltage—if leakage is neglected.

Example 4.2
An elementary reluctance machine is shown in Figure 4.6. The machine is singly excited, that is, it carries only one winding on the stator. The exciting winding is

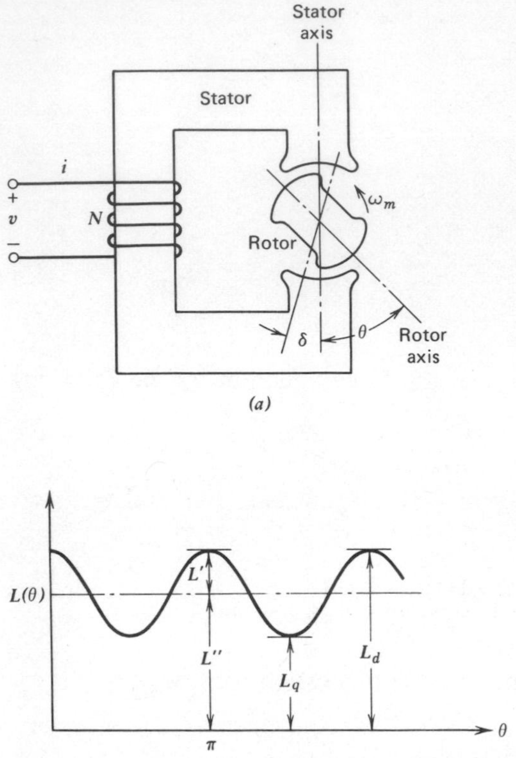

(b)

Figure 4.6. (*a*) A reluctance machine. (*b*) Inductance variation.

wound on the stator and the rotor is free to rotate. The rotor and the stator are shaped so that the variation of the inductance of the windings is sinusoidal with respect to the rotor position. The space variation of the inductance is of double frequency, that is,

$$L(\theta) = L'' + L' \cos 2\theta$$

where the symbols are defined as in Figure 4.6. For an excitation,

$$i = I_m \sin \omega t$$

determine the instantaneous and average torques.

The magnetic energy stored is

$$W_m = \frac{1}{2} L(\theta) i^2$$

and the flux linkage

$$\lambda(\theta) = L(\theta) i$$

where i is the independent variable.

Therefore, from (4.11)

$$T_e = -\frac{\partial W_m}{\partial \theta} + i\frac{\partial \lambda}{\partial \theta}$$

which, expressed in terms of inductance and current, becomes

$$T_e = -\frac{1}{2}i^2\frac{\partial L}{\partial \theta} + i^2\frac{\partial L}{\partial \theta} = \frac{1}{2}i^2\frac{\partial L}{\partial \theta}$$

For given current and inductance variations,

$$T_e = -I_m^2 L' \sin 2\theta \sin^2 \omega t$$

If the rotor is now allowed to rotate at an angular velocity ω_m, so that at any instant

$$\theta = \omega_m t - \delta$$

(where δ is the rotor position at $t=0$, when the current i is also zero), then in terms of ω and ω_m the expression for instantaneous torque becomes

$$T_e = -\frac{1}{2}I_m^2 L' \left\{ \sin 2(\omega_m t - \delta) - \frac{1}{2}[\sin 2(\omega_m t + \omega t - \delta) \right.$$
$$\left. + \sin 2(\omega_m t - \omega t - \delta)] \right\}$$

To obtain the above final form, we have used the following trigonometric identities:

$$\sin^2 A = \frac{1}{2}(1 - \cos 2A)$$

and

$$\sin C \cos D = \frac{1}{2}\sin(C+D) + \frac{1}{2}\sin(C-D)$$

From the above expression it can be concluded that the time-average torque is zero, since the value of each term integrated over a period is zero. The only case for which the average torque is nonzero is when $\omega = \omega_m$. At this particular frequency, the magnitude of the average torque becomes

$$T_{av} = \frac{1}{4}I_m^2 L' \sin 2\delta$$

or from Figure 4.6b

$$T_{av} = \frac{1}{8} I_m^2 (L_d - L_q) \sin 2\delta$$

Thus, for example, at $I_m = 4$ A, $L_d = 0.2$ H (henry) and $L_q = 0.1$ H, the maximum average torque is 0.2 N-m (newton meter).

A number of conclusions can be drawn from the preceding analysis. The machine develops an average torque only at one particular speed corresponding to the frequency $\omega = \omega_m$, which is known as the synchronous speed. The reluctance machine is therefore a synchronous machine. The torque developed by the machine is called the reluctance torque, which will be zero if $L_d = L_q$. The torque varies sinusoidally with the angle δ, called the torque angle. Angle δ is a measure of the torque. The inductances L_d and L_q are the maximum and minimum values of inductance and are called the direct-axis inductance and quadrature-axis inductance, respectively. The maximum torque occurs at the point at which $\delta = 45°$, called the pull-out torque. Any load requiring a torque greater than the maximum torque results in unstable operation of the machine.

Example 4.3

In this example we would like to study the effect of including in our analysis the resistance of a voltage-excited coil, such as that in Figure 4.5. Let r be the resistance of the N-turn coil. Neglecting saturation, find the instantaneous and average values of the electrical force if the terminal voltage $v = V_m \sin \omega t$ and the reluctance of the entire magnetic circuit may be expressed as $R = a + bx$, where a and b are constants.

The stored energy in the magnetic field may be expressed as

$$W_m = \frac{1}{2} R\phi^2$$

where $N\phi = \lambda =$ flux linking the coil. Consequently, we have

$$F_e = -\frac{1}{2}\phi^2 \frac{\partial R}{\partial x} = -\frac{1}{2} b\phi^2$$

since $R = a + bx$.

The problem now reduces to relating ϕ and v as follows:

$$v = ri + N\frac{d\phi}{dt}, \qquad \text{where } \phi = \frac{Ni}{R}$$

Therefore, under steady state conditions

$$\phi = \phi_m \sin(\omega t + \Psi)$$

where

$$\phi_m = \frac{NV_m}{\sqrt{\left[(rR)^2 + (\omega N^2)^2\right]}}$$

and

$$\Psi = \tan^{-1}\frac{\omega N^2}{rR}.$$

The instantaneous force is, therefore,

$$F_e = -\frac{1}{2}\frac{b\ N^2V_m^2}{(rR)^2+(\omega N^2)^2}\sin^2\left(\omega t + \tan^{-1}\frac{\omega N^2}{rR}\right)$$

The time-average value of the force is found from

$$(F_e)_{av} = \frac{1}{T}\int_o^T F_e\,dt = -\frac{b}{4}\frac{N^2V_m^2}{(rR)^2+(\omega N^2)^2}$$

Notice that the average value of $\sin^2$ (or $\cos^2$) is $1/2$.

4.4 CURRENT AND FLUX VARIATIONS

In a voltage-excited system it is interesting to investigate the variation of the input current as a function of time. The following discussion is meant to be of a qualitative nature only. We shall consider a numerical example later.

Consider the system shown in Figure 4.7. With no applied voltage the movable iron (armature) is, let us say, at a distance x_o for the core. The corresponding inductance is L_o—the minimum value of the inductance—and the time constant τ_o is L_o/r. If the iron is held in the original position and a step voltage is applied, the circuit behaves like an rL-circuit, with time constant τ_o. Final current is V/r, where V is the applied voltage. If, however, the iron is allowed to move and its final position is x_f, the inductance of the circuit

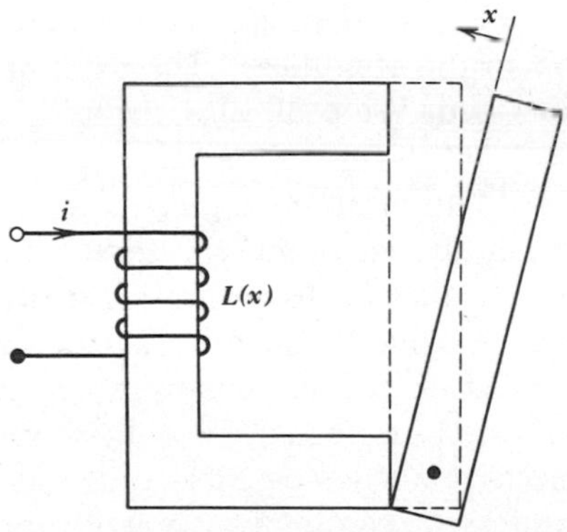

Figure 4.7. Model of an electromagnetic relay.

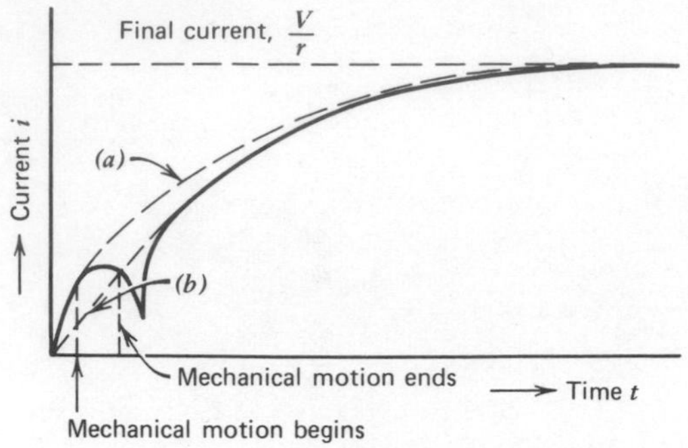

Figure 4.8. $i(t)$ for a step input voltage.

increases to L_f and the corresponding time constant is $\tau_f = L_f/r$. It is evident that $\tau_f > \tau_o$. For the initial and final positions of the movable iron the currents are shown, respectively by curves (a) and (b) in Figure 4.8. However, the transition from (a) and (b) is not smooth because as soon as the iron starts moving, the mechanical time constant τ_m of the system comes into play. It should be recognized that $\tau_m > \tau_f > \tau_o$. This explains the nature of the current variation when the movable iron is allowed to move at a constant voltage.

In case of constant-current excitation, the variations of flux are of interest. For initial position x_o the reluctance is maximum and the corresponding flux ϕ_o is a minimum. The flux reaches its maximum value ϕ_f when the motion is complete. The variation of the flux is governed by the mechanical as well as by the electrical time constants, as in the preceding case. It may be of interest to obtain the variation of flux as a function of time. (See Problem 4.12.)

4.5 DYNAMICS OF ELECTROMECHANICAL SYSTEMS

In the preceding discussions we have assumed electrical and mechanical motion during the electromechanical energy conversion process. In this section, we shall study electromechanical dynamics by deriving the pertinent equations of motion and then solving these equations for various operating conditions. The dynamical behavior of a given system will thus be obtained. We shall also consider a few illustrative examples.

In obtaining the necessary quantitative information about a given energy-conversion device, several difficulties are generally encountered. First, the parameters involved in the equations of motion are difficult to evaluate. Liberal approximations have to be made to obtain models amenable to analysis. To take a specific example, in calculating the inductances of various windings of an electromagnetic device, saturation effects are often neglected as a first approximation. Similarly, in rotary energy converters, realistic considerations such as harmonics, leakages, and effects of slots and teeth cannot conveniently be taken into account, although each of these factors contributes to the performance of the device. In such cases the actual device is replaced by an idealized model. Simplifying assumptions are made and, if necessary for accuracy, the second-order effects are included in the solution by special techniques such as numerical and graphical methods. Once parameters in equations of motion are determined—a step so important to the study of energy conversion devices—formulation of the equations themselves (discussed later) is a routine matter.

A second difficult step in the study of an energy-conversion device is obtaining solutions for the resulting equations of motion. These equations are most often nonlinear differential equations with time-variable coefficients. No general methods are available for solving these nonlinear equations and approximations have to be made to obtain the end results. For example in the case of transducers, incremental motion is assumed (often not too unrealistically) and the resulting nonlinear equations of motion are linearized about a quiescent operating point.

After they are linearized, the equations may be solved by a standard method. For rotary energy converters such as motors and generators, the procedure for formulating the equations of motion is similar to that for a transducer. The solution of these equations is facilitated either by linear transformations or by some other technique, such as by numerical methods or by using analog or digital computers, which may also be used to solve the equations of motion without linearizing them.

The analysis and study of electromagnetic energy converters, therefore, include the following aspects:

1. **Topological considerations and physical descriptions:** These give the locations of input and output terminals, identify the fixed and moving elements, and specify the structure of the magnetic circuit, winding data, and various physical dimensions.
2. **Choice of a model and simplifying assumptions:** These generally depend on the problem at hand and the degree of refinement desired in the solution. For example, in a magnetic device, as a first approximation, fringing, saturation, and hysteresis are neglected. The permeability of the magnetic material is ideally assumed to be infinitely greater than that of free space; consequently, the magnetic energy is assumed to be stored in the air gap alone.
3. **Determination of the system parameters:** This includes the evaluation of resistances, inductances, and capacitances for the electrical portion of the system; and evaluation of the mass (or moment of inertia), stiffness, and friction coefficient for the mechanical portion. The parameters are usually obtainable from the physical description of the system and from the simplifying assumptions and choice of model. In most cases of practical importance, parameter determination consists of calculation of various inductances, a difficult but important step upon which the performance of an electromagnetic device depends.

4. **Formulation of the electrical and mechanical equations of motion:** Respectively, these turn out to be the volt-ampere equations and the force-balance (or torque-balance) equations. The equations of motion can be derived by one of the methods discussed later.

5. **Solution of the equations of motion:** This step is performed after the equations of motion have been formulated. Almost invariably, the resulting differential equations are nonlinear. In simple cases, such as in the case of transducers for small-signal applications, the equations of motion are first linearized and then solved by treating them as linear differential equations with constant coefficients. For complicated cases of rotating electrical machines, certain types of linear transformations become necessary. No sufficiently general method is available which covers all cases. With the availability of digital computers, numerical solutions may be quite easily obtained, even without linearizing the equations of motion.

4.5.1 FORMULATION OF THE EQUATIONS OF MOTION

The electrical and mechanical characteristics of an electromechanical system are given, respectively, by the electrical and mechanical equations of motion, which are the voltage- (or current-) balance equation and the force- (or torque-) balance equations. These equations may be obtained by equating the "applied forces" to the "restoring forces." In the electrical equation, electrical forces of mechanical origin, such as voltage induced by mechanical motion, are obtained from Faraday's law. Mechanical forces of electrical origin are determined by using the force equation derived earlier. In formulating the mechanical equation, the mechanical force resulting from magnetic field effect may be considered as an externally applied force. We shall term this force as an electrical force, denoted by F_e. We now illustrate the formulation of the equations of motion by the following example.

An electromagnetic relay may be modeled by the lumped-parameter system shown in Figure 4.9. There is no externally applied mechanical force. First of all, we wish to formulate the dynamical equations of motion in the following steps.

1. *Assumptions.* We neglect saturation of the magnetic circuit, which we assume to be infinitely permeable, and ignore leakage and fringing fluxes. Also we assume the friction force to be linearly proportional to velocity and the spring force linearly proportional to the elongation.
2. *Parameters.* The mechanical parameters are mass, M; friction coefficient, b; and spring stiffness, k. The parameters for the electrical circuit are resistance, r; and inductance, L; which may also be expressed in terms of the dimensions shown in Figure 4.9 as follows:

$$L(x)=\frac{\mu_o aN^2}{l_1-x}=\frac{A}{C+x} \tag{4.19}$$

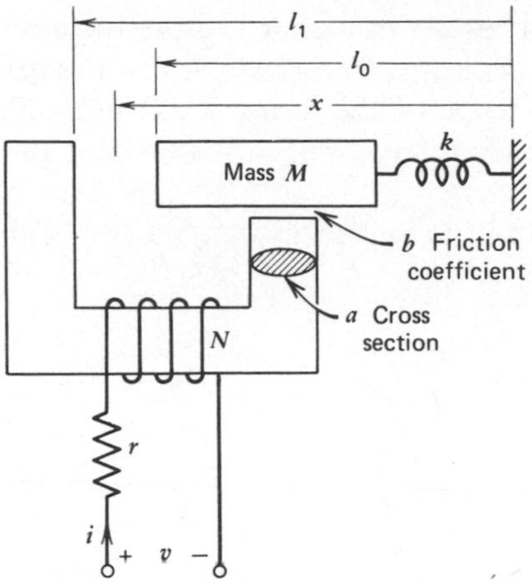

Figure 4.9. An electromechanical system.

where A and C are constants. (Notice that $A = -\mu_o a N^2$ and $C = -l_1$.)
3. *Equations of motion.* We can now identify the different "forces" acting on the system.
(a) Electrical:

$$ri + \frac{d}{dt}(Li) = v \tag{4.20}$$

where the terms on the left-hand side denote the "restoring forces" or voltage drops.
(b) Mechanical:

$$M\ddot{x} + b\dot{x} + k(x - l_o) = F_e = \frac{1}{2}i^2\frac{\partial L}{\partial x} \tag{4.21}$$

where the left-hand side is the sum of the restoring forces and where F_e the electrical force, considered as an external force, is given by (4.11) or (4.18).

4.5.2 A RECONSIDERATION OF (4.20) AND (4.21)

Because the equations of motion are significant in defining the behavior of a system they are reconsidered here. Of particular importance in the electrical equation is the term arising from mechanical motion, and in the mechanical equation the term attributable to the "electrical motion." Thus, in (4.20) we have $d(Li)/dt$, which may also be written as

$$\frac{d}{dt}(Li) = L\frac{di}{dt} + i\frac{dL}{dt} = L\frac{di}{dt} + i\frac{dL}{dx}\dot{x} \tag{4.22}$$

where $\dot{x} = dx/dt =$ mechanical velocity, which results in a change in inductance with time. Notice that the second term in (4.22) arises from mechanical motion and is called a *motional voltage*. Also the presence of the terms (4.22) in (4.20) make it a nonlinear differential equation. We shall have more to say about this later.

We now consider the right-hand side of (4.21), which in conjunction with (4.19) yields

$$\frac{1}{2} i^2 \frac{\partial L}{\partial x} = \frac{1}{2} i^2 \frac{\partial}{\partial x}\left(\frac{A}{C+x}\right) = -\frac{A}{2} \frac{i^2}{(C+x)^2} \tag{4.23}$$

which is also a nonlinear term making (4.21) a nonlinear differential equation. Recall that $A = -\mu_o a N^2$ and $C = -l_1$ implying that the electrical force is positive and acts in the positive x-direction.

4.5.3 ANALYTICAL SOLUTIONS TO EQUATIONS OF MOTION

For convenience we rewrite the equations of motion (4.20) and (4.21) as follows:

$$L\frac{di}{dt} + i\frac{dL}{dx}\dot{x} + ri = v \tag{4.24}$$

$$M\ddot{x} + b\dot{x} + k(x - l_o) = \frac{\mu_o a N^2 i^2}{2(x - l_1)^2} \tag{4.25}$$

As mentioned in the last section, these equations are nonlinear and explicit analytical solutions cannot be obtained. Nevertheless, for small signals and incremental motion, useful information about the system can be obtained by solving the corresponding linearized differential equations of motion.

In addition to the small-signal constraint, linearization of the equations for a given physical system requires that a stable equilibrium point exists. The small signals (or motion) are then excursions about this equilibrium point. Let (V_o, I_o, X_o) denote the steady state stable equilibrium point such that

$$\begin{aligned} v(t) &= V_o + v_1(t) \\ i(t) &= I_o + i_1(t) \\ x(t) &= X_o + x_1(t) \end{aligned} \tag{4.26}$$

where (v, i, x) are the original variables and (v_1, i_1, x_1) are small perturbations about (V_o, I_o, X_o). The smallness is measured by the fact that product-type terms such as i_1^2, $i_1 x_1$, and so forth, are negligible. With these constraints in mind, let us examine the nonlinear terms in (4.24) and (4.25). For instance, by substituting $x = X_o + x_1$ in (4.19) we obtain

$$L = \frac{\mu_o a N^2}{l_1 - x} = \frac{\mu_o a N^2}{l_1 - X_o - x_1} = L_o\left(\frac{1}{1 - x_1/(l_1 - X_o)}\right)$$
$$= L_o\left(1 - \frac{x_1}{l_1 - X_o}\right)^{-1} \tag{4.27}$$

where $L_o = [\mu_o a N^2/(l_1 - X_o)]$. For small values of x_1, (4.27) may be expanded in the binomial series as

$$L = L_o\left[1 + \frac{x_1}{l_1 - X_o} + \left(\frac{x_1}{l_1 - X_o}\right)^2 + \left(\frac{x_1}{l_1 - X_o}\right)^3 + \cdots\right] \tag{4.28}$$

Also, from (4.28) we get

$$\frac{\partial L}{\partial x_1} = \frac{L_o}{l_1 - X_o}\left[1 + \frac{2x_1}{l_1 - X_o} + 3\left(\frac{x_1}{l_1 - X_o}\right)^2 + \cdots\right] \tag{4.29}$$

The linearized forms of (4.28) and (4.29) would then be

$$L \simeq L_o\left(1 + \frac{x_1}{l_1 - X_o}\right) \tag{4.30}$$

and

$$\frac{\partial L}{\partial x_1} \simeq \frac{L_o}{l_1 - X_o}\left(1 + \frac{2x_1}{l_1 - X_o}\right) \tag{4.31}$$

where $x_1/(l_1 - X_o) \ll 1$ is assumed and all terms of second order and higher in (4.28) and (4.29) are neglected.

The flux-linkage λ then becomes

$$\lambda = Li$$
$$\simeq L_o\left(1 + \frac{x_1}{l_1 - X_o}\right)(I_o + i_1)$$
$$\simeq L_o I_o + L_o i_1 + \frac{L_o I_o}{l_1 - X_o} x_1 \qquad \text{(neglecting } x_1 i_1\text{)}$$

and

$$\frac{d}{dt}(Li) \simeq L_o \frac{di_1}{dt} + \frac{L_o I_o}{l_1 - X_o} \dot{x}_1 \tag{4.32}$$

$$ri = rI_o + ri \tag{4.33}$$

Substituting (4.26), (4.32) and (4.33) in (4.20) yields

$$L_o \frac{di_1}{dt} + ri_1 + \frac{L_o I_o}{l_1 - X_o} \dot{x}_1 + rI_o = v_1 + V_o \tag{4.34}$$

The steady state or dc operating point is then given by

$$rI_o = V_o \tag{4.35}$$

and the remaining terms in (4.34) give the electrical dynamics (about this operating point) as

$$L_o \frac{di_1}{dt} + ri_1 + \frac{L_o I_o}{l_1 - X_o} \dot{x}_1 = v_1 \tag{4.36}$$

This is the linearized electrical equation of motion.

Next, we consider the right-hand side of (4.21) in which we substitute (4.31) and

$$i^2 = I_o^2 + 2I_o i_1 + i_1^2 \simeq I_o^2 + 2I_o i_1 \qquad \text{(neglecting } i_1^2\text{)}$$

to obtain

$$\frac{1}{2} i^2 \frac{\partial L}{\partial x} \simeq \frac{1}{2}(I_o^2 + 2I_o i_1) \frac{L_o}{l_1 - X_o}\left(1 + \frac{2x_1}{l_1 - X_o}\right)$$

$$= \frac{L_o I_o^2}{2(l_1 - X_o)} + \frac{L_o I_o}{l_1 - X_o} i_1 + \frac{L_o I_o^2}{(l_1 - X_o)^2} x_1 \tag{4.37}$$

(neglecting the $i_1 x_1$ term). Substituting (4.26) and (4.37) in (4.21) yields

$$M\ddot{x}_1 + b\dot{x}_1 + kx_1 + k(X_o - l_o)$$

$$= \frac{L_o I_o^2}{2(l_1 - X_o)} + \frac{L_o I_o}{l_1 - X_o} i_1 + \frac{L_o I_o^2}{(l_1 - X_o)^2} x_1 \tag{4.38}$$

The steady state mechanical equilibrium is given by

$$k(X_o - l_o) = \frac{L_o I_o^2}{2(l_1 - X_o)} \tag{4.39}$$

whereas the mechanical dynamics, as obtained from the remaining terms of (4.38), are given by the linearized equation:

$$M\ddot{x}_1 + b\dot{x}_1 + \left[k - \frac{L_o I_o^2}{(l_1 - X_o)^2}\right]x_1 = \frac{L_o I_o}{l_1 - X_o} i_1 \tag{4.40}$$

The electromechanical dynamics of the system may then be obtained by solving the linearized equations, (4.36) and (4.40), by one of the standard methods. The solutions are discussed in Section 4.5.4.

4.5.4 EXISTENCE OF THE EQUILIBRIUM POINT

We recall that (4.35) and (4.39) give the steady state operating points. Clearly, (4.35) is a statement of Ohm's law for a resistive circuit; and for a given r, (V_o, I_o) can be determined. Knowing I_o, we now can proceed to solve (4.39). The solution is illustrated graphically in Figure 4.10, where the spring force and the electrical force (for two different currents) are plotted as functions of X_o. An equilibrium point exists if the line denoting the spring force cuts the curve for the electrical force. Thus, for a current greater than I_s (2.45 A) there is no equilibrium point and the electrical force is always greater than the spring force pulling the iron mass to the extreme position ($X_o = l_1$). For the current I_o' at the point 3 ($X_o = 2.3$ cm) the spring force is equal to the electrical force. However, a slight disturbance about this point would result in the electrical force exceeding the restoring force of the spring. Consequently, for $I_o' = 2.45$ A, $X_o = 2.3$ cm is not a stable operating point.

Next, if the current is decreased to I_o'' ($= 2.0$ A), we obtain the points 1 and 2 as two equilibrium points. At the point 2, if the mass is moved to the left (X_o increased), the electrical force becomes always greater than the spring force, resulting in a continuous motion to the left until $X_o = l_1$. On the other hand, if the mass is moved to the right (X_o decreased), the spring force is always greater than the electrical force and the motion will continue up to the point 1 ($X_o = 2.13$ cm), which is the stable equilibrium point. A disturbance to the left or right of the point 1 would result in moving the mass back to 1. Hence, the quiescent stable operating for the system is ($V_o = 6$ V; $I_o = 2$ A; $X_o = 2.13$ cm).

4.5.5 NUMERICAL SOLUTIONS

The use of state variables in linear systems theory is a standard practice and a formulation of equations in terms of state variables facilitates the study of

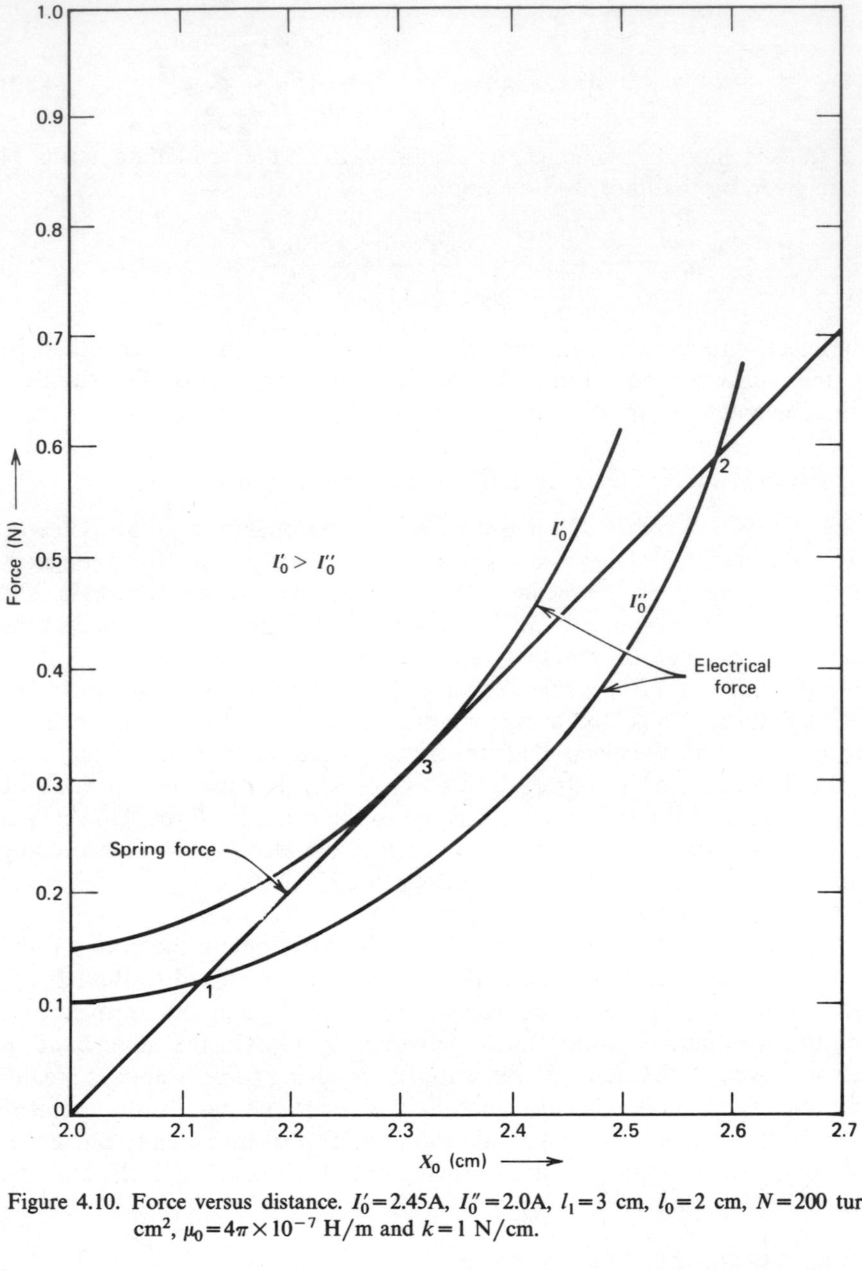

Figure 4.10. Force versus distance. $I_0'=2.45$A, $I_0''=2.0$A, $l_1=3$ cm, $l_0=2$ cm, $N=200$ turns, $a=1$ cm^2, $\mu_0=4\pi\times10^{-7}$ H/m and $k=1$ N/cm.

dynamical systems. Electromechanical devices, such as transducers and rotating electric machines, may be viewed as dynamic systems. With the aid of standard computer subroutines, the solution to the state equation can be obtained and the dynamical characteristics of the system may be studied. It is no more difficult to program the nonlinear equation than the linear system and the linearization step is avoided. Analysis of the nonlinear system is actually simpler than analysis of the linearized approximation of the system.

The equations that describe the behavior of an electromechanical system can be reduced to the state formulation

$$\dot{\mathbf{y}} = f(\mathbf{y}, \mathbf{v}, t)$$

where $\mathbf{y}$ is the state vector and $\mathbf{v}$ is the input vector. With the assumption of small motion about some quiescent point, the state equations can be linearized to

$$\dot{\mathbf{y}} = \mathbf{A}\mathbf{y} + \mathbf{B}\mathbf{v}$$

where the $\mathbf{A}$ and $\mathbf{B}$ matrices are composed of system parameters. As the example shows, when the decision has been made to generate the solution on a computer, it is easier to work with the nonlinear equations.

For the electromagnetic relay let us find the current, armature position, and velocity as functions of time for a step voltage input. In the equations of motion (4.20) and (4.21), with L given by (4.19), let the state variables by $y_1 = i$, $y_2 = x$, $y_3 = \dot{x}$. The state equations become

$$\begin{aligned}
\dot{y}_1 &= \frac{(C+y_2)}{A}\left[v_1 - ry_1 + \frac{Ay_1y_3}{(C+y_2)^2}\right] \\
\dot{y}_2 &= y_3 \\
\dot{y}_3 &= \frac{-1}{M}\left[\frac{Ay_1^2}{2(C+y_2)^2} + k(y_2 - l_o) + by_3\right]
\end{aligned} \tag{4.41}$$

The linearized equations (4.36) and (4.40) may now be rewritten as

$$\begin{aligned}
L_o\frac{di_1}{dt} + \gamma\dot{x}_1 + ri_1 &= v_1 \\
M\ddot{x}_1 + b\dot{x}_1 + \beta x_1 - \gamma i_1 &= 0
\end{aligned} \tag{4.42}$$

where

$$\gamma = \frac{L_o I_o}{(l_1 - X_o)}$$

$$\beta = k - \frac{L_o I_o^2}{(l_1 - X_o)^2}$$

In state form these equations are

$$\dot{y}_1 = \frac{1}{L_o}(ry_1 - \gamma y_3 + v_1)$$

$$\dot{y}_2 = y_3 \tag{4.43}$$

$$\dot{y}_3 = \frac{1}{M}(\gamma y_1 - \beta y_2 - by_3)$$

```
   SUBROUTINE FCT (X, Y, DERY, NDIM)
   REAL*4Y(NDIM), DERY(NDIM), SOL(100,4)
   COMMON SOL,N,V,AM,AK,R,OL,B,C,A
   IF(Y(2)) 1,1,2
1  DERY (1)=(V-Y(1)*R)*C/A
   DERY (2)=0.
   DERY(3)=0.
   RETURN
2  DERY (1)=(V-Y(1)*R)*(Y(2)+C)/A+Y(1)*Y(3)/(Y(2)+C)
   DERY (2)-Y(3)
   DERY (3)=-B*Y(3)/AM-AK*(Y(2)-OL)/AM-A*Y(1)*Y(1)/(2.*AM*(Y(2)+C)*
   1(Y(2)+C))
   RETURN
   END
```

Figure 4.11. Program for solution of the nonlinear system.

Even with such a simple system of equations as (4.43) one can use the computer to determine and plot the solution. Once it has been decided to use the computer, it becomes evident that it is no more difficult to program (4.41) than to program (4.43). By using the nonlinear equation, no additional work for linearization has to be done.

The relay of Figure 4.9 may be analyzed using the subroutine RKGS and the results plotted. Figure 4.11 shows the programmed equations for the nonlinear system. These equations are fed into RKGS by the external subroutine FCT. Figure 4.12 shows the current and position for an applied step voltage. The constants used for this example are $M = 10$ g, $b = 0.0001$ N sec/cm, $k = 1$ N/cm, $R = 1\ \Omega$, $X_o = 2.13$, and other data are shown in Figure 4.10.

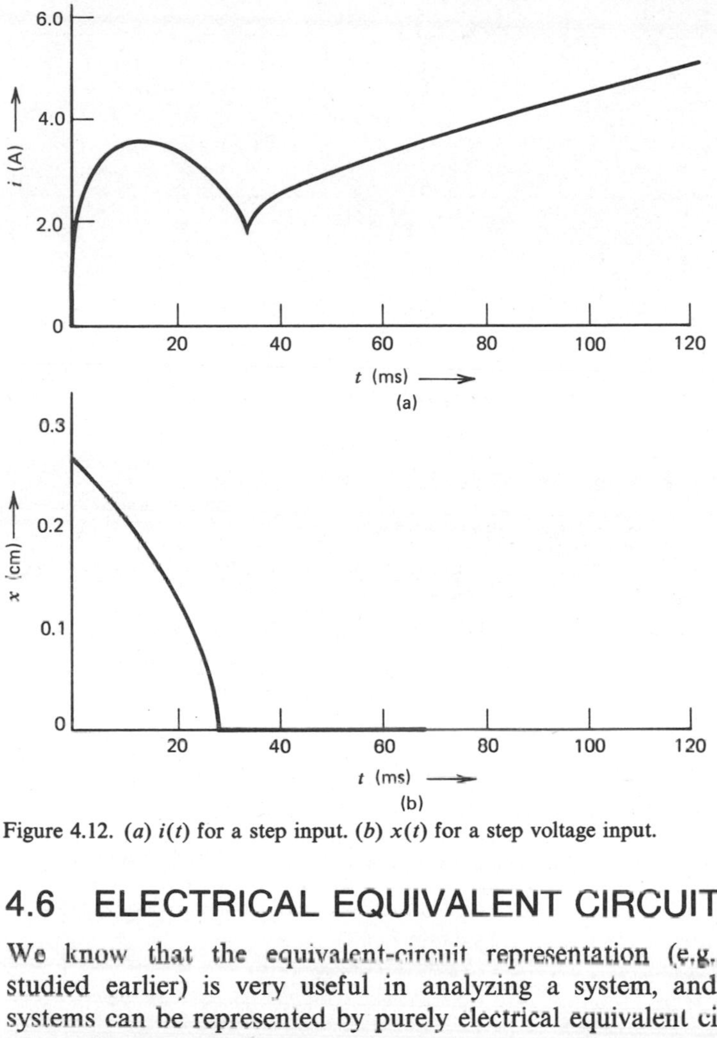

Figure 4.12. (*a*) $i(t)$ for a step input. (*b*) $x(t)$ for a step voltage input.

4.6 ELECTRICAL EQUIVALENT CIRCUITS

We know that the equivalent-circuit representation (e.g., for transformers studied earlier) is very useful in analyzing a system, and electromechanical systems can be represented by purely electrical equivalent circuits. The method of developing an equivalent circuit for a mechanical system is illustrated by means of the following example.

A purely mechanical system consisting of a spring, a mass, and a damper is considered in Figure 4.13 *a*, in which the damping force is assumed to be directly proportional to the velocity. If F_{ext} is an applied external force, then from Newton's law the force-balance equation is written as

$$M\ddot{x} + b\dot{x} + kx = F_{ext}$$

If the force F_{ext} is assumed to be analogous to voltage v, then the network of

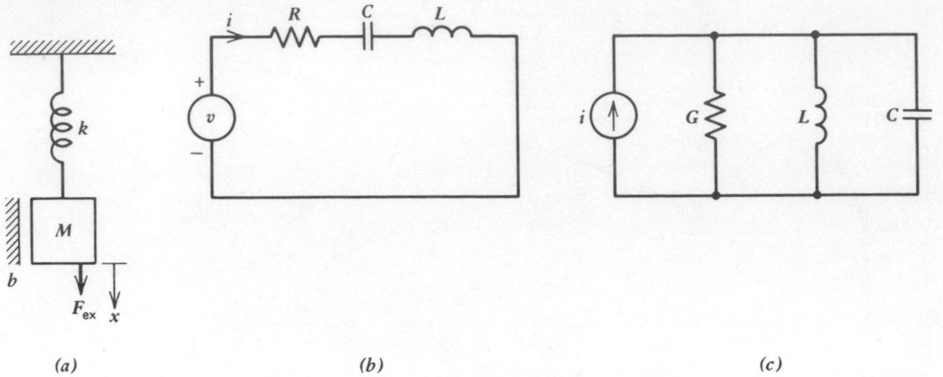

Figure 4.13. (*a*) A mechanical system. (*b*) Force-voltage electrical equivalent circuit. (*c*) Force-current electrical equivalent circuit.

Figure 4.13 can be drawn from the following correspondences:

Force-Voltage Analogy

Force F	Voltage v
Velocity $\dot{x}$	Current i
Damping b	Resistance R
Mass M	Inductance L
Spring constant k	Elastance $1/C$ = Reciprocal of capacitance

On the basis of the principle of duality, if force is taken to be similar to current, the following analogy can also be drawn:

Force-Current Analogy

Force F	Current i
Velocity $\dot{x}$	Voltage v
Damping b	Conductance G
Mass M	Capacitance C
Spring constant k	Reciprocal of inductance $= 1/L$

This results in the circuit shown in 4.13(*b*) and (*c*).

The preceding example involved a very simple, purely mechanical system. But the method is applicable to more complicated electromechanical systems. Based on the preceding analysis we may now develop an equivalent circuit for the "relay" studied in the last section. For this purpose, we recall (4.42)

$$L_o\frac{di_1}{dt}+ri_1+\gamma\dot{x}_1=v_1 \tag{4.44}$$

$$M\ddot{x}_1+b\dot{x}_1+\beta x_1-\gamma i_1=0 \tag{4.45}$$

We let $\gamma\dot{x}_1 = v_2$ so that (4.44) becomes

$$L_o\frac{di_1}{dt} + ri_1 = v_1 - v_2 \tag{4.46}$$

Substituting $\gamma\dot{x}_1 = v_2$ in (4.45), we have

$$M\ddot{x}_1 = \frac{M}{\gamma}\dot{v}_2, \quad b\dot{x}_1 = \frac{b}{\gamma}v_2, \quad \text{and } \beta x_1 = \frac{\beta}{\gamma}\int v_2\,dt$$

so that (4.45) becomes

$$C_2\dot{v}_2 + G_2 v_2 + \frac{1}{L_2}\int v_2 = i_1 \tag{4.47}$$

where $C_2 = M/\gamma^2$, $L_2 = \gamma^2/\beta$, and $G_2 = b/\gamma^2$. By the use of these constants, (4.46) and (4.47) may be represented by the equivalent circuit given in Figure 4.14.

On the other hand, (4.44) and (4.45) can also be written as

$$L_o\frac{di_1}{dt} + ri_1 + r_o i_2 = v_1 \tag{4.48}$$

$$L_2\frac{di_2}{dt} + r_2 i_2 + \frac{1}{C_2}\int i_2\,dt - r_o i_1 = 0 \tag{4.49}$$

where $b = r_2$, and $C_2 = 1/\beta$. Notice that reciprocity of mutual interaction does not hold because the coefficients of the coupling terms in the two equations do not appear with the same sign. This can be taken into account by the use of a gyrator, and the equivalent circuit then becomes as shown in Figure 4.15.

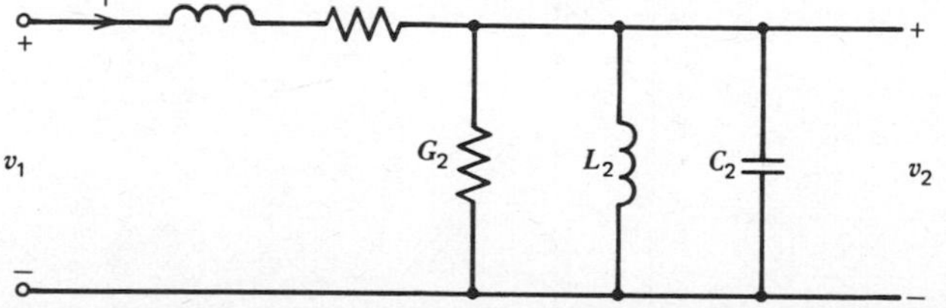

Figure 4.14. An equivalent circuit for the electromechanical system.

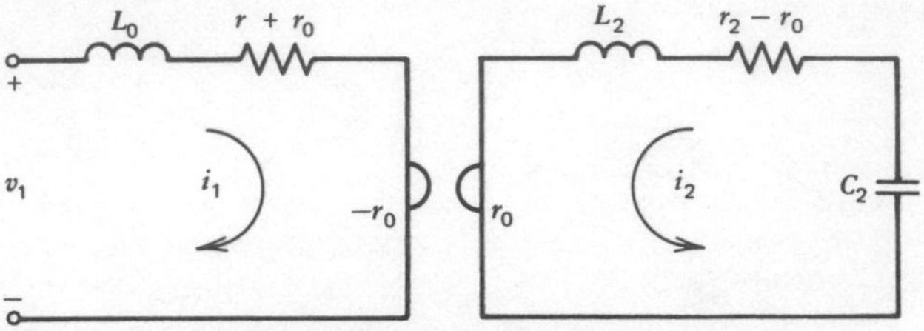

Figure 4.15. An alternate equivalent circuit.

4.7 DOUBLY-AND MULTI EXCITED SYSTEMS

In our discussions so far we have considered systems having only one coil (i.e., single electrical input). But the general principles are equally applicable to multicoil (or multiexcited) systems. To find the electrical force, for instance, we need to determine the energy stored. Thus for a doubly excited (or two-coil) system (Figure 4.16), we can express the flux linkages in terms of the inductances as follows:

$$\lambda_1 = L_{11} i_1 + L_{12} i_2$$

$$\lambda_2 = L_{12} i_1 + L_{22} i_2$$

Suppose that there are incremental changes in the flux linkages so that

$$d\lambda_1 = L_{11} di_1 + L_{12} di_2$$

$$d\lambda_2 = L_{12} di_1 + L_{22} di_2$$

To find the change in the stored magnetic energy, we multiply the $d\lambda$'s above

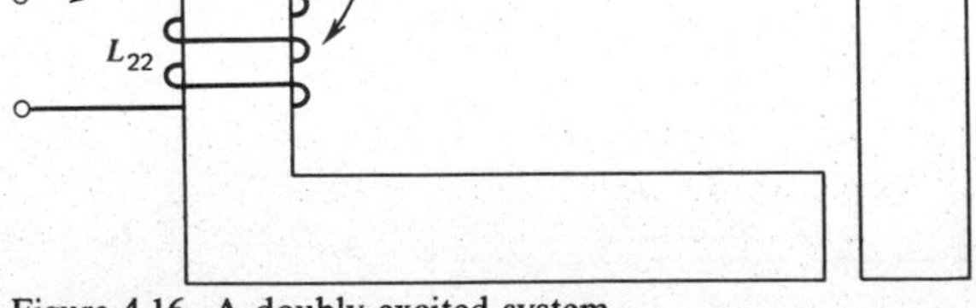

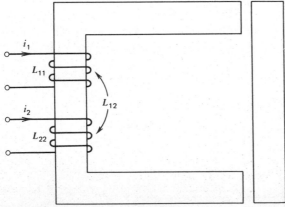

Figure 4.16. A doubly excited system.

by the appropriate currents. Therefore,

$$\begin{aligned} dW_m &= i_1 d\lambda_1 + i_2 d\lambda_2 \\ &= L_{11} i_1 di_1 + L_{12} i_1 di_2 + L_{12} i_2 di_1 + L_{22} i_2 di_2 \\ &= L_{11} i_1 di_1 + L_{12} d(i_1 i_2) + L_{22} i_2 di_2 \end{aligned}$$

The total stored energy is calculated by integrating this equation to get

$$\begin{aligned} W_m &= \frac{1}{2} L_{11} i_1^2 + L_{12} i_1 i_2 + \frac{1}{2} L_{22} i_2^2 \\ &= \frac{1}{2}\left(L_{11} i_1^2 + L_{12} i_1 i_2 + L_{12} i_1 i_2 + L_{22} i_2^2\right) \end{aligned} \tag{4.50}$$

Also, (4.50) can be generalized to

$$W_m = \frac{1}{2} \sum_k \sum_m L_{km} i_k i_m \tag{4.51}$$

or, in matrix notation we have

$$W_m = \frac{1}{2} \tilde{\mathbf{i}} \mathbf{L} \mathbf{i} \tag{4.52}$$

where $\mathbf{i}$ is a column matrix, $\tilde{\mathbf{i}}$ is the transpose of $\mathbf{i}$ (i.e., $\tilde{\mathbf{i}}$ is a row matrix) and $\mathbf{L}$ is the inductance matrix of the system.

Once the stored energy has been found, the rest of the procedure in the study of a multiexcited system is similar to that presented earlier.

4.8 INSTANTANEOUS, AVERAGE, AND RMS VALUE PROBLEMS

The instantaneous and average value of a force of electrical origin are important to the study of electromechanical systems. For instance, to start the mechanical motion the instantaneous force must be nonzero. Similarly, for continuous motion the average force must be nonzero. The average force is defined as

$$F_{av} = \frac{1}{T} \int_0^T F_{inst}\, dt \tag{4.53}$$

where T is a suitable period and F_{inst} is the instantaneous force.

From the definition of the average force a new interpretation of the rms (root-mean-square) value emerges. This is illustrated by the following equations.

For a current-excited inductive system the instantaneous force is

$$F_i = \frac{1}{2} i^2 \frac{\partial L}{\partial x} \tag{4.54}$$

where the current i may have any waveform. The average force is, from (4.53) and (4.54),

$$F_{av} = \frac{1}{2} \frac{\partial L}{\partial x} \frac{1}{T} \int_0^T i^2 \, dt \tag{4.55}$$

If a direct current I_{dc} excites the system, the average force is

$$F_{av} = \frac{1}{2} \frac{\partial L}{\partial x} I_{dc}^2 \tag{4.56}$$

The value of the direct current that produces the same average force is, therefore, from (4.55) and (4.56)

$$I_{dc} = \left(\frac{1}{T} \int_0^T i^2 \, dt \right)^{\frac{1}{2}} \tag{4.57}$$

Recall that (4.57) is identical to the definition of the rms or effective value. Thus, in order that the same average force be exerted for both ac and dc excitations, it is necessary that the rms value of the alternating current be equal to the magnitude of the direct current.

In summary, in this chapter we have studied the fundamental principles governing the process of electromechanical energy conversion. We have applied these principles to study the dynamics of a number of electromechanical systems.

PROBLEMS

4.1 The singly excited electromechanical system shown in Figure 4.17 is constrained to move only horizontally. The pertinent dimensions are shown in the diagram. Determine the electrical force exerted on the movable iron member for

(a) Current excitation:

$$i = I \cos \omega t$$

(b) Voltage excitation:

$$v = V \cos \omega t$$

For parts (a) and (b) neglect the winding resistance, leakage fields, and fringing. Assume all energy to be stored in the air gaps, that is, the

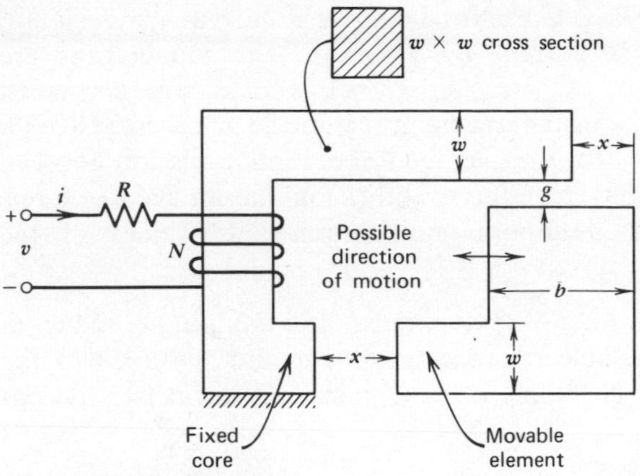

Figure 4.17. Problem 4.1.

permeability of iron is very large as compared with that of free space. What modifications have to be made if the winding resistance is not negligible?

4.2 An electromagnetic structure is characterized by the following Θ-dependent inductances

$$L_{11} = 2 + \cos 2\Theta = L_{22}$$

$$L_{12} = 1 + 0.5\cos\Theta = L_{21}$$

Assume winding resistance to be zero. Find the torque (as a function of Θ) when both windings are connected to the same source, so that $v_1 = v_2 = 155\sin 377\,t$.

4.3 The system shown in Figure 4.18 carries two coils, having self- and mutual

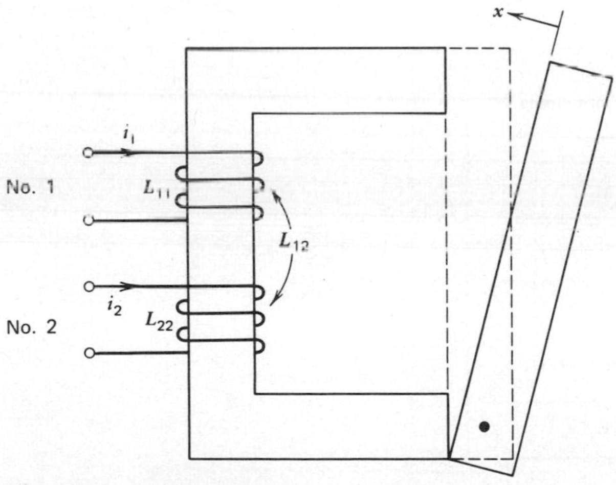

Figure 4.18. Problem 4.3.

inductances L_{11}, L_{12}, and L_{22}. Coil No. 1 carries a current $i_1 = I_1 \sin\omega_1 t$, while coil No. 2 carries a current $i_2 = I_2 \sin\omega_2 t$. The inductances are $L_{11} = k_1/x$, $L_{22} = k_2/x$, and $L_{12} = k_3/x$; k_1, k_2, and k_3 are constants. Derive an expression for the instantaneous force on the armature. Give an expression (in integral form) for the average force. Find a relation between ω_1 and ω_2 for (*i*) maximum average force, and (*ii*) minimum average force. Determine the maximum, minimum, and intermediate values of the average force.

4.4. A mechanical system consisting of spring, mass, and damping is shown in Figure 4.19. Choose a suitable coordinate system and write down the equations of motion from Newton's laws. Obtain its electrical equivalent circuit using
(a) The force-voltage analogy.
(b) The force-current analogy.

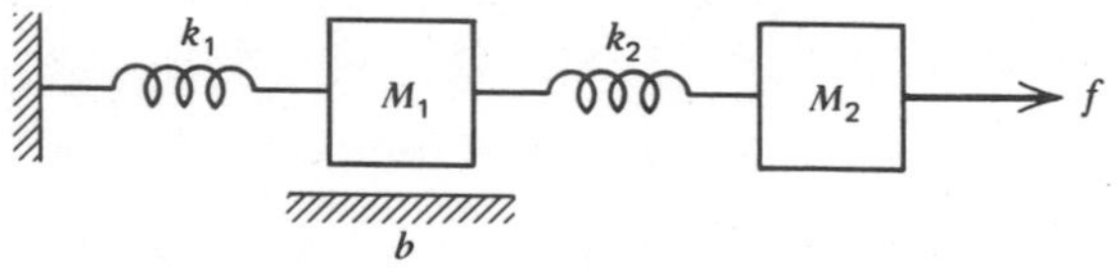

Figure 4.19. Problem 4.4.

4.5 The following data relate to the system shown in Figure 4.20: r_1, r_2 = resistances of the two coils; i_1, i_2 = specified currents: n_1, n_2 = number of turns; μ_o = permeability of free space; A = area of cross-section of the center limb and of the outer limbs; M = mass of the armature; k = spring constant; b = friction coefficient; l_o = length of the spring when $i = 0$. Assume all magnetic energy to be stored in the air gap.

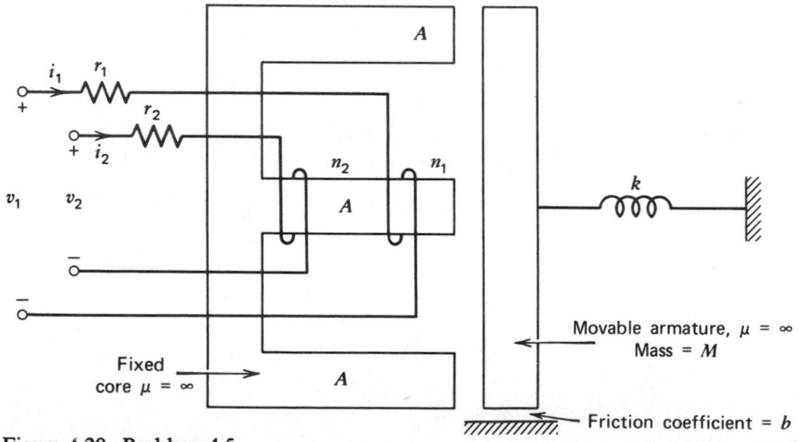

Figure 4.20. Problem 4.5.

(a) If $i_1 = I_{dc}$ and $i_2 = I_m \sin t$, write
   (i) The electrical equations of motion in matrix notation
   (ii) The mechanical equation of motion.
   (iii) If the above equations are nonlinear, identify the nonlinear terms.

(b) If $i_1 = I_{dc}$ and $i_2 = 0$, obtain the quiescent operating point and linearize the equation about this point.

4.6 Two coils have self- and mutual inductances (in henries) as functions of a displacement x (in meter) as follows:

$$L_{11} = L_{22} = 3 + \frac{2}{3x}; \quad \text{and } L_{12} = L_{21} = \frac{1}{3x}$$

The resistances are negligible. Both coils are excited by the same voltage source $v = v_1 = v_2 = 100 \cos 50t$ volt.

(a) Find an expression for the electrical force (i.e., the force of electrical origin).

(b) Calculate the time-average value of the force at $x = 1$.

(c) Does the force tend to increase or decrease x?

4.7 An electromechanical system is shown in Figure 4.21. The core structure is cylindrical. Other data are as shown. The coil has negligible resistance, and has a voltage v across it.

(a) Write down the general electrical and mechanical equations of motion.

(b) Given: $a = 1$ cm; $h = 5$ cm; $l_1 = 4$ cm; $w = 0.1$ cm; and $N = 1000$ turns. Calculate the inductance of the coil, at $x = 0.5$ cm.

(c) For the data in (b) calculate the force on the slug at $x = 2$ cm; and $i = 10$ A dc.

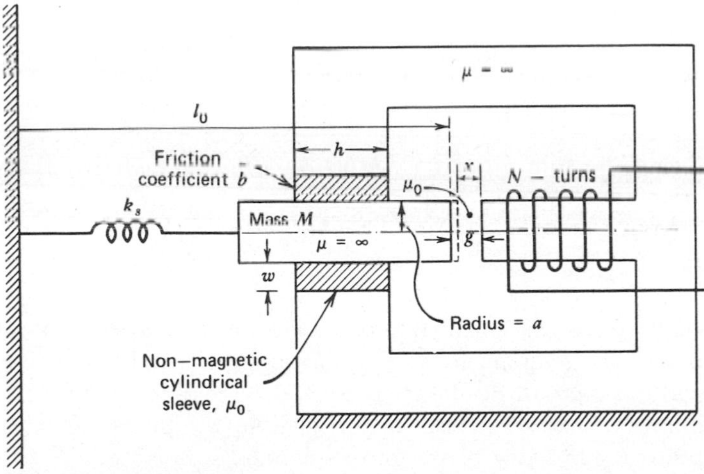

Figure 4.21. Problem 4.7.

4.8 In an electromagnet the i-λ relationship is

$$i=\lambda^3+\lambda x$$

where x is some arbitrary air gap. If λ is the independent variable, (*a*) find an expression for the electromagnetic force at $x=0.5$ and at $x=1$. (*b*) Does the force change with x? Explain.

4.9 The electromagnet of Problem 4.8 is current-excited with 3 ampere direct current in an N-turn coil. If the magnet is required to develop a force of 12 newton at $x=1.0$ cm, calculate N, in terms of effective permeance P.

4.10 Given the electromechanical system shown in Figure 4.22, derive an expression for the electrical force developed. State all the simplifying assumptions made in this derivation. Write down the electrical and mechanical equations of motion. Find the equilibrium position. What is the direction of the electrical force? Next assume that the spring is detached and friction ignored. An external force is applied to pull the plunger through a distance d. Determine the change in the stored energy in the coupling field. If there is a change in the stored energy, where does it go? Or, where does it come from, as the case may be?

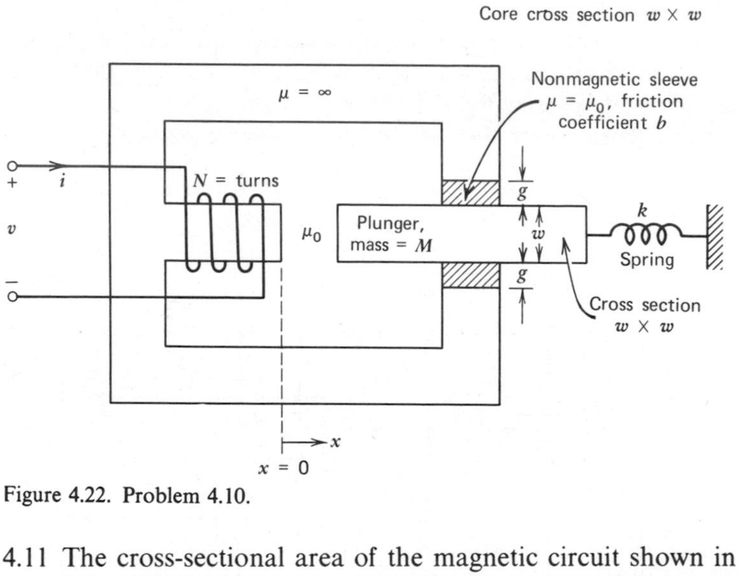

Figure 4.22. Problem 4.10.

4.11 The cross-sectional area of the magnetic circuit shown in Figure 4.23 is a. The slide has a viscous damping force $F_{\text{fric}}=b\dot{x}$ opposing motion of the slide and acting at the support. Position x is equal to x_o is obtained with no electrical excitation. With $v=V_m\cos\omega t$ and negligible winding resistance, determine the following in terms of the given symbols:

(a) The magnetic force acting on the slide.

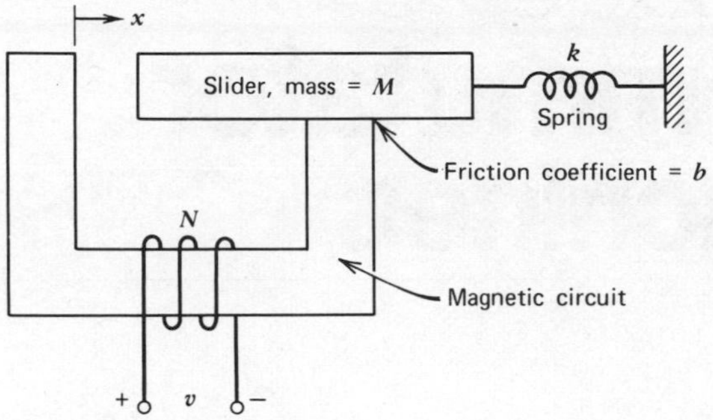

Figure 4.23. Problem 4.11.

(b) The distance x as a function of time in the steady state, and the condition for the validity of this expression.

(c) The average power supplied by the voltage source if the magnetic material is lossless.

4.12 For the system shown in Figure 4.9 with the parameters defined in the text (Section 4.5.5), determine the flux variation as a function of time for a 2-A input current.

Chapter 5
DC Commutator Machines

In this chapter, a family of rotating electromagnetic devices, dc commutator machines are introduced. While analyzing dc commutator machines we shall make use of many of the principles developed in earlier chapters, particularly the magnetic circuit concepts of Chapter 2 and the force relationships developed in Chapter 4. The material that we shall present in this chapter would usually be entitled only "dc machines." We have inserted the word *commutator* for a number of reasons. First of all, the commutator is the distinguishing characteristic of the devices discussed in this chapter. Without the commutator, these machines would be indistinguishable from many other types of machines. The commutator is a mechanical rectifier/inverter that permits connection to a dc source, and, in the process, provides this machine configuration with some most useful characteristics as a motor, as a generator, and as a control device.

Second, the commutator type is not the *only* type of dc machine. In fact, it is not a truly dc machine at all—if, by "dc" we mean a device whose currents and voltages are undirectional for a given condition of speed and torque. Electronics-oriented readers would probably object to the intrusion of a "mechanical" recitifier/inverter as part of a device used so frequently in electronics and control systems and ask, "Why not supply the rectifier/inverter function with solid state devices?" Such devices are fairly common today and are known as *brushless dc machines*. And, if one is still wondering what *is* the truly dc machine, it is a device known as the *homopolar machine*, which is a grandchild of the Faraday disc generator, developed by Michael Faraday in the 1830s. A liquid metal version of this configuration, has been used in various aerospace applications. Therefore, we use "dc commutator machines" in order to typify a specific configuration of dc machines.

The term "dc" is used to describe this generic class of machines to indicate that they are conventionally energized from a source of dc electrical energy. There are several types of machines with commutators that are normally operated from an ac source: these include *repulsion motors* and *polyphase commutator machines*. The latter machines are used in Europe for many of the applications for which dc commutator machines are used in the United States. Finally, there is a very common type of dc commutator motor, known as the *universal motor*, that is operated from low-frequency ac sources.

The dc commutator machine is a most versatile rotating device and is built in a wide range of sizes, from small control devices in the 1-W power rating up to very large motors of 10,000 hp or more used in rolling mill applications. Their principal applications today are as industrial drive motors, especially where large magnitude, precisely controlled torque is required. Such motors are used in steel and aluminum rolling mills, in traction motors, in overhead cranes, fork-lift trucks, electric trains, electric vehicles and golf carts. Commutator machines are used in portable tools supplied from batteries, in automotive vehicles as starter motors, blower motors, and in many control applications as actuators and as speed or position sensing devices.

There are almost no modern uses of commutator machines as power generators, although in the early stages of electrical power distribution the Edison three-wire distribution system and the dc commutator generator were the principal means of supplying electrical energy to both residential and industrial customers. Vestiges of these systems still remain in the central-city districts of some large cities. However, it should be noted that the dc commutator machine is a bilateral device and many dc commutator motors often operate as generators in a "regenerative" or "dynamic braking" mode, especially in traction applications.

5.1 DESCRIPTION OF A DC COMMUTATOR MACHINE

Figures 5.1 to 5.4 show examples of various types of dc commutator machines. Physically, the dc commutator machine consists of a rotating member, called the *rotor* or the *armature*, and a stationary member referred to as the *stator*. There are two types of geometry that are universal in all types of rotating machines and which are designated by the air-gap dimensional coordinate, namely, the *radial air gap*—which is by far the most common. The second geometry, the *axial air gap*, is often associated with what are called "pancake motors." In radial air-gap geometries, the rotating element is generally cylindrical in shape and fits inside of the stator, which is a hollow cylinder or is annular in shape and concentric with the rotor. The air gap is the radial distance, measured on a radius with the origin on the axis of rotation of the rotor, between the outer diameter of the rotor and the inner diameter of the stator. In axial air-gap geometries, both rotor and stator are disclike in shape and the air gap is a

Figure 5.1. Cutaway of a 4000-hp 700-V dc motor. (Courtesy General Electric Company.)

General Electric
"JB" traction motors
. . . lightweight • long life
• versatile performance
2) Terminal Studs are sized for speedy, dependable connection of external leads.
6) Brush Interface Systems are optimized for speed, currents, voltage and load conditions, providing simple brush replacement and greater life expectancy.
4) Shielded Ball Bearings, with high radial and axial thrust capacity, are prelubricated for long-life operation without regreasing.
5) Commutators utilizing clamp ring construction, coupled with silver-bearing copper and premium selected mica, contribute to a solid, stable commutator unit.
1) Magnetic Steel Tubing for the motor frame, coupled with one-piece rugged end shields are machined with rabbets to assure accurate alignment and provide rigid support for the bearings.
3) High-Temperature Insulation—Both armature and field coils feature bar type conductors insulated with materials that have been carefully selected and tested to assure long, dependable motor life.
1
2
3
4
5
6

Figure 5.2. Direct current commutator motor for traction application. (Courtesy General Electric Company.)

Figure 5.3. Typical control motors and generators. (Courtesy Servo-Tek Products Company.)

distance parallel to the axis of the concentric discs between two adjacent surfaces of the discs.

The theory and analysis developed in this text apply equally to either geometry. However, where geometry does enter into the formulation of an equation, the development will be in terms of the more common radial geometry. In either geometry, there is a bearing and lubrication system to support the

Figure 5.4. Rotor body of a 3500 hp 220 V 40/80-rpm dc motor—one half with stampings. (Courtesy Brown Boveri Company.)

rotor and to permit its rotation with minimum frictional losses and minimum eccentricity. The bearings are supported by or fixed to the housing, the outer structural portion of the stator. The housing configuration varies with the type of ventilation system used to cool the machine and with the means used to prevent dirt and other foreign particles from entering the machine. Small machines are often hermetically sealed, and internally produced heat is transferred to the outer machine surface by conduction only. Some of these structural features may be observed in the photographs of Figures 5.1 to 5.4.

Electromagnetically, the dc commutator machine consists of two or more sources of magnetic excitation coupled magnetically by means of a magnetic circuit. The two principal excitation systems are the *field* or exciting system, which may be an electrical winding or a permanent magnet and which is located on the stator, and the *armature winding*, located on the rotor (Figure 5.5). Thus, the dc commutator machine is a *doubly excited* machine. The magnetic circuit consists of the armature magnetic material, the air gap, the field poles (soft magnetic material or permanent magnets), and the yoke which is often an integral part of the machine housing. Additional sources of excitation result from the use of interpoles and compensating windings (Figure 5.6), which will be described later in this chapter in connection with commutation problems.

Figure 5.5. Direct current armature with double commutator for a 2000-kW 450-rpm generator. (Courtesy Brown Boveri Company.)

The function of the field system obviously is to supply energy to establish a magnetic field in the magnetic circuit. The use of an electrical field winding (Figure 5.7) gives the great diversity and variety of performance characteristics that typify dc commutator machines. Permanent magnet excitation of the field system is often less costly and occupies less space than electric excitation and eliminates the need for a separate electrical source of energy.

The armature winding is often referred to as the power winding of a dc commutator machine, since the machine's electromagnetic torque is a function of the armature winding current, and the armature winding terminals are connected to the external power source through the *commutator/brush* system. As has already been noted, the commutator/brush system serves as the electrical connection. The system acts as a mechanical switching device between the external armature circuit and the armature windings within the machine, in which currents and induced voltages are time-varying and reversing in polarity. As a bilateral device, the commutator/brush system is somewhat analogous to the action of an antiparallel pair of rectifiers.

The rotor or armature of a dc commutator machine with radial air gap is cylindrical in shape and is mounted on a shaft that is supported on the bearings (Figure 5.5). One or both ends of the shaft are brought out beyond the machine's housing. This is the mechanical input/output "terminal" of the

Figure 5.6. Stator of a 1030-kW dc motor showing compensating bars and interpoles. (Courtesy Brown Boveri Company.)

Figure 5.7. Magnet pole of a 2550-kW dc motor for rolling mills. (Courtesy Brown Boveri Company.)

machine and may be coupled mechanically to a loading device or source of mechanical energy, depending upon whether the machine is to be operated as a motor or generator. In many machines, the "active" portion of the rotor, that is, the portion serving as a magnetic conductor in the magnetic circuit, is a hollow cylinder of soft magnetic material, supported on the shaft by a purely structural member known as a web or spider. This form of construction is used to reduce the weight over a construction using a solid cylinder.

On the surface of the rotor are a number of parallel, axial slots—usually evenly spaced around the circumference of the rotor, in which are laid the armature coils that make up the armature winding. In larger machines, these coils are preformed and laid in the slots according to winding rules, which will be discussed in Section 5.2. Various types of electrical insulation are used to insulate the coils from the conducting surfaces of the steel slots, depending upon the voltages developed in the coils, the type of cooling or heat transfer technique used to remove heat from the coils, and the type of environment in which the coils will be operated.

For mechanical support, protection from abrasion, and for further electrical insulation, nonconducting slot liners are often wedged in between the coils and the slot walls. The magnetic material between the slots are called "teeth." Several cross-sectional views of common slot/teeth geometry are shown in Figure 5.8. The slot cross-section, although it may appear to be a trivial design detail, has a significant influence upon the ultimate performance characteristic of the machine, and affects such things as the armature inductance, magnetic saturation in the teeth, induced eddy current losses in the stator poles, as well as the cost and complexity of laying the coils in the slots.

The mechanical and structural design of rotating machines is a very challenging subject in itself and an area that is continuously changing with the advent of improved magnetic, electrical, and insulating materials, the use of improved heat transfer techniques, and the development of new manufacturing processes. Reference 1 gives a good treatment of many of the more detailed structural features of both ac and dc rotating machines. Our interest, however, is more in the development of the electromagnetic principles that make machines work.

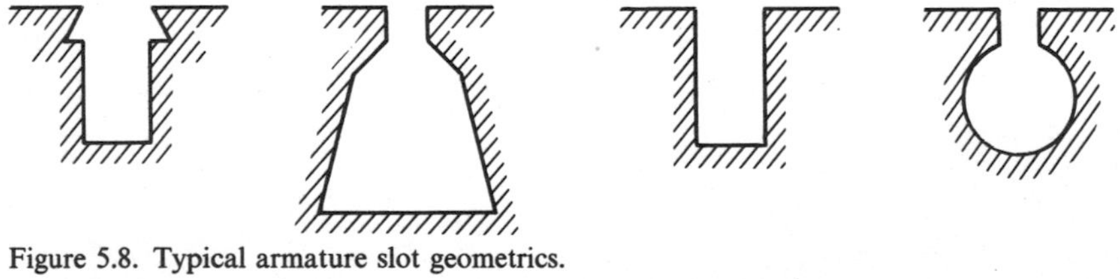

Figure 5.8. Typical armature slot geometrics.

5.2 ARMATURE WINDINGS

Many types of armature windings have been used in dc commutator machines, the earliest configuration being the Gramme ring type, which was wound on rotors that were toroidal in shape. In present-day radial air gap machines, most armature windings are either of the *lap* or *wave* type. Small rotors are usually machine-wound by turning the rotor axis end-over-end while wire is fed into the slots one turn at a time. In larger rotors, the winding is composed of preformed coils which are laid in the slots (Figure 5.5). References 1 and 2 provide further descriptions of machine windings.

An armature winding is a *continuous winding*, that is, it has no beginning or end. It is composed of a number of *coils* in series, one end of which is electrically connected to one commutator bar. A coil may consist of one *turn* or loop of wire or may be one of a number of turns in series (Figure 5.9). Only the beginning of the first turn and the end of the last turn of a series of turns making up a coil are connected to the commutator bars. The manner in which the coil ends are connected to the commutator bars determines whether the winding is lap or wave.

The coil shapes of the two types of windings are generally the same, and it is not possible to determine the type of winding from a visual inspection of fully wound armature and commutator. The differences between the two types is illustrated in Figures 5.10 and 5.11. To understand these winding layouts, visualize that an axial "slit" has been made on the surface of the rotor and on the commutator, and the two cylindrical surfaces (rotor and commutator) have been spread out flat. The resulting surfaces would appear as in Figures 5.10 and 5.11 if we ignore for the moment the fact that another set of axial lines should appear on these diagrams to indicate the slot sides. The term *lap* arises from the

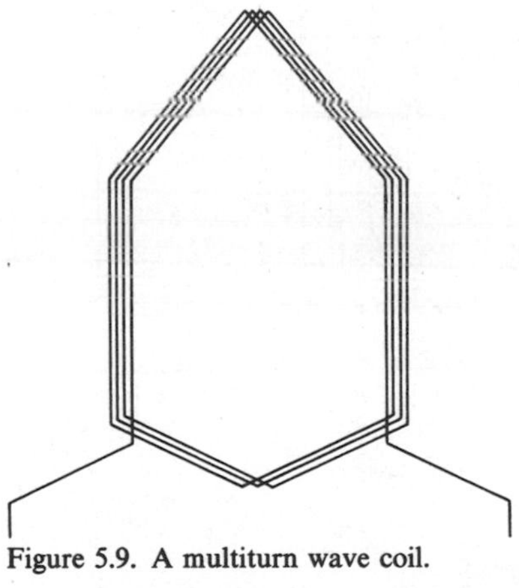

Figure 5.9. A multiturn wave coil.

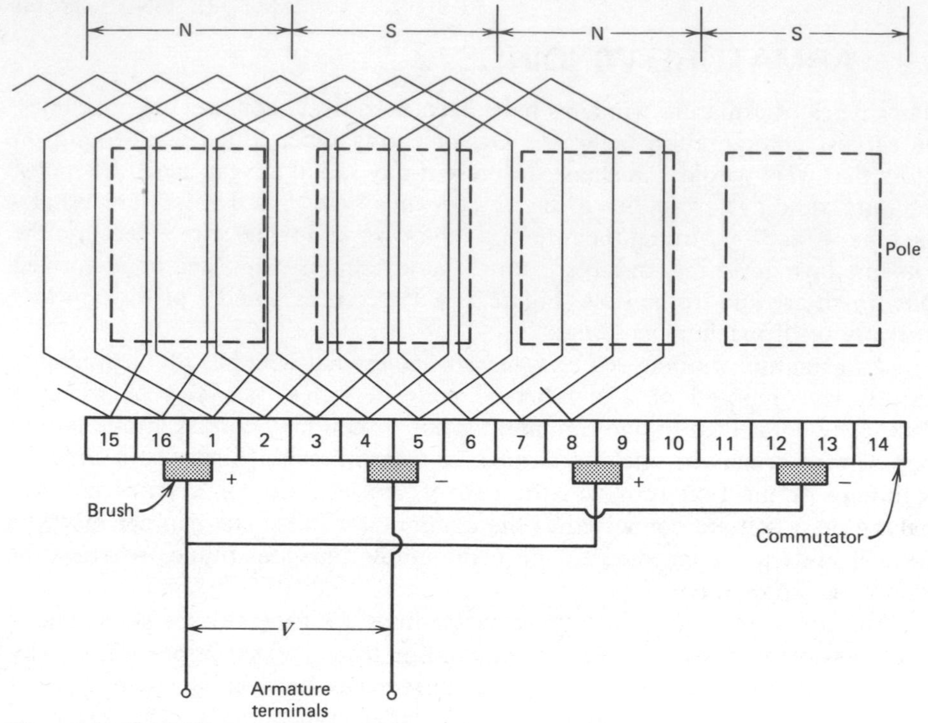

Figure 5.10. Partial layout of a lap armature.

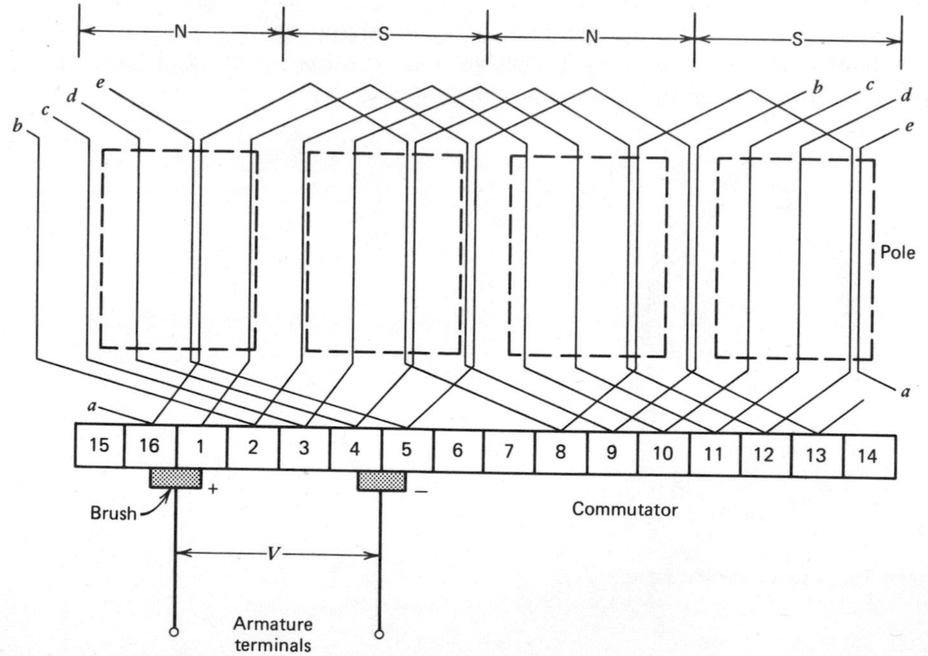

Figure 5.11. Partial layout of a wave armature winding.

fact that as the winding progresses around the rotor, the coil end *laps back* on itself and is connected to the next adjacent commutator bar from the bar to which the coil beginning is connected. The origin of the term *wave* is more obscure, except that when the winding is drawn as in Figure 5.11 there is a general pattern of a wave.

The two sides of a coil are generally laid in slots that are one *pole pitch* apart. Pole pitch is a means of stating the distance between the centers of the field poles and may be expressed in units of slots (total number of slots divided by total number of poles) or in degrees,

$$\alpha = \frac{360}{p} \tag{5.1}$$

where p = number of poles. A winding in which coil sides are a pole-pitch apart is known as a full-pitch winding. For various reasons, it is sometimes desirable to have the coil sides less than a pole-pitch apart. Such a winding is known as a short-pitch or fractional-pitch winding. This practice is much more common in the stator windings of ac machines than in dc armature windings.

The external connections to the armature windings are made through the brushes resting on the commutator, which are held in stationary, fixed positions as the commutator and rotor turn. The brush positions are likewise located one pole-pitch apart. In a lap winding, a number of brush positions equal to the number of poles is required. Half of the positions are at a positive polarity, half at a negative, and the positive and negative groups are paralleled through external electrical connections.

With a wave winding, only two brush positions are required, one positive and one negative, although more positions are often used in order to decrease the current density in the brushes. The principal difference between the two windings, as far as the electrical performance of the machine is concerned, is in the number of parallel electrical paths through the winding between the positive and negative terminals of the armature. Designating the symbol "a" for number of parallel paths between the armature winding terminals, we have

$$\begin{aligned} &\text{For lap winding} \quad && a = p \\ &\text{For wave winding} \quad && a = 2 \end{aligned} \tag{5.2}$$

In a problem at the end of this chapter the reader can verify this conclusion. Since the lap winding provides more parallel paths in the armature winding than the wave winding in all but two-pole machines, the lap winding is generally more suited for low-voltage, high-current applications. The wave winding is also used more frequently on high-voltage applications. A common form of winding in an axial airgap machine is shown in Figure 5.12.

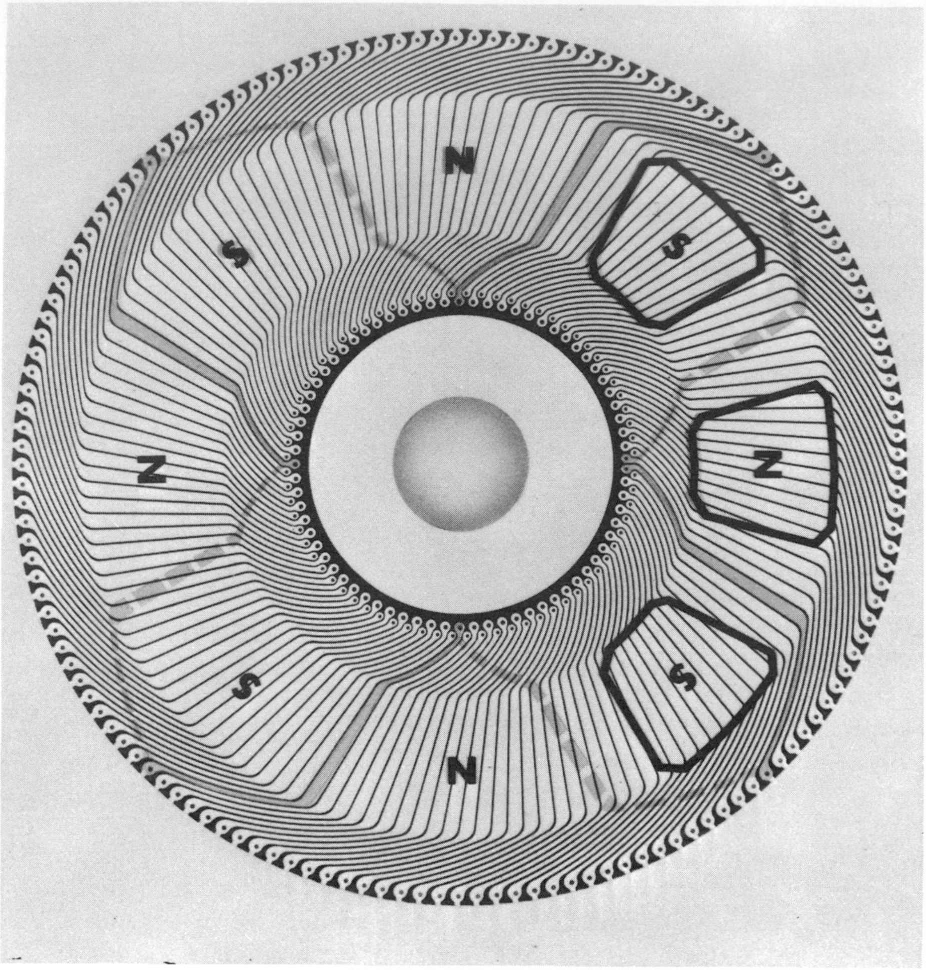

Figure 5.12. Printed circuit winding layout for a disc-type motor. (Courtesy PMI Motors, Division of Kollmorgen Corporation.)

5.3 ARMATURE VOLTAGE

In this section, we shall develop expressions for the voltage induced in the armature winding as the armature rotates in a magnetic field that is produced by the stator poles. There are several possible forms in which this voltage can be expressed, such as in terms of self- and mutual inductances of the field and armature windings, as discussed in Section 4.7. In analysis of dc commutator machines it has been customary to develop armature voltage expressions in terms of the field flux rather than in terms of self- and mutual inductances of Section 4.7. This emphasis relates armature voltage to the physical dimensions of the field system. Since a great many dc commutator machines are excited by permanent magnets rather than by electrical winding, it is almost impossible to use inductance expressions in machine performance calculations.

Therefore, we shall use the conventional approach in terms of magnetic flux and note that this component can usually be calculated in a straightforward manner and related to the physical dimensions of the field poles and to the electrical winding by means of the principles developed in Chapter 2. When approaching a problem involving rotating machines, the general guidelines for choosing the appropriate model given in Section 4.5 should be kept in mind.

We shall use a very simple form of a dc machine for the following development, Figure 5.13, which is a cross-sectional view of a simple two-pole machine. Two conductor cross sections are shown, which are the sides of a single one-turn coil. The reference axis for the coil position, θ_m, is the vertical axis. If ϕ_p is the total flux emanating from a pole and is assumed to be perfectly horizontal, the flux linking the coil as a function of coil (rotor) position is

$$\phi = \phi_p \cos\theta_m \tag{5.3}$$

If the rotor is turning at an angular velocity, ω_m rad/s, the position as a function of time is

$$\theta_m = \omega_m t \tag{5.4}$$

From Faraday's law, for a single turn, (2.25), the voltage induced in the coil is

$$e = \phi_p \omega_m \sin\omega_m t \tag{5.5}$$

If there are "p" poles on the stator instead of the two shown in Figure 5.13, note that frequency of the induced voltage in the coil is increased by $p/2$ times. It is

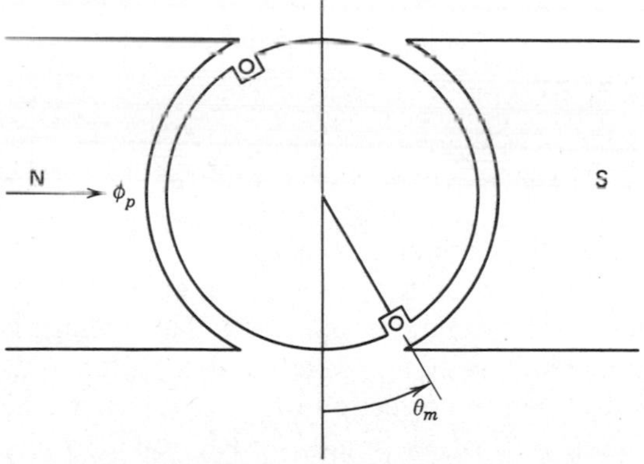

Figure 5.13. Simplified axial cross section of a two-pole machine showing only one armature coil.

just as if the speed of rotation had been increased by $p/2$. We can redefine the angle, θ_m, shown in Figure 5.13 as an equivalent "electrical" angle

$$\theta_e = \frac{p}{2}\theta_m \tag{5.6}$$

The units of the electrical angle are often referred to as "electrical radians" or "electrical degrees" and are related to the true mechanical angle by the number of pole-pairs, $p/2$. Replacing θ_m by θ_e in the above analysis results in an expression for induced voltage in an armature coil

$$e = \phi_p \frac{p}{2}\omega_m \sin \omega_m \frac{p}{2} t \tag{5.7}$$

For N turns in series, (5.7) becomes

$$e = N\phi_p \frac{p}{2}\omega_m \sin \frac{p}{2}\omega_m t \tag{5.8}$$

The voltage appearing at the terminals of the armature is the half-wave average of the maximum value of instantaneous voltage because of the action of the commutator.

$$E = \frac{p}{2\pi}\int_0^{2\pi/p} e \, d(\omega_m t) = \frac{2}{\pi}\left(\phi_p N \frac{p}{2}\omega_m\right) = \frac{Np\phi_p\omega_m}{\pi} \tag{5.9}$$

It has been the custom in dc commutator machine practice to express series turns in terms of "conductors" Z, (coil sides), and parallel paths, a or

$$N = \frac{Z}{2a} \tag{5.10}$$

Making this substitution for N in (5.9) gives

$$E = K_a \phi_p \omega_m \tag{5.11}$$

where $K_a = (Zp/2\pi a)$ and is often called the armature constant.

The above expressions give the armature-induced voltage as a function of motor speed, field pole magnetic flux, and winding configuration. This voltage is entirely independent of whether the machine is being operated as a motor or as a generator; that is, it is not dependent upon the direction of current flow in the armature. In motor operation, this voltage is often termed "back emf." In any event (5.11) is known as the emf equation.

5.3.1. AN ALTERNATIVE DERIVATION

The emf equation may also be derived, perhaps more simply, by considering only one conductor of Figure 5.13. This conductor rotates in the field system such that the total flux, ϕ, cut by the conductor in n revolutions is given by

$$\phi = p\phi_p n \tag{5.12}$$

Thus, the flux "cut" per unit time giving the voltage induced in the conductor is simply

$$e = \frac{p\phi_p n}{60} \tag{5.13}$$

where n = speed in rpm. Finally, the total voltage induced in the winding can be expressed as

$$E = \phi_p \frac{nZ}{60}\frac{p}{a} = \frac{Zp}{2\pi a}\phi_p \omega_m \tag{5.14}$$

where the symbols have been defined earlier.

The graphical representation of (5.11) is one of the chief tools in the analysis of electrically excited dc commutator machines and is known as the *magnetization or excitation curve*. The curve is obtained by driving the machine at a constant mechanical speed, ω_m and varying the field current (which varies ϕ_p) while measuring the induced voltage, E, across the open-circuited armature. Typical magnetization curves at two different mechanical speeds are shown in Figure 5.14. This experimental procedure is termed the "no-load" test, and is described in Section 5.10.

If certain physical dimensions of the machine such as air-gap length and pole face area are known, the units of the ordinates in Figure 5.14 can be converted to the units of B and H, and machine magnetization curves are often expressed in these units, as well as in units of ϕ_p versus I_f. If appropriate flux sensing instruments are available, such as Hall probes or an integrating fluxmeter, the variables B or ϕ_p can be measured directly in the machine air gap at zero speed in a manner similar to that used in any other magnetic circuit. However, the open-circuit voltage measurement is much simpler than the typical flux or flux-density measurement and a voltmeter is generally more available than a flux measuring instrument. Therefore, the voltage versus ampere turns (or just field current) are commonly used to describe the magnetic characteristics of the machine magnetic circuit.

The form of (5.11) in terms of flux is not always convenient. For this reason, (5.11) is often expressed in terms of field current rather than in terms of field flux as

$$E = K_t I_f \omega_m \tag{5.15}$$

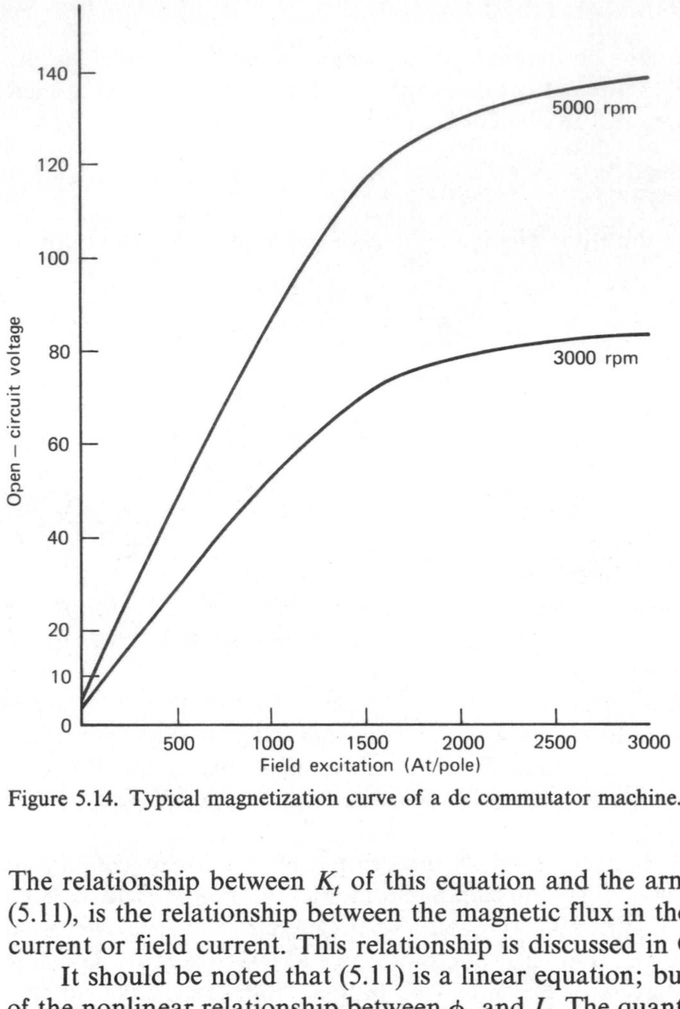

Figure 5.14. Typical magnetization curve of a dc commutator machine.

The relationship between K_t of this equation and the armature constant, K_a, of (5.11), is the relationship between the magnetic flux in the pole and the exciting current or field current. This relationship is discussed in Chapter 2.

It should be noted that (5.11) is a linear equation; but (5.15) is not, because of the nonlinear relationship between ϕ_p and I_f. The quantity K_t is nonlinear and varies with the slope of the magnetization curve of Figure 5.14. Since this characteristic can be experimentally obtained for a dc commutator machine, the constant K_t is relatively easy to obtain as compared to the K_a in (5.11). Also, many machines, especially control-type machines, are operated in the linear region of the magnetization curve and, in such cases, the nonlinearities of (5.15) can be ignored.

5.4 ELECTROMAGNETIC TORQUE

Let us now return briefly to semantics and examine the two words that compose the title to this section, since these terms will appear frequently in discussions of rotating machines. The second of these words, torque, is probably familiar to the reader and may already invoke a physical image. Torque, often termed *moment* is the product of a force and a lever arm. It is sometimes represented as a vector

$$\mathbf{T}=\mathbf{r}\times\mathbf{F} \tag{5.16}$$

where **F** represents the force and **r** the lever arm or radius about an axis of rotation with a direction outward from the axis of rotation.

The interesting aspect of (5.16) is the direction associated with torque. In electromagnetic rotating machines, **r** is generally the radius of the rotating element of the machine, the rotor, and **F** is the force acting upon a current-carrying conductor at or near the surface of the rotor. In dc commutator machines and many other classes of rotating machines, these two vectors are orthogonal. From the vector algebra of the cross product in (5.16), it is seen that this situation results in a direction for the torque vector along the axis of rotation, that is, parallel to the shaft of the machine. It is seen that there are only two *possible* directions for **T** in this case, depending upon the relative directions of **r** and **F**, both of which are parallel to the axis of rotation. This rather pedantic point is raised, since a common mistake is to associate the direction of torque with the direction of the force, **F**.

Torque is more commonly treated as a scalar quantity. Its units are those of energy or work, as can be seen from (5.16), which are scalar quantities. In SI units, torque units are newton-meters; in English units, pound-foot. The treatment of torque as a scalar quantity is also more amenable to the derivation of electromagnetic torque from variations in stored magnetic energy, as described in Example 4.2.

The second word in the title of this section, *electromagnetic*, implies an interaction between electric and magnetic fields, which is at the core of the energy conversion process in rotating machines. This is best observed from (5.8) above and (4.15) in Chapter 4. These equations describe the mechanism by which electrical energy is converted to mechanical energy, or vice versa, in both motors and generators.

This energy conversion process, although it is the fundamental "raison d'être" of rotating machines is described by a very simple relationship

$$EI_a = T_d\omega_m \tag{5.17}$$

where T_d is the electromagnetically developed torque and I_a is the armature current. The nomenclature here is that used for a dc commutator machine, but the significance of the two sides of (5.17) is the same for all types of rotating machines. The left side of this equation represents electrical power, and the right side represents mechanical power. Notice that (5.17) is bilateral, and describes either motor or generator action.

Equation (5.17) introduces a simple means of deriving an expression for the electromagnetically developed torque of a dc commutator machine. Solving for

T_d and substituting for E from (5.11), we obtain

$$T_d = K_a \phi_p I_a \qquad N-m \tag{5.18}$$

The above derivation may still leave the reader with the feeling that electromagnetic torque is a rather obscure quantity with little physical significance. Therefore, let us derive equation (5.18) from (5.16). This will also give us a chance to introduce the Lorentz force, (2.4). The reader should first observe that the conditions that led to the form of the Lorentz force expressed in (2.6) are applicable to dc commutator machines. Substituting this equation into (5.16) and assuming an orthogonal relationship between r and F, which results in a scalar form for (5.16), we obtain

$$T_d = rBlI_a \qquad N-m \tag{5.19}$$

which is the torque on one armature conductor of length, l_a, and at a radius, r, and carrying a current, I_a. For Z armature conductors connected so as to have "a" paths, (5.19) becomes

$$T_d = \frac{ZBlrI_a}{a} \qquad N-m \tag{5.20}$$

Assuming that the flux density is uniform over the armature area opposite a field pole, ϕ_p, the flux per pole, can be expressed as

$$\phi_p = B \times A_p = B \times \left(2\pi r \frac{l}{p}\right) \tag{5.21}$$

Substituting for B from (5.21) into (5.19) gives

$$T_d = \frac{Zp}{2\pi a} \phi_p I_a \qquad N-m$$

which agrees with (5.18). As in the case of (5.11), it is often desirable to express the torque in terms of field current rather than field flux, such that,

$$T = K_t I_f I_a \tag{5.22}$$

5.5 DIRECT-CURRENT COMMUTATOR MACHINE CONFIGURATIONS

The performance characteristics of a dc commutator machine vary considerably as a function of the manner in which the field and armature circuits are connected to each other. Before describing these connections, let us summarize the three principal equations for dc commutator machines that will be of value in observing the effects of these interconnections:

$$E = K_a \phi_p \omega_m \tag{5.11}$$

$$T = K_a \phi_p I_a \tag{5.18}$$

$$V = E \pm I_a R_a \tag{5.23}$$

where R_a represents the armature circuit resistance.

The last equation relates the armature terminal voltage, V, to the armature induced voltage, E. The sign on the right side of this equation depends upon whether the machine is operated as a motor or generator, the positive sign used with motor operation. The speed relationships of a machine can be observed by substituting (5.11) into (5.23) and solving for speed, which gives

$$\omega_m = \frac{V \pm I_a R_a}{K_a \phi_p} \text{ rad/sec} \tag{5.24}$$

Figure 5.15 illustrates schematically the different methods of connection of the field and armature circuits in the dc commutator machines. The circular symbol represents only the *active* portion of the armature circuit, that is, the generated voltage described by (5.11). It should be realized that there are also passive circuit elements in the armature circuit representing the resistance and inductance of the armature winding and of other windings connected in series with the armature for the purpose of improving commutation. The squares at the sides of the circle in the armature symbol indicate that the connection to the armature is through the brush commutator system.

There is another aspect of a symbolic nature in the circuits shown in Figure 5.15. The field and armature circuits are always drawn at 90° with respect to each other. This represents the actual spacial orientation of the magnetic fields produced by these circuits in a dc commutator machine. It should be noted that although the angle between the two circuits is shown as a physical angle of 90°, this represents, in the general multipolar machine, an *electrical* angle of 90°.

The direction of the arrows in Figures 5.15*d* and 5.15*e* is also symbolic of the direction of the magnetic fields associated with the two field windings. The last four of the circuits shown represent only two machine configurations, the differential-compound and the cumulative-compound configurations. The long-shunt and short-shunt circuits are shown to indicate the two possible means of connecting the shunt field winding in a compound machine. There is only a negligible difference in machine performance characteristics between the two types of connections.

To the five basic configurations shown in Figure 5.15 should be added a sixth: the permanent magnet-excited dc commutator machine. This might be

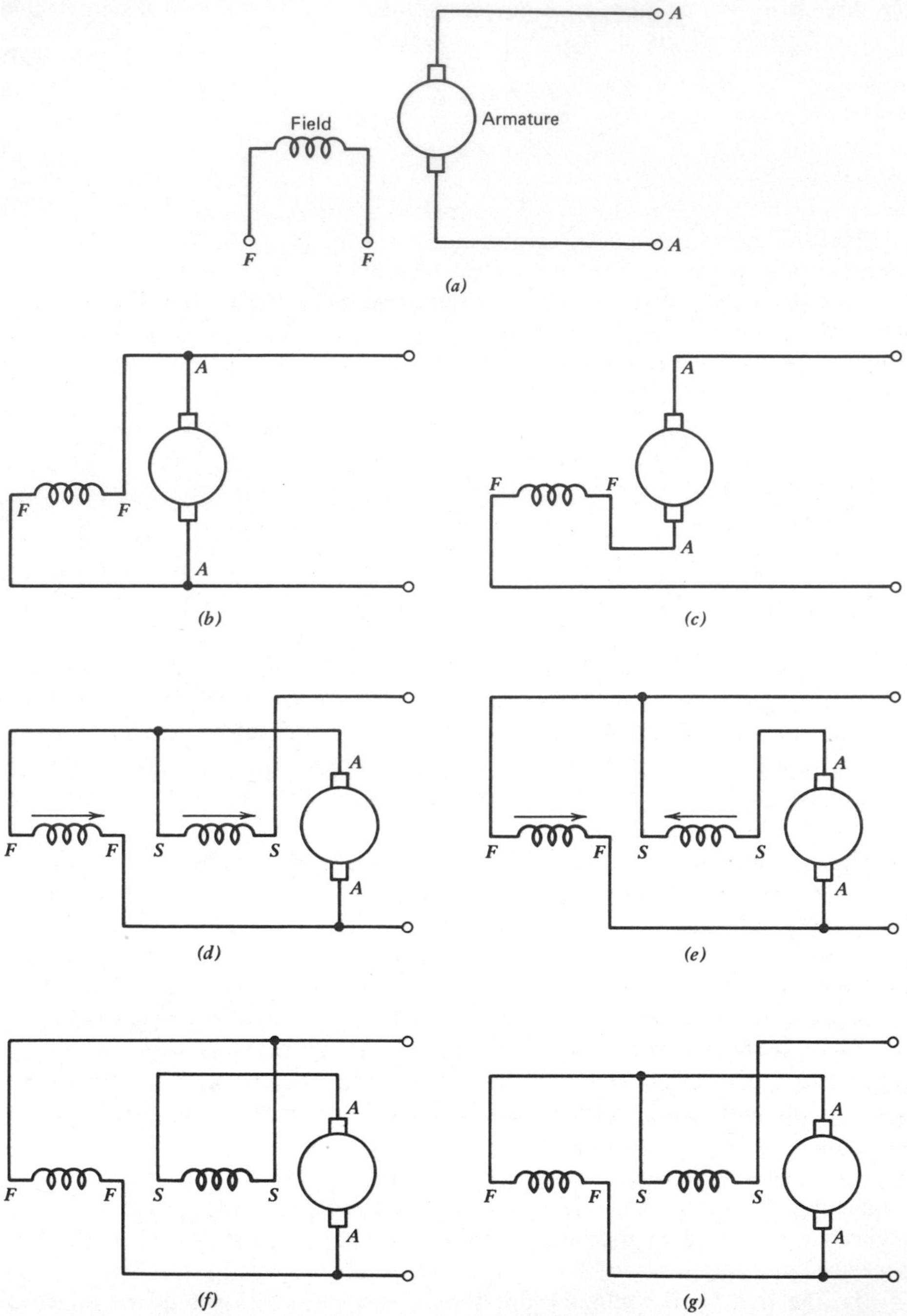

Figure 5.15. Classification of dc machines. (*a*) Separately excited. (*b*) Shunt. (*c*) Series. (*d*) Cumulative compound. (*e*) Differential compound. (*f*) Long-shunt. (*g*) Short-shunt.

considered as a form of the separately excited machine, Figure 5.15*a*, with a separate but constant excitation produced by a permanent magnet rather than the electrically energized winding shown. All of the six configurations are capable of operation as a motor or generator but, as noted at the beginning of this chapter, the motoring mode is the more common application of the dc commutator machine today. Of these six configurations, the first—the separately excited machine—is the most flexible, since full control of both armature and field circuits is possible. The separately excited machine is also the most versatile in that a wide variation of performance characteristics as a motor or as a generator is possible.

A widely used configuration of dc commutator machine is the series field motor, Figure 5.15*c*, which is the principal traction motor in use today and which also finds many other applications as a "universal" motor with either dc or ac excitation.

The shunt configuration, Figure 5.15*b*, is used primarily in low-power, fixed-speed applications, where additional circuitry is inserted into the field circuit. This configuration becomes a variation of the separately excited motor configuration, Figure 5.15*a*. As a generator, the shunt configuration is known as a selfexcited generator and has an interesting characteristic known as voltage "build-up." This phenomenon is the process of the armature supplying its own field excitation and is useful in generator applications where no external source of excitation is available, such as on isolated farms or camping areas.

Voltage build-up requires that some residual magnetism exist in the magnetic circuit of the generator to get the process started. Also, the field circuit must be connected to the armature circuit in such a manner that, for the given direction of armature rotation, the armature voltage causes a current to flow in the field circuit in a particular direction. The direction of current flow is such that the magnetic field resulting from this current *aids* the residual magnetism. The reader may observe the process with the aid of Figure 5.16.

The magnetization curve of a typical machine is shown along with a "field resistance line," which is merely the plot of voltage versus current for the resistance of the field circuit of the selfexcited shunt generator. Assuming that the resistance of the shunt field is linear, this plot is a straight line. The build-up process is as follows: Assume that the field circuit is initially disconnected from the armature circuit. The armature rotates at a given speed, which results in the residual induced voltage E_r. When the shunt field circuit is connected to the armature circuit, there is initially zero current in the field circuit as a result of the effect of the field circuit inductance.

The voltage, E_r, appearing across the armature circuit, eventually causes the current, I_1, to flow in the field circuit. The rate of rise of I_1 depends upon the time constant of the field circuit. However, with I_1 exciting the field, the armature voltage—according to the magnetization curve—builds up to E_1. The increased armature voltage eventually increases the field current to I_2, which, in turn, builds up the armature voltage to E_2.

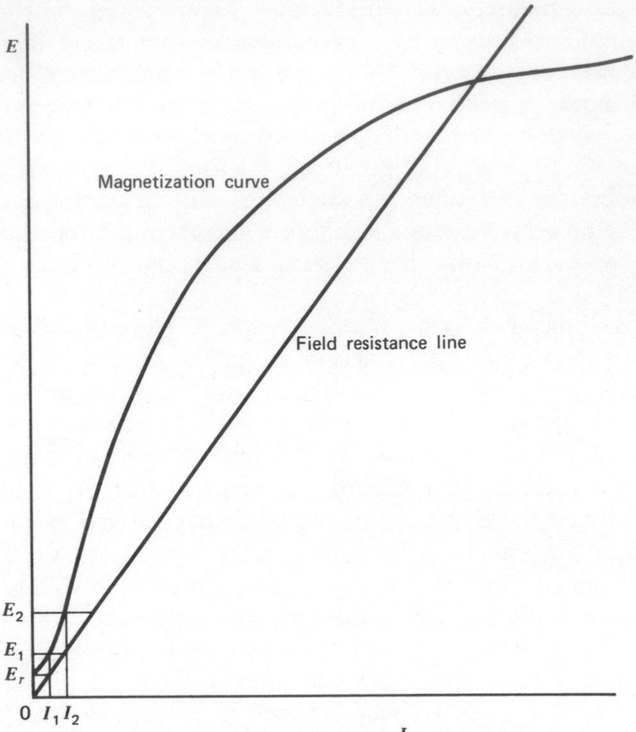

Figure 5.16. Voltage buildup in a self-excited generator.

This process is continued until the magnetization curve and resistance line cross, at which point voltage buildup ceases. Now stable operation as a shunt generator supplying load current can continue. It should be noted that, as the resistance of the field circuit is increased, the field resistance line in Figure 5.16 moves counterclockwise. At some value of resistance, the resistance line is approximately coincident with the linear portion of the magnetization curve. This results in an unstable voltage condition as far as supplying power is concerned. The coincidence has some useful characteristics, however, as a control device. It is sometimes called a "tuned generator" and is marketed under the trade name of Rototrol. If the resistance is increased beyond this value, voltage buildup is not possible.

The compound configurations also have some interesting characteristics. The cumulative compound configuration, Figure 5.15*d*, implies that the magnetic fields of the two field windings are additive. Cumulative compounding of both motors and generators is frequently used to capture the merits of both shunt and series excitation, although, with the advent of electronic control techniques for both armature and field circuits, this method is less frequently applied.

The differential compound machine has rather infrequent application, except, perhaps in the laboratory where the neophyte student of machinery may inadvertently set it up. In this configuration, the fluxes of the shunt and series fields oppose each other and it is possible that, at some state of operation, there may be zero flux in the magnetic field. If this occurs in a motoring mode, (5.24) shows that the motor speed will become dangerously high. This situation is to be avoided. The differential connection has some useful characteristics in control applications but, in general, it results in unstable operation that may damage the machine.

Example 5.1
A six-pole, 30-slot dc commutator generator has a lap-wound armature winding. Determine the armature constant, K_a, for this machine from (5.12).

A lap-wound armature winding has a number of parallel paths equal to the number of poles, in this case, $a=6$. The number of active conductors in a double-layer winding is equal to twice the number of slots (two conductors per slot). Therefore, $Z=2\times 30=60$ conductors. Substituting into (5.11) yields

$$K_a=\frac{60\times 6}{2\pi\times 6}=\frac{30}{\pi}=9.55$$

Example 5.2
The generator described in Example 5.1 is operated with a separately excited field such that the field flux per pole is 0.04 weber/pole. If the generator is driven at a speed of 3000 rpm, determine the no-load armature voltage.

Noting that 3000 rpm is equal to 314.16 rad/s, the no-load voltage is

$$E=9.55\times .04\times 314.16=120 \text{ V}$$

Example 5.3
The generator of Example 5.1 is operated as a motor. The field circuit is adjusted as in Example 5.2 so that the flux is 0.04 weber/pole. It is desired to supply a load requiring a torque of 50 n-m at 4000 rpm. What is the required armature current and voltage to supply this load? Armature circuit resistance is 0.075 ohm.

The armature current can be found directly from (5.18):

$$I_a=50/(9.55\times 0.04)=130.9 \text{ A}$$

The back emf is

$$E=9.55\times 0.04\times 418.9=160 \text{ V}$$

The necessary armature voltage is, from (5.23)

$$V=160+130.9\times 0.075=160+9.8=169.8 \text{ V}$$

Example 5.4
A certain separately excited generator has the magnetization characteristic shown in Figure 5.14. The generator is driven at 5000 rpm with a field excitation of 2000 ampere-turns and supplies a load current of 150 A. The total armature circuit resistance is 0.08 ohm. Determine the terminal voltage, the power output, the electromagnetic power, and the electromagnetic torque input required at the stated load condition.

From Figure 5.14, the no-load voltage at 2000 ampere-turns is 131 V. At the given load current, the terminal voltage (neglecting armature reaction is)

$$V = 131 - 150 \times 0.08 = 119 \text{ V}$$

The output power is $P = 119 \times 150 = 17{,}850$ W. The electromagnetic power is $P_e = EI = 131 \times 150 = 19{,}650$ W. The electromagnetic torque input is $19{,}650/523.6 = 37.5$ N-m.

5.6 ARMATURE AND FIELD CIRCUIT MAGNETIC CHARACTERISTICS

There are several types of diagrams that are of value for illustrating the spacial relationships among magnetic fields produced by different sources of excitation. Figure 5.17 illustrates in a schematic manner the cross section (perpendicular to the axis of rotation) of a two-pole dc commutator machine. In this simplified representation, the commutator is not shown, but the brushes that connect to the external armature circuit are shown in proper physical location to the "active" portions of the armature conductors, that is, the portions located in the armature slots.

The reader should relate this type of diagram to the winding layout diagrams of Figures 5.10 and 5.11. The two-pole machine is used to simplify the diagram and the subsequent explanation. In a multipole machine with "p" poles, the pattern shown in Figure 5.17 would be repeated $p/2$ times around the circumference of the armature. This form of machine representation is independent of whether the winding is lap-wound or wave-wound. It is most nearly realized in a practical machine by the printed-circuit type of winding (Figure 5.12) in which there is no commutator and the brushes make electrical contact with the conductors. The reader should verify that field flux, and armature rotation for the directions of armature currents, shown in (Figure 5.17), represent motor operation.

The field poles and their associated magnetic flux, ϕ_f, can typify either electric excitation or permanent magnet excitation and thus, requires little further discussion at this point. The magnetic flux resulting from the armature current, ϕ_a, is seen to be orthogonal to the field flux and colinear with the brush axis. In this diagram, the directional arrows representing magnetic flux are not to be interpreted as vectors. Rather, they represent the center lines or axes of the

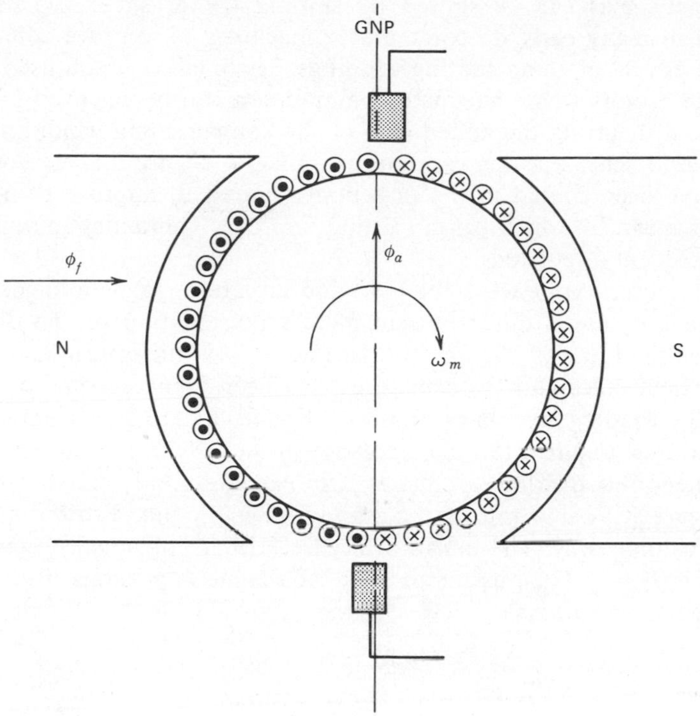

Figure 5.17. Brushes on the geometric neutral plan (GNP).

magnetic flow lines or flux lines produced by the field and armature mmfs. The flux directional arrows are often replaced by flux density vectors, but this type of vector representation can describe flux density at only one point, namely, along the horizontal and vertical axes of Figure 5.17.

The important point to observe is that the direction of the armature flux lines or flux density resulting from armature current along the horizontal axis is *stationary in space* and *independent of the armature velocity of rotation.* This situation results from the action of the brushes, for, as a conductor passes a brush location, the current in this conductor reverses direction. It is this process that is termed *commutation*. Referring to Figure 5.17, all of the conductors in the left half of the armature always have current flow in the direction shown. Those in the right half, also, always have current in the opposite direction even though the conductors are continuously rotating in the direction shown. This condition exists, assuming no change in the direction of field excitation, in the direction of rotation, or in the direction of armature current in the external armature circuit. The interaction of the stationary magnetic field, which results from the flow of armature current, with the magnetic field produced by the field winding, is termed *armature reaction*.

Since the brush axis determines the axis of the magnetic field of the armature, the magnetic axis can be shifted by shifting the brush axis. This technique was used in many early dc commutator machines to improve commutation before the advent of compensating windings. The method is still used in automotive starter motors where adequate commutation can be achieved by this technique without incurring the added cost of the compensating windings. Figure 5.18 illustrates a schematic representation similar to Figure 5.17 except that the brushes have been shifted in the direction required to improve commutation. The mechanism by which brush-shifting improves commutation will be explained in subsequent discussions.

To observe the interaction between the field and armature magnetic fields, it is more convenient to use a different schematic representation of the dc commutator machine than that of Figure 5.17. In Figure 5.19, the circumferential surfaces of the field poles and the armature have been "unwrapped" in a manner similar to the winding layouts of Figures 5.10 and 5.11 to give a linear diagram of these surfaces. Figure 5.19*a* illustrates the linearized air-gap surfaces. The mmf and magnetic flux distribution due to field excitation only are shown in Figure 5.19*b*. Figure 5.19*c* illustrates the mmf and magnetic flux distribution due to armature excitation only. The mmf's in Figure 5.19 are obtained by the application of Ampere's law, (2.2), assuming that the magnetic portions of the machine are ideal magnetic materials.

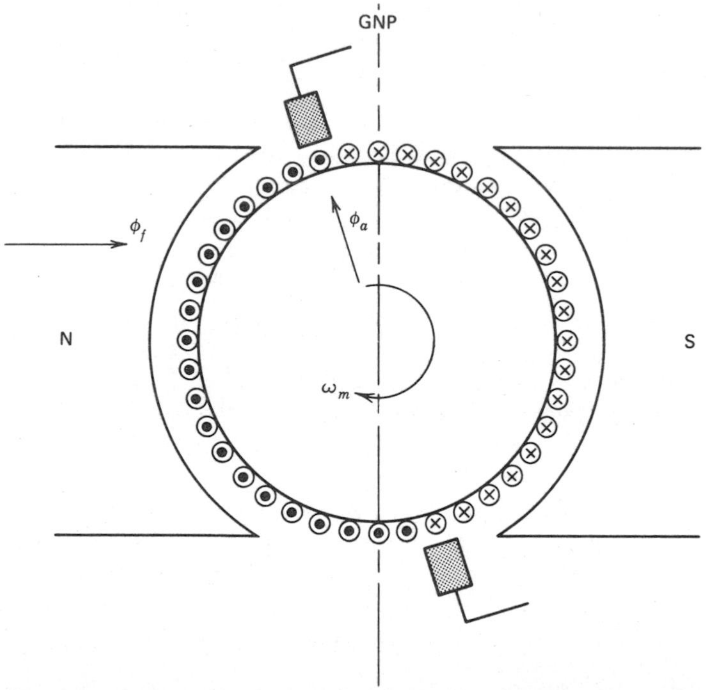

Figure 5.18. Brushes shifted from the geometric neutral plan (GNP).

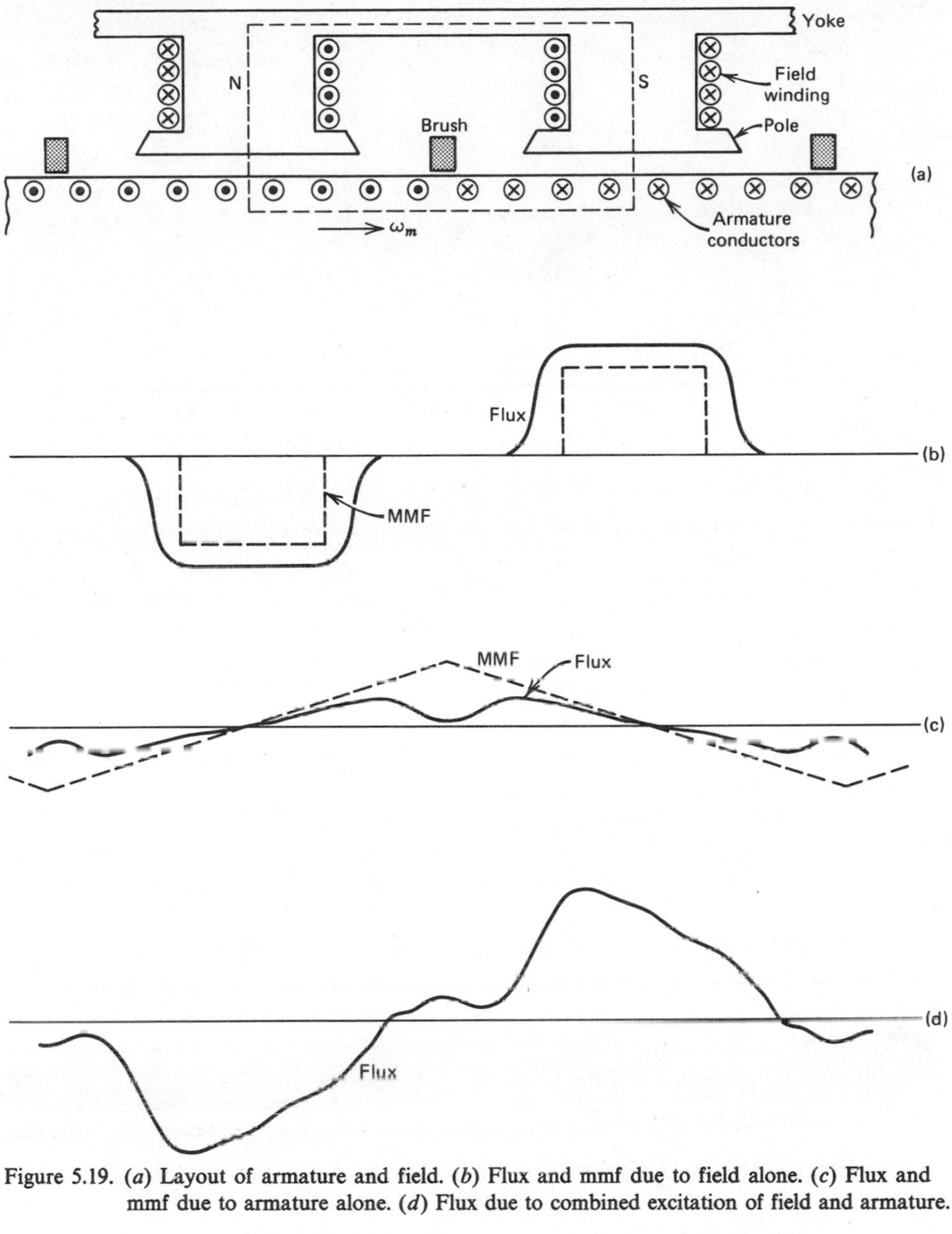

Figure 5.19. (*a*) Layout of armature and field. (*b*) Flux and mmf due to field alone. (*c*) Flux and mmf due to armature alone. (*d*) Flux due to combined excitation of field and armature.

To apply this relationship, the closed path of integration should be chosen to include all magnetic sources within a pole pitch, (5.1). Such a path is shown in Figure 5.19*a* through the center lines of the two magnetic poles. The mmf due to the field windings alone, Figure 5.19*b*, is a fairly straightforward application of (2.2) and should not require further explanation. The magnitude of the field mmf obtained in this manner is $N_f I_f$.

The armature mmf distribution is more difficult to visualize. For the path illustrated in Figure 5.19*a*, the mmf due to the armature conductors is zero, since an equal number of conductors carrying positive currents and negative currents is included in the closed path. As this path is moved to the right, an excess of positive current-carrying conductors is included within the path, resulting in a positive mmf. If the geometry of the closed path is held constant as the path is moved, the mmf increases in a step function pattern as conductor boundaries are passed. The triangular function shown in Figure 5.19*c* is an averaging of these step changes in mmf.

Strictly speaking, this smooth, triangular function represents the mmf of a "current sheet." The function occurs when a uniform circumferential distribution of armature current of one polarity between brush positions is followed by a uniform armature current distribution of the opposite polarity between the next set of brush position. The smooth triangular function is approached in a machine with discrete conductors and slots if there are a large number of narrow, deep slots between brush positions.

The meaning of these mmf diagrams can be interpreted from (2.2). The mmf described by these diagrams is the mmf that occurs within one pole pitch as a function of the circumferential nature of the location of the closed path as it crosses the air gap. Since there is no reluctance drop in the magnetic members traversed by the closed path (under the assumption of ideal magnetic materials) the entire reluctance drop of $\oint \mathbf{H} \cdot d\mathbf{l}$ on the left-hand side of (2.2) occurs in the air gap.

The mmf diagrams also represent the magnitude of the magnetic field intensity, **H**, about the circumference of the air gap. Using the mmf plots as representing air gap field intensities and considering the reluctance of the closed path, we can qualitatively estimate the magnetic flux density as a function of circumferential position because of both the field and armature mmf's. These approximate flux plots are shown in Figures 5.19*b*-5.19*d*.

It is seen that the flux density distribution of the field mmf generally corresponds to the mmf plot, with some allowance for fringing at the pole sides. The correspondence results from the low reluctance of the path of integration when this path includes the two poles. In contrast, when the armature mmf is maximum, the path of integration traverses the high reluctance air spaces between the poles. Under these circumstances the flux density distribution does not correspond to the mmf plot. It should be noted again that the flux density plots of Figure 5.19 are based upon the assumption of ideal magnetic materials; they ignore possible saturation effects within actual magnetic materials.

The above discussions and Figures 5.17 to 5.19 illustrate the nature of the armature magnetic field or armature reaction. The actual effect of armature reaction upon the performance of a given machine depends upon many factors besides the basic relationships described above. These factors are the relative magnitudes of field and armature mmf's under normal operating conditions, the length of the air gap, the geometry of the armature slots, the type of field excitation (electrical or permanent magnet), and the various corrective measures used to improve commutation. The following considerations of the effects of armature reaction are general and may vary considerably from one machine to another, depending upon the machine's construction and application.

1. *Armature reaction* is basically a *cross-magnetizing* field in a dc commutator machine, since it is orthogonal to the field of the main excitation source. This spacial relationship is observed clearly in Figure 5.17.
2. *The cross-magnetizing effects* of armature reaction are mainly a *distortion* of the magnetic field of the main source of excitation. The two principal distortions of the main magnetic field are:
   a. Creating a magnetic field in the interpole regions, the regions where armature conductors are undergoing commutation.
   b. Distorting the field under the main field poles by adding mmf to one side of the pole and subtracting mmf from the other side. As seen from Figure 5.19*d*, this tends to increase the flux density under one tip of a main field pole.

   Both of these distorting effects of armature reaction tend to have adverse effects upon the commutation process. The flux in the interpole region means that the coil undergoing commutation is moving through a magnetic field during the time that it is shorted by the brushes. The high flux level at one tip of a main field pole implies that coils passing through this flux have higher than normal voltages induced in them. This situation may cause a voltage breakdown across the commutator bars to which the coil is connected. Both of these phenomena tend to cause sparking at the brush/commutator region.
3. There may be some *demagnetizing* effects from the cross-magnetizing field of armature reaction:
   a. In the region under the main poles, the addition of mmf to one side and subtraction of mmf from the other side, caused by armature reaction, is an *algebraic addition of mmf's*. The flux resulting from this addition depends upon the degree of saturation of the magnetic circuit. If the nominal field excitation is designed to cause the magnetic circuit to be excited at about the "knee" of the magnetic material's saturation curve, the addition of mmf on one side of the pole will result in a smaller increase in flux density than the subtraction of mmf on the other side of the pole. Therefore, there will be a

net reduction in the total flux emanating from under a main pole. In a typical dc commutator machine, this phenomenon generally results in a reduction in the full-load generator voltage (or back emf) of from 2 to 5%.

b. Shifting the brushes produces a demagnetizing (or magnetizing) mmf. It is seen from Figure 5.18 that the armature mmf is no longer orthogonal to the main field mmf. It has both cross and demagnetizing components. If the brushes are shifted in such a manner as to improve commutation, the noncrossmagnetizing field will always be a demagnetizing field.

4. The effects of armature reaction upon the main field mmf are of the opposite algebraic sign depending upon whether the machine is being operated as a motor or as a generator. This is because of the fact that the difference between motor and generator operation is governed by the sign of the armature current.

5.7 COMMUTATION

Commutation has been mentioned in several sections of this Chapter without any specific definition of the term. In dc commutator machines, commutation refers to the process of reversing the current in an armature coil. This process occurs as the commutator bars to which the coil is connected pass under a stationary brush connected to the external armature circuit. Commutation in the broader sense is often associated with the concept of "turning off" a current, such as in a thyristor (SCR). The term is not imcompatible with the use of this term in dc commutator machines, since the armature current must first be "turned off" and then be caused to flow in the opposite polarity. *Ideal commutation*, in a dc commutator machine, is the linear variation of armature current from $+I_a$ to $-I_a$, as depicted in Figure 5.20. Linear commutation is a goal to be sought but one that can seldom be achieved in the design or operation of a dc

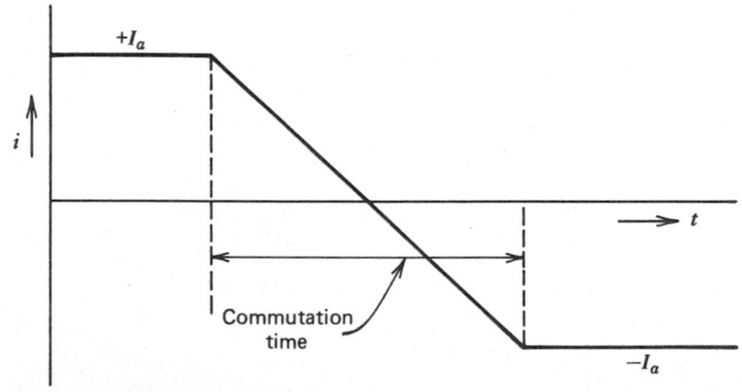

Figure 5.20. Linear or ideal commutation.

commutator machine. The principal impediments to linear commutation, several of which have already been described, are:

1. **Magnetic flux** in the "spacial zone of commutation," that is, in the area surrounding the conductors undergoing commutation. This flux is due to the distorting effects of armature reaction.
2. **Self-"reactance voltage"**, the voltage equal to $L(di/dt)$ associated with the inductance of the coil undergoing commutation.
3. **Mutual "reactance voltage,"** the voltage $M(di/dt)$, associated with adjacent coils undergoing commutation.
4. **High "bar-to-bar" voltages** induced by the excessive flux at a pole tip. This is caused by the armature reaction fluxes, which aid the main field flux on one side of the pole, as discussed in the preceding section.

There are several solutions that have been developed to compensate for these armature reaction effects, as follows.

1. Interpoles: These are auxiliary poles located between the main field poles, that is, roughly in the region opposite the brushes. Interpoles are designed to supply a flux in the polarity opposite to the flux existing in the interpolar region caused by armature reaction as illustrated in Figure 5.19*d* plus an additional flux designed to counteract the effects of the self and mutual reactance voltages.

The mmf required to counteract the flux of armature reaction can be obtained with reference to Figure 5.19*c*: the crossmagnetizing mmf per brush position is

$$NI_b = \left(\text{conductors in } \tfrac{1}{2} \text{ pole span}\right) \times \frac{I_a}{a} \tag{5.25}$$

The mmf of the interpole must be larger than the value obtained by application of (5.25) because of the need to compensate for the effects of self and mutual reactance voltages. This value varies with the construction of each machine, but is generally about 1.25 times the mmf obtained from (5.25).

The winding producing the interpole mmf is connected in series with the armature circuit, since the armature reaction that is to be compensated by the interpole varies linearly with the armature current. Figure 5.21 illustrates the role of interpoles in reducing the interpolar flux and overcompensating for reactance voltage.

2. Compensating Windings: These windings are located in the pole faces of the main poles and are designed to compensate for the distortive effects of armature reaction under the main poles. Figure 5.22 illustrates the location of compensating windings. By applying (2.2), it is seen that the compensating

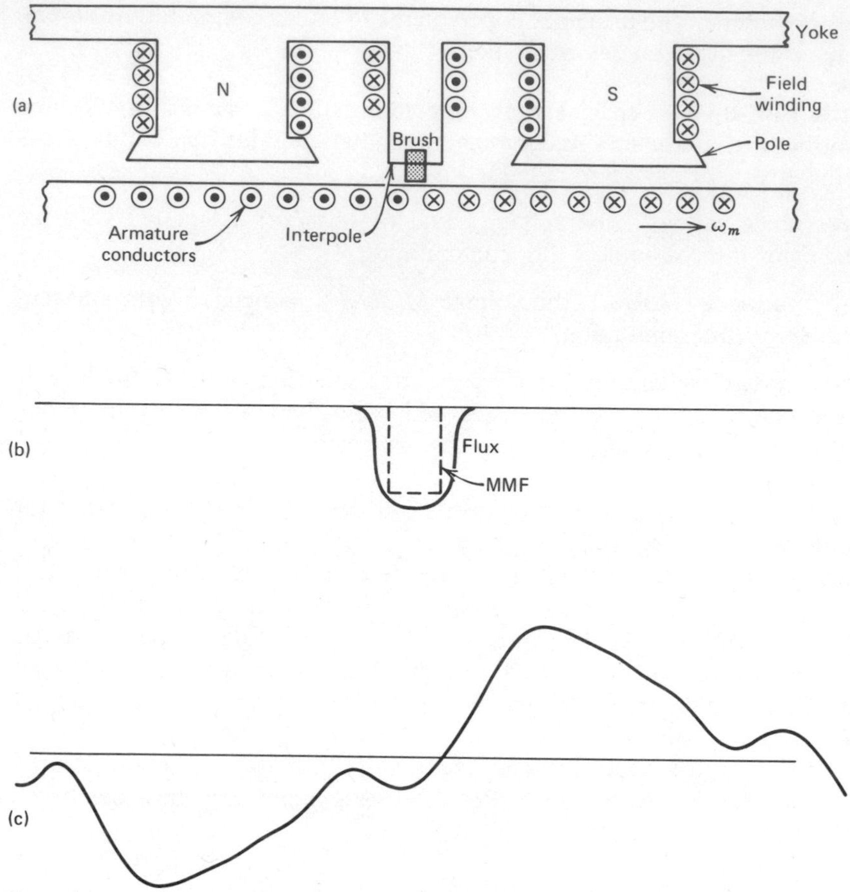

Figure 5.21. (*a*) Layout showing interpole location. (*b*) Flux and mmf due to interpole. (*c*) Resultant flux of main field, armature, and interpole excitation. (Compare with Figure 5.19*d*.)

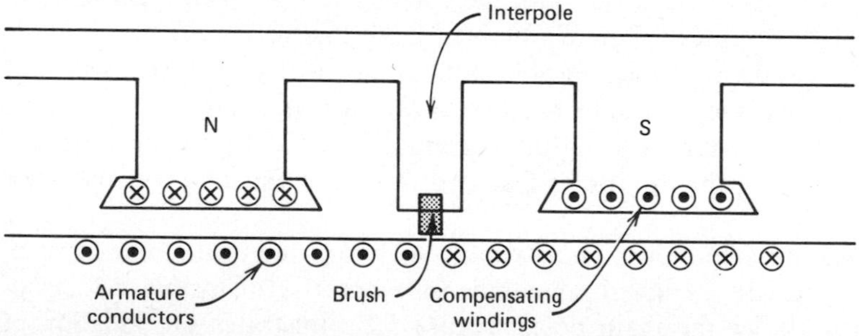

Figure 5.22. Compensating windings.

windings "compensate", that is, eliminate the armature reaction mmf under the main poles. Compensating windings, like interpole windings, are connected in series with the armature winding.

3. Brush Shifting: This is an archaic method of compensating for the effects of armature reaction. Figure 5.18 illustrates how the armature reaction mmf is changed from a purely crossmagnetizing mmf to an mmf with both crossmagnetizing and demagnetizing components by means of brush shifting. This technique was used to compensate for the effects of armature reaction before the introduction of interpoles and compensating windings. It is still used for this purpose in machines where the cost of interpoles or compensating windings is prohibitive, such as in automotive starter motors. Brush shifting is applicable for only one type of energy conversion, either for motor or for generator action. Also, the effects of brush shifting are fixed in magnitude as a function of armature current—in contrast to interpole or compensating winding correction.

The various methods of compensation for the effects of armature reaction are used in dc commutator machines, depending upon the machine environment in terms of cost, application, and lifetime of operation. Improper commutation generally results in sparking or ring-fire at the commutator and shortens the useful operating life of the machine. Inadequate commutation compensation results in increased maintenance of the machine due to commutator wear and deterioration. The use of the various compensating techniques described above is a trade-off in initial cost of the machine versus operating and maintenance cost.

5.8 LOSSES AND EFFICIENCY

One of the important characteristics of any machine is its power or energy efficiency, defined in Chapter 1. Energy efficiency is useful in applications of varying speed and torque, such as traction applications. The power efficiency of dc commutator machines varies greatly with torque. Therefore, the power efficiency of the machine at one condition of speed and torque may not be at all indicative of the average efficiency over a range of speed and torque conditions. For this reason, energy efficiency during a specific operating cycle or "duty cycle" is often more meaningful in specifying the performance of a machine that does not operate at constant conditions of speed and torque. In the following development, loss components and efficiencies are expressed in terms of power quantities. Energy terms can be obtained from the time integral of power quantities.

The mechanical power at the shaft of a rotating electromagnetic machine is

$$P_m = T_s\omega_m \text{ watts} \tag{5.26}$$

This power, in a motor, is the output power available, that is, the power that can be supplied to an external mechanical load, such as a pump, generator, or overhead crane. In a generator, this is the required input power. The electrical power at the terminals of a dc commutator machine is

$$P_e = VI + P_f \tag{5.27}$$

where V and I are the voltage and current, respectively, of the external electrical system connected to the machine and P_f is the power supplied to a separately excited field winding, if the machine is operated with such a field. In a motor, P_e is the *input power*; in a generator, the VI portion of P_e is the *output power*. The ratio of P_m and P_e is the machine power efficiency.

The difference between the input and output consists of various components of the internal machine losses, described with the aid of the energy-flow diagram in Figure 5.23. In this diagram, power (or energy) flow can be in either direction, depending upon whether the machine is being operated as a motor or

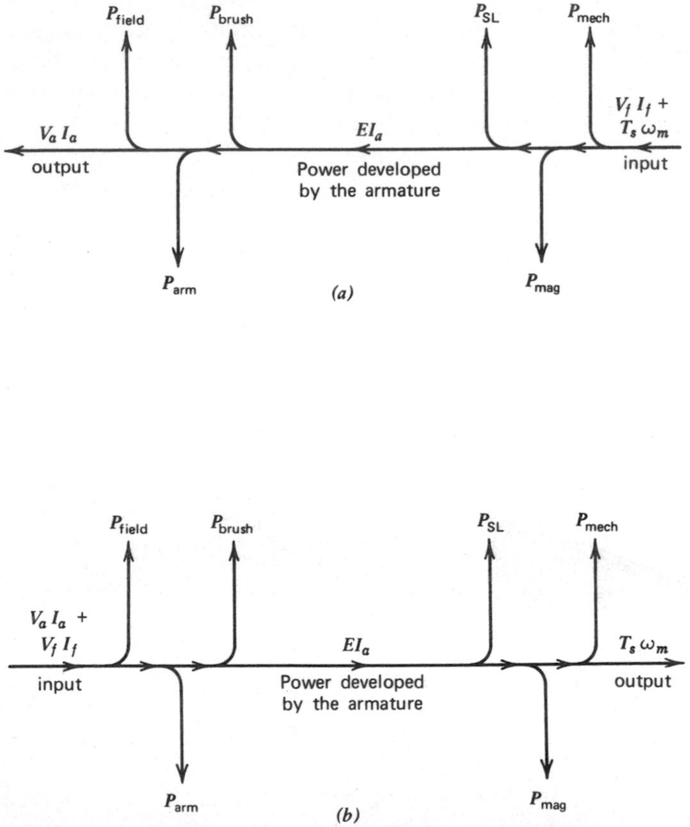

Figure 5.23. Power flow in a dc commutator machine. (*a*) Generator. (*b*) Motor.

as a generator. Many of the loss components are typical of any electromagnetic rotating machine, but the descriptions given next are in terms of the dc commutating machine.

1. **Mechanical Losses:** There are three components of the mechanical loss in a dc commutator machine.

a. **Bearing friction:** This component varies with the type and quality of bearing used in the machine, with the torque-loading (axial and radial) on the bearing, and with the speed of rotation. For an accurate evaluation of bearing friction, the bearing manufacturer should be consulted. An approximate relationship of the loss as a function of machine rotor speed, however, is given by[3]

$$P_b = K_b \times 10^{-6}(\text{rpm})^{5/3} \text{ watts} \tag{5.28}$$

where K_b is a constant depending upon bearing size, type, and method of lubrication. For two similar bearings per machine and oil mist lubrication, K_b is between 1.1 and 1.7.

b. **Windage:** This component is due to the air motion in the air-gap and interpolar regions caused by the rotation of the armature. It is a function of many of the detailed construction features of the machine but is most influenced by whether turbulent or laminar air flow is induced in the air-gap regions. The windage loss of an axial air-gap configuration is generally less than that of a radial air-gap configuration of equivalent power rating. A useful formula for estimating the windage loss of a radial air-gap machine assuming turbulent flow is[4]

$$P_w = \pi C_d \rho R^4 \omega_m^3 l \text{ watts} \tag{5.29}$$

where R and l are the radius and length, respectively, of the cylindrical rotor, ρ is the air density, and C_d is a skin friction coefficient evaluated from

$$\frac{1}{\sqrt{C_d}} = 2.04 + 1.768 \ln\left(Re\sqrt{C_d}\right) \tag{5.30}$$

where Re is the Reynolds number for the radial air gap.

c. **Brush friction:** This component is unique to rotating machines which require a brush/commutator system. It is often the largest component of mechanical loss and is one of the undesirable features of this class of machines. Brush friction loss depends upon the speed of rotation, brush pressure, and the composition of the brushes. The latter feature is chosen on the basis of a rather complex "engineering trade-off" of a number of parameters including brush friction, brush electrical resistance, brush lifetime, brush current density, brush pressure, cost of the brush assembly and so on. Brushes are constructed of a base carbon (graphite) material intermixed with copper or some other conductive material. In general, the conductive material, while increasing the electrical conductivity, decreases the lubricating features of the graphite. It creates additional friction loss, commutator wear and scarring, and audible noise due to brush "chatter." The magnitude of this loss component is difficult to calculate and must be obtained through experimental measurements.

2. **Magnetic losses** or core losses occur in magnetic sections of dc commutator machines in which there is a time-varying magnetic field. The nature of these losses has been described in Chapter 2. Magnetic losses occur primarily in the armature of dc commutator machines, since the magnetic field in the pole and yoke sections is relatively constant. The armature magnetic material is continusouly rotating through the main field flux and armature field flux and therefore is usually constructed of laminations to minimize eddy currents. Pole and yoke sections are often constructed of solid magnetic material. However, there is a source of eddy currents and associated loss located in the regions of the poles adjacent to the air gap. This area is called the "pole face."

3. **Winding losses,** often called "copper losses," are ohmic losses in the armature, field, interpole and compensating windings. These losses are readily calculated if the winding resistances are known and can be measured by the test methods discussed in Section 5.9. The armature, interpole, and compensating pole resistances are usually lumped together as the total armature circuit resistance. Winding resistances are usually measured at room temperature, which is lower than the temperatures at which the windings are operated. When using measured resistances to calculate losses or voltage drops, measured values must be corrected to correspond with actual operating temperatures.

4. **Brush electrical loss:** The carbon graphite material used as the principal material in brushes has a different resistance characteristic than that of metallic conductors, namely, a negative resistance-temperature characteristic similar to that of a semiconductor. Also, the brush loss results in part from the contact resistance between brush and commutator, which also cannot be described as a conventional ohmic resistance. For these reasons, the brush voltage drop and power loss are often treated separately from the armature circuit voltage and loss. The brush loss depends upon many factors, including the composition of the brushes, the contact pressure, the temperature of the brushes, the condition of the commutator surface, and so forth. For pure graphite brushes, it is often assumed that the voltage across the brush-contact resistance is constant as a function of armature current with a value of 2.0 V. The brush loss is therefore $2I_a$ W. However, in practice, the voltage drop value varies considerably with the composition of the brushes and must be determined for each machine. In small machines, the brush drop and loss are usually included with the total armature circuit parameters.

5. **Stray-load loss** is a somewhat anomalous loss used to account for loss increase caused by loading the machine. It is attributable mainly to increase in magnetic losses caused by the changes in flux distribution due to the armature reaction field. This loss is more pronounced in large machines. The stray-load

loss is difficult to measure accurately (see Section 5.9), and a value of 1% of the machine output in large machines is often assigned to account for this loss.

5.9 NAMEPLATE RATINGS

The "rating" of a machine refers to the conditions of voltage, current, speed and power at which the machine is normally operated. There are several ratings associated with dc commutator machines, the principal rating being known as the "continuous" rating or one-hour rating, which is the rating described on the nameplate of a machine. The continuous power rating of a machine is a *thermal* rating. At this power the machine can be operated for long periods of time without a rise in temperature beyond the limitations of the conductor insulating materials, bearings, and other components, which are critically affected by temperature. In addition, there are several short-time ratings which are not stated on the nameplate but which can be obtained from the manufacturer.

For short periods of time, most types of dc commutator machines can be operated at conditions of power and/or torque that greatly exceed the continuous rating. For example, industrial motors can usually be operated intermittently or for one-minute intervals at torques of from three to five times the continuous torque.[5] The nameplate power refers to an *output* power, that is, to output shaft power for a motor and to output terminal power for a generator.

The speed rating of dc commutator machines is more complex than its power rating, since most dc machines are variable speed devices. In small machines or control machines, a speed range is often given. In shunt or separately excited motors, one speed rating refers to the maximum speed at which full field excitation can be sustained. It is seen from (5.11) and (5.24) that this speed is determined by the terminal voltage and the size of the machine's magnetic circuit (which determines the value of ϕ before saturation). The machine can be operated above this speed by "field weakening," that is, by reducing the value of ϕ.

A second speed is often given on the nameplate, which is an upper operating speed limit in the field weakening range of operation. In a series motor, the nameplate speed refers to the speed that is possible with rated voltage and rated power output on the motor. Added to these rated or nameplate speeds, is a mechanically safe operating speed limit—the upper speed at which the machine can be operated before mechanical damage occurs. In dc commutator machines, this speed limit is determined by the commutator/brush assembly design and cannot be exceeded. It can be obtained from the manufacturer.

A *power* motor nameplate includes at least the following information: power outout (usually in horsepower), speed (rpm), terminal voltage, armature current at rated output, field current, temperature rise under rated conditions of operation, and service factor. The service factor is a multiplier that can be applied to the machine's output rating to permit operation at increased output under particular conditions specified by the manufacturer. For a more detailed discussion of machine nameplate ratings, consult References 6 and 7.

Torque is seldom stated on the nameplates of power motors but can be obtained from nameplate values by the relationships,

$$
\begin{aligned}
P\,(\text{in horsepower}) &= \frac{2\pi NT}{33{,}000} \\
P\,(\text{in watts}) &= \frac{NT}{7.04}
\end{aligned}
\tag{5.31}
$$

where T is in pound-feet (lb-ft) and N in rpm. We observe that pound-foot are still very much in commercial use, prompting us to write (5.31).

The nameplates of *control* motors usually include voltage, current, and torque, and often rotor inertia. The nameplate listings of power generators are similar to those of power motors except that power output is listed in the units of kilowatts.

5.10 EXPERIMENTAL DETERMINATION OF MACHINE PARAMETERS

To predict the performance of a specific machine or to match the capabilities of a machine to a certain application, it is necessary to know the values of some of the machines' internal parameters. Of particular importance, especially in control applications, are the inductance and resistance of the armature and of field circuits. These can generally be obtained by conventional bridge methods or voltage-current measurements as used for any other type of circuit. It is often desirable to measure resistance and inductance of the armature circuit exclisive of the brushes. This can be done by locating the instrument probes underneath the brushes on the commutator circuit.

The armature circuit resistance is also frequently obtained from a standard machine test known as the *blocked rotor test*. In this test, the machine's rotor is mechanically blocked and prevented from rotating. This is done by fastening a brake arm to the rotor shaft. A scale can then be used to measure the torque developed in the machine. In a shunt or separately exicted machine, the field should be energized from a source of variable voltage in order to vary the field current. The armature circuit is supplied from a source of relatively low voltage, since the only impedance in the armature circuit at standstill is the armature circuit resistance.

The test procedure is as follows: With the rotor blocked and the field current (if shunt) set at a prescribed value, increase the armature voltage from zero to a value giving the desired armature current. Measure the armature voltage, the armature current, the force on the lever arm (which can be converted to torque if the length of the lever arm is known), and the field current. This test is usually repeated for other values of field and armature current. The armature circuit resistance can be obtained from the ratio of armature voltage to armature current, and is usually more indicative of the resistance existing under actual operating conditions than of measurements made at low current levels such as bridge measurements. From the torque, armature, and field current measurements, the machine's torque constant, (5.22), is obtained. The block-rotor test requires relatively low power levels (as compared to actual load testing) and generally simulates the effects of armature reaction under load conditions as they influence motor torque production.

Another useful machine test is the *no-load test*. The machine to be evaluated is driven as a generator from an external source at a constant speed and at zero armature current (armature circuit open-circuited). The measurement of armature voltage as a function of field current produces the magnetization curve data, as shown in Figure 5.14. The torque constant, K_t, can also be obtained from this data, as seen from (5.15). If the machine is driven from a source whose output torque can be measured, such as a motoring dynamometer or calibrated dc machine, this torque should be measured at each value of field current. The product of the input torque (in newton-meters) and the shaft speed (in radians/s) is the input power supplied to the test machine. The product is a measure of the *no-load loss* of the machine at the particular values of speed and field current. The no-load loss is the sum of mechanical losses and magnetic losses. The no-load test is repeated for several values of speed that encompass the operating speed range of the test machine. The resulting family of curves describes the variation of mechanical and magnetic losses over the range of speeds and excitations (field currents) used in the test.

In some series machines and many low-power machines, the field winding terminals are not accessible and the test as described above cannot be performed. Also, many laboratories and test facilities are not equipped with an appropriate drive machine to operate the test machine as a generator. In situations where the no-load test cannot be performed, some of the same information can be obtained by operating the test machine as an un-loaded motor. Magnetization curves can be obtained by varying the armature voltage as field current is varied in order to maintain constant speed and measuring armature current and voltage, field current, and speed.

The range of field current values that can be obtained at a constant speed in this manner is smaller than for the no-load generator method. Since the motor is unloaded, the armature current will be small and the effects of armature reaction are negligible. The induced voltage, E, is found by subtracting the small armature resitance voltage drop from the measured armature terminal voltage. The no-load loss is found by subtracting the armature I^2r loss from the armature power input, which is equal to the product of armature current and armature terminal voltage. Care must be exercised, when operating a motor with no external load, that the maximum safe operating speed of the motor is not exceeded during testing at low values of field current.

5.10.1 CALIBRATION OF THE MACHINE

The shunt or separately excited dc commutator machine is a very versatile device because its torque can be controlled over a wide range of speeds during both motor and generator operation. Because of this capability, it is widely used in the laboratory as a loading device for testing rotating electric machines, internal combustion engines, and many other types of machines. The principal embodiment of the separately excited dc commutator as a test instrument is the dynamometer. In the dynamometer, stator is "cradled" and free to rotate through a small angle in order to measure the developed electromagnetic torque. However, an uncradled conventional separately excited machine is also very useful as a loading or driving device for testing other rotating machines or engines. For this application, the machine must be experimentally "calibrated," that is, the relationship between input and output power over a wide range of speed and torque conditions must be determined.

To calibrate a dc commutator machine, which involves determining the power losses as a function of speed and torque, the following tests are required:

1. The machine to be calibrated is mechanically coupled to and driven by another machine. It is preferable that this machine be calibrated or of known characteristics, but if accurate instrumentation is available, any dc commutator machine capable of driving the test machine is suitable. The driving machine must have a power rating larger than the *expected losses* of the test machine (usually, at least 15% of the rated power output of the test machine). This machine will be referred to as the "drive" machine in the following discussion.

2. The drive machine is operated as a motor, uncoupled to the test machine, over the total speed range to be used for the test machine. The input power at selected speed point is measured.

3. The drive motor and test machine are mechanically coupled. The test machine is driven over its speed range, and the input to the drive motor is measured at each speed point. The difference between the power input to drive motor at each speed point determined in Steps 2 and 3 is the mechanical loss of the test motor. Plotting these points as a function of speed gives the mechanical loss of the test motor as a function of speed. The mechanical loss can be separated into two components, if desired, by performing this test twice: first, with the brushes of the test machine lifted, which gives the windage and bearing friction losses of the test machine; and, second, with the brushes riding on the commutator, which gives the total mechanical loss. The difference between these two tests is the mechanical brush friction loss.

4. The test described in Step 3 above is repeated with the field current of the test machine set at a prescribed value. In addition to the power input to the

drive motor, field current and armature voltage on the test machine must now be measured. The difference between the power input to the drive motor measured in this step and the power measured in Step 3, at each speed point, is the magnetic loss of the test machine for the prescribed value of field current. This step should be repeated for several different values of field current—enough values of field current to cover the range of operation of the test machine, since it is used as a "calibrated" machine.

5. An accurate value of the armature resistance of the test machine must be obtained. This may or may not include the brush-contact resistance, depending upon the factors discussed in earlier sections of this chapter. The resistance versus temperature characteristics of the armature and brush-contact resistance must be known. This is a fairly simple matter for the armature winding itself if the conductor material is known. If the brush electrical loss is to be treated independently from the armature winding loss (which is frequently done for power machines with graphite brushes), the brush contact voltage drop as a function of armature current should be measured. In small machines or in machines with metal graphite brushes, the brush-contact resistance can usually be lumped in with the armature winding resistance and the resistance as measured at the armature terminals can be used in calculating the armature electrical loss. Temperature variations of resistance are made, in this case assuming the characteristics of the winding conductor material.

6. The final step involves the measurement of stray-load loss. As noted above, in small machines, this loss is often ignored. The following test can be used to determine whether or not this loss is significant: The test machine is driven at a given speed by the drive motor as in previous steps. With *zero* field excitation, the armature circuit is short-circuited. A small armature current will usually appear because of residual magnetism. Slowly and carefully increase the field excitation from zero to a value that causes rated armature current to flow. Measure armature current, field current, and drive motor input power. Calculate the armature I^2r loss and brush loss for rated armature current from the resistance value determined in Step 5, making corrections for armature temperature rise as indicated. Determine the mechanical loss and magnetic loss of the test motor for the speed and field current used from the data taken in Steps 2, 3, and 4. Subtract the armature I^2r loss, brush loss, mechanical loss, and magnetic loss from the measured loss. The difference is the stray load loss at this speed and armature current.

The test should be repeated for other values of armature current and, if the test machine is to be operated at varying speeds, at other speeds. If this difference comes out as a negative value or so small as to be within the errors attributable to instrument inaccuracies, the stray-load loss can be neglected. (*Note*: the total loss in this test is the measured power input to the drive motor minus the input measured in Step 2. If the field or armature currents are significantly different from those measured in Step 2, corrections for variations in armature and field I^2r losses should be made in the drive motor inputs to account for these changes.)

The data obtained from the above tests can be used to predict the performance of the machine for either motor or generator operation under almost any operating condition. This process is made simpler if the results are presented graphically.

a. Mechanical losses versus speed from Step 3.
b. Magnetic loss and induced armature voltage versus speed for the chosen values of field current from Step 4 (a family of curves).
c. Stray-load loss (if significant) versus armature current for the chosen values of speed from Step 6 (a family of curves).

In using these data it will often be necessary to interpolate among the curves for magnetic loss, induced voltage, and stray-load loss.

5.11 PREDICTION OF PERFORMANCE CHARACTERISTICS

For the steady state performance prediction, the graphical data (items *a*, *b*, and *c* from the previous section) should be available. This method can be used for the prediction of both motor and generator characteristics. It is most adaptable to shunt and separately excited configurations but can, with some modification, also be applied to series and compound configurations. We shall illustrate the method for a separately excited motor. The power-loss diagram of Figure 5.23 will be an aid in keeping track of the loss components during the power summations used in the method. To start the performance prediction, it is necessary to assume either the power input or output. This choice will generally be made upon the basis of how the machine is used and controlled. For a motor with both field and armature control, it is common practice to assume the output parameters.

1. Assume output torque, T_s, and shaft speed, ω_m; the product of T_s (in N-m) and ω_m (in rad/s) gives the mechanical output power.

2. For the particular speed assumed, ω_m, look up the mechanical loss, P_{mech}, from the curve of item *a*.

3. Assume a value of field current, I_f. Since the method is going to be repeated for a number of difference values of field current, it is not too critical which value is assumed. However, the field current should be within the range of field currents used in the test data and one that is expected to be used during the actual operation of the machine. For the assumed values of field current and speed, look up the induced voltage, E, and the magnetic loss, P_{mag}, from the curves of item *b*.

4. Estimate or guess a value of stray-load loss, P_{SL}, from the curves of item *c*.

5. The electromagnetic or developed power of the machine is

$$P_e = T_s\omega_m + P_{\text{mech}} + P_{\text{mag}} + P_{\text{SL}} \tag{5.32}$$

6. The armature current for the assumed output and field current is

$$I_a = \frac{P_e}{E} \tag{5.33}$$

7. At this point, an iterative step can be inserted for accuracy. Now that the armature current is known, the stray-load loss can be obtained from the curve of item *c*. Look up this value and compare it with the estimate made in Step 4. If the value is considerably different, insert the new value into (5.32) and determine a new value of armature current in (5.33). This process can be repeated several times until the value of P_{SL} and I_a are consistent with the measured data in item *c*. If this method is being performed on a computer, this iteration step can be inserted as a simple routine. Some criteria should be used in the routine, based upon the desired difference between the estimated P_{SL} and the value from item *c*, which is used to stop the iteration process.

8. Calculate the armature circuit loss, P_{Cu}, and the brush loss, P_b (if separately treated).

9. Calculate the field winding loss, P_f.

10. The motor efficiency is

$$\eta = \frac{T_s\omega_m}{T_s\omega_m + P_{\text{mech}} + P_{\text{mag}} + P_{\text{SL}} + P_{Cu} + P_b + P_f} \tag{5.34}$$

11. The required power input for the assumed output conditions is, of course, the denominator of (5.34).

12. The required armature terminal voltage is, neglecting armature reaction,

$$V = E + I_a R_a \tag{5.35}$$

where R_a has been corrected for an assumed temperature rise and includes brush drop. Brush drop may be handled independently as discussed in Section 5.8.

13. It is possible to include the demagnetizing effects of armature reaction in lieu of using (5.35) if the demagnetizing armature ampere-turns are known. This is a difficult parameter to define numerically, since it varies with saturation in the machine's magnetic magnetic circuit. Because demagnetizing effects of armature reaction can be obtained only through extensive load testing, this parameter tends to defeat the intent of the technique being described, which is aimed at predicting performance with the use of only no-load and short-circuit test data. Therefore, if armature reaction demagnetizing effects are to be

illustrated along with the method under discussion, estimated values of this parameter should be used. Otherwise, complete load testing can be used.

To introduce the effects of armature reaction into this method, the abscissa of the curves of item b must be expressed in "ampere-turns." Then, in Step 3, the assumed field current can be expressed as an assumed value of ampere-turns. Add to this the assumed value of armature reaction demagnetizing ampere-turns, AR. Look up a new value of E from the curves of item b, which we shall call E'. Use E' in (5.35) to determine V. Note that E is still used to determine magnetic losses in Step 3 and armature current in Step 6, since the magnetic loss and induced voltage are a function of flux (not mmf). However, the true field current required now must be

$$I_f' = \frac{N_f I_f + AR}{N_f} \tag{5.36}$$

This new value of field current, I_f', must be used in calculating the field winding loss in Step 9.

14. Repeat Steps 3 through 13 for other values of field current (or field ampere-turns if armature reaction is to be included).

15. Repeat Steps 1 through 13 for other values of output torque and speed.

This simple performance prediction technique has taken many words to describe but actually requires a relatively short time to calculate, even by hand if the curves of items a to c are readily available. It is still desirable to program the above process on a computer whenever feasible, however, since it is readily amenable to computerized calculation.

Example 5.5

A separately excited motor is rated 1 HP, 220 V, 4.6 A, 1750 rpm, and 350 mA field current; the armature resistance is 4.8 ohm including temperature correction; and the field resistance is 630 ohm. Laboratory tests have resulted in the loss information and magnetization curve shown in Figures 5.24 to 5.26. Determine the required input armature voltage, input power, and efficiency when the motor is operated to supply rated torque at rated speed.

The rated torque, although not required for the solutions to this problem, can be found from (5.30) or (5.31) as 3.0 lb-ft. or 4.06 N-m. The power output in watts is 746. Follow the procedure outlined in Steps 1 to 15 above. From 5.24 (Item a), the mechanical loss at rated speed is 26 W. From Figure 5.25 (Item b), the magnetic loss at rated field current of 350 mA is 22 W. From Figure 5.26 (Item c), assume that the armature current will be at rated value or 4.6 A, giving a stray-load loss at rated speed of 17 W. The electromagnetic power is therefore $746+26+22+17=811$ W. From Figure 5.25, the induced voltage at 350 mA is 192 V, giving an armature current of $811/192=4.25$ A. This is close enough to the assumed value used in determining the stray-load loss so that an iteration is

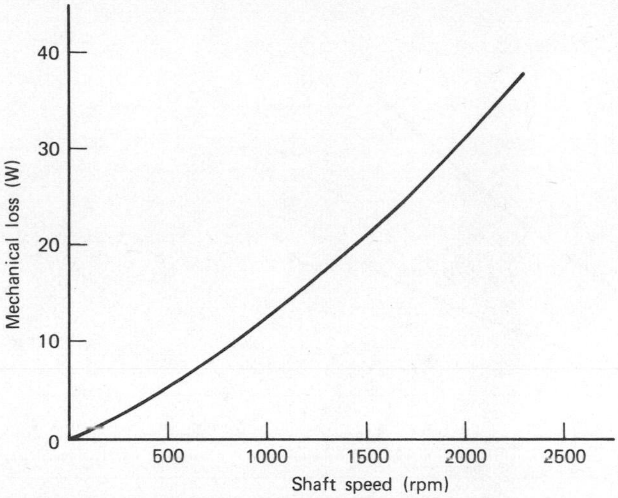

Figure 5.24. Mechanical loss for a 1-hp machine.

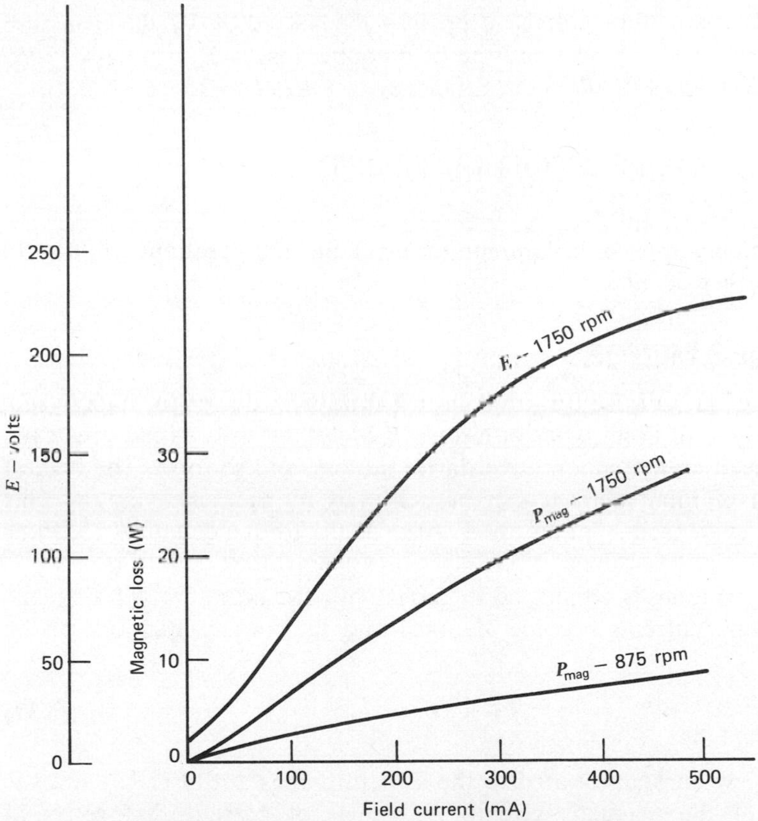

Figure 5.25. Magnetic loss and induced voltage for a 1-hp machine.

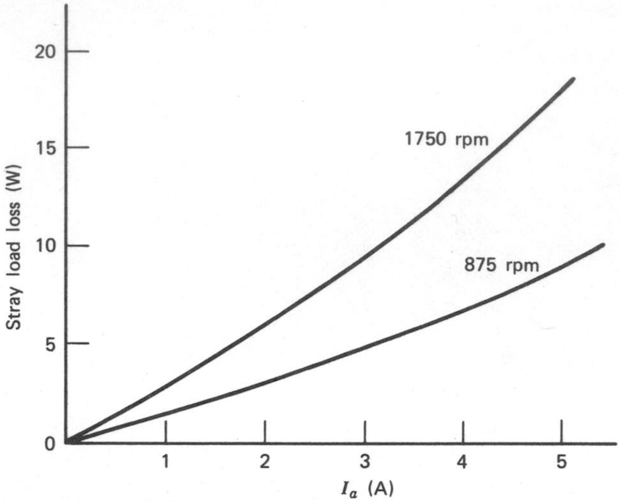

Figure 5.26. Stray-load loss for a 1-hp machine.

not required. The armature voltage drop is $4.25 \times 4.8 = 20.4$ V; the armature copper loss is $4.25^2 \times 4.8 = 87$ W; and the field copper loss is $0.35^2 \times 630 = 77$ W. Power input is $811 + 87 + 77 = 975$ W. Efficiency is $746/975 = 0.77$ or 77%.

5.12 DC CUMMUTATOR MACHINE CHARACTERISTICS

The most commonly applied dc commutator machine configurations will now be presented in further detail.

5.12.1 SERIES MOTORS

The series motor has been the principal electrical traction device for many years and is still widely used in all types of electric vehicles, electric trains, streetcars, industrial overhead cranes, automotive starter motors, and so forth. The highest torque per unit of input current can be achieved by the series motor. This characteristic is evidenced from an observation of the torque equation, (5.18) or (5.22).

If the field winding is connected in series with the armature winding, the field and armature currents become identical and the torque equation can be expressed as

$$T = K_m I_a^2 \tag{5.37}$$

where K_m is a new constant related to the armature constant of (5.18) and the characteristics of the magnetic circuit. At values of current that result in

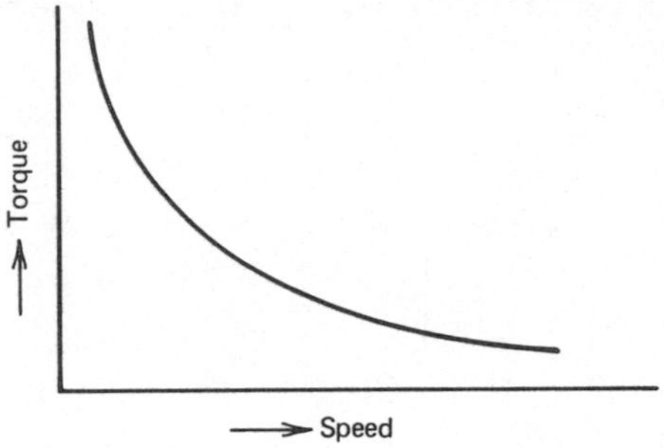

Figure 5.27. Speed-torque characteristic of a series motor.

unsaturated conditions of the magnetic circuit, the torque of a series motor varies with the square of the armature current. Above saturation, the torque varies with the first power of armature current as in a shunt or separately excited machine. Equation (5.37) is applicable in the unsaturated region of operation of the magnetic circuit. The general characteristics of a series motor excited from a constant voltage source are shown in Figure 5.27.

An undesirable characteristic of the series motor is its tendency toward excessive speeds at light loads. This can best be observed from (5.24). If the machine is lightly loaded, the armature current is small. In a series machine, the series field flux is therefore small, which results in a small value for the denominator of (5.24), and calls for a large value of mechanical speed, ω_m. This is a common cause of "machine runaways" in the typical college laboratory.

5.12.2 SEPARATELY EXCITED MOTORS

This is the most versatile form of the dc commutator motor, as previously mentioned in this chapter. The separately excited machine has the capability of both independent armature control and field control, which—in terms of the operating parameters of a machine—imply independent speed and torque control. An early approach to exploiting this characteristic was the Ward-Leonard system, which is still widely used in high-power applications requiring variable speed and variable torque control, such as in steel and aluminum rolling mills. A schematic representation of the Ward-Leonard system is shown in Figure 5.28. The separately excited configuration is especially adaptable to feedback control. For example, the effects of demagnetizing armature reaction and armature and brush resistance drop can be compensated for by increased field excitation.

A common speed/torque characteristic used in many variable-speed applications is shown in Figure 5.29. This characteristic consists of two sections: a constant torque section at low speeds up to the "break speed", N_b, and a constant power section at speeds above the break speed. The separately excited machine is frequently operated in this mode using Ward-Leonard or various types of electronic control systems. In the constant torque region, operation is

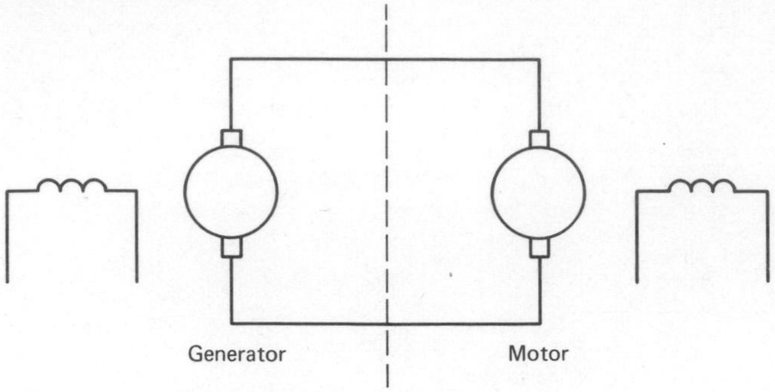

Figure 5.28. Schematic diagram of a Ward-Leonard system.

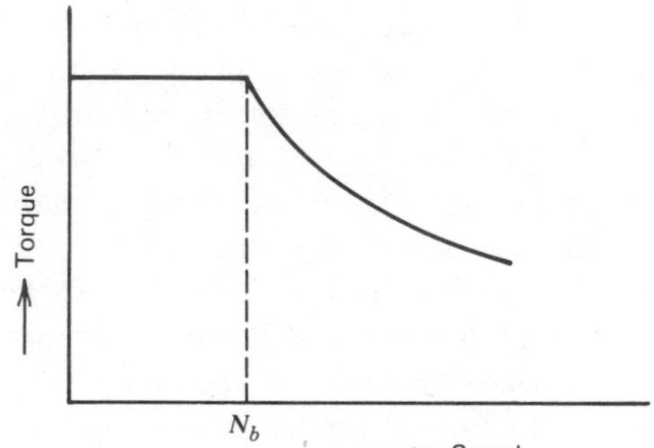

Figure 5.29. Torque-speed characteristic of a traction motor.

usually at a constant value of field excitation (the maximum excitation in this region determines the maximum torque capability). Control of torque is achieved by controlling the armature voltage (or current).

In the constant power regions at higher speeds, armature current is held constant and torque and speed are varied by varying the field current, or by "field weakening." These two modes of operation are understood by reference to (5.15) and (5.21). For example, constant electromagnetic power implies

$$P = EI_a = K_t I_f \omega_m I_a = \text{constant} \tag{5.38}$$

If the armature current is held constant in (5.38) the product, $I_f \omega_m$, must also be maintained constant indicating an inverse relationship between speed and field current. This is the actual situation in the constant power region of Figure 5.29 and results in the parabolic characteristic above the break speed since, as seen from (5.22), torque decreases with I_f (or inversely with ω_m). The speed, torque, and required excitation at the break speed determine the size of a motor designed to meet this type of characteristic.

A comparison of the series and separately excited motor characteristics in Figures 5.27 and 5.29, shows a similarity in general shapes. The separately excited motor has the capability of controling the two extremes of the characteristic—at very low and very high speeds—which are relatively uncontrollable in the straight series configuration. It should be added, however, that with variable armature voltage control, the series motor speed-torque characteristic can be made similar to that of Figure 5.29.

5.12.3 SEPARATELY EXCITED GENERATORS

The principal applications today are as excitation sources for large synchronous alternators in power generating stations, as the control generator in Ward-Leonard systems, and as auxiliary and emergency power supplies. The separately excited generator has all of the control flexibility of the separately excited motor. One important characteristic used in specifying a generator and its excitation control is *voltage regulation*. Voltage regulation is a measure of the variation in output voltage of a generator with load and is defined as

$$\text{percent regulation} = \frac{V(\text{no-load}) - V(\text{load})}{V(\text{load})} \times 100 \tag{5.39}$$

In a separately excited generator, load is synonomous with armature current.

Voltage regulation is usually specified in terms of the "full-load" or rated load condition, although the above formula is applicable to any other load. The variation of voltage with load in a separately excited generator arises from two causes: armature resistance voltage drop and the demagnetizing effects of armature reaction. With a fixed value of field excitation, a typical voltage characteristic might resemble that shown in Figure 5.30. The drop in voltage with increasing load can be eliminated by appropriately increasing the field excitation. This is usually done by means of a feedback control system using armature voltage as the feedback signal.

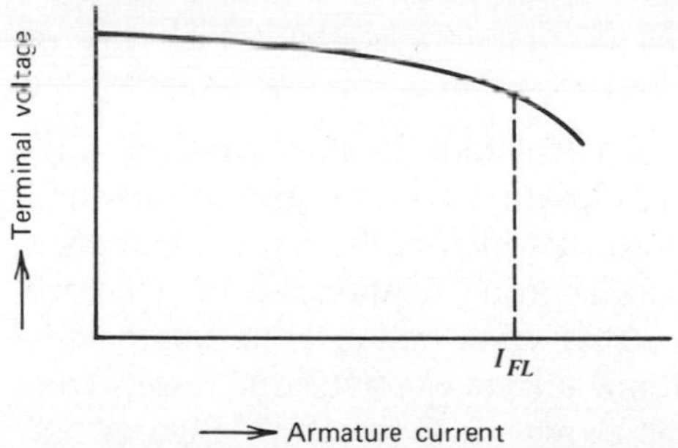

Figure 5.30. Voltage characteristic of a generator.

5.12.4 SHUNT MACHINES

The terms *shunt field* and *separately excited field* are used interchangeably in practice. However, in this text we have distinguished between these two field arrangements and we have treated them as two separate machine configurations. The designation, "shunt machine" refers to the configuration in which the field is permanently connected across (or "in shunt with") the armature, as shown in Figure 5.15*b*, with no variable resistance in series with the field. The shunt machine obviously does not have the control flexibility of the separately excited machine discussed in the previous two sections. The shunt generator, or self-excited generator, has been briefly discussed in Section 5.5. Shunt motors are used in industrial and automotive applications where precise control of speed and torque are not required.

The speed-torque relationship for a typical shunt motor is shown in Figure 5.31. The declining speed versus torque characteristic is caused by the armature resistance voltage drop and armature reaction. At some value of torque, usually a value about 2.5 times the rated torque of the motor, armature reaction becomes excessive, causing a rapid decrease in field flux and a rapid decline in developed torque down to a stall condition. This abrupt drop in developed torque, which often leads to unstable operation at high overloads, can be smoothed out considerably by adding a series field winding connected in series with the armature circuit. Such a configuration is called a stabilizing winding, although contrary to its name, it results in a greater variation in speed in the normal operating torque range. This characteristic is shown in Figure 5.31, also. Such a configuration is termed a *compound motor*. Shunt and compound motors are used where a reasonably constant speed versus torque characteristic is required and the cost of field control is not justified.

5.12.5 CONTROL MOTORS

Control motors are motors of relatively low power rating (usually less than a few hundred watts) with fixed field excitation. Fixed field excitation adds another variation to the basic torque equation, (5.22),

$$T = K_c I_a \tag{5.40}$$

where K_c is a new constant in N-m/A. Control motors have relatively high armature current to diminish the effects of armature reaction and to provide a linear relationship between voltage and torque, which is the basis for applications as a control device. The desired speed-torque relationship of a control motor can be shown to be a degeneration of the shunt motor torque characteristic illustrated in Figure 5.31. The shunt motor torque characteristic results from a large increase in armature resistance, as shown in Figure 5.32. Figures 5.32 and 5.33 illustrate the physical characteristics of several types of control motors.

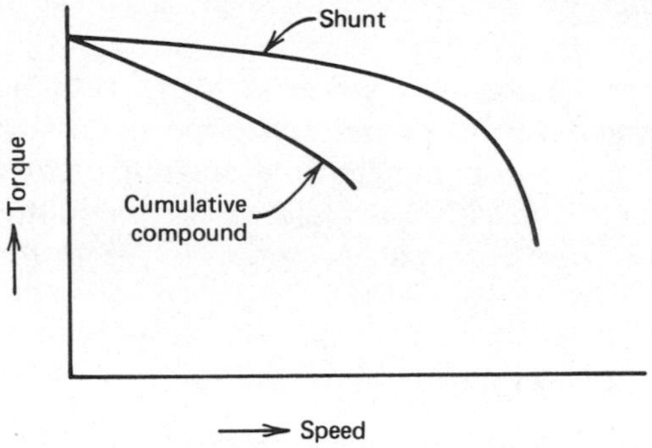

Figure 5.31. Torque-speed characteristics of shunt and compound motors.

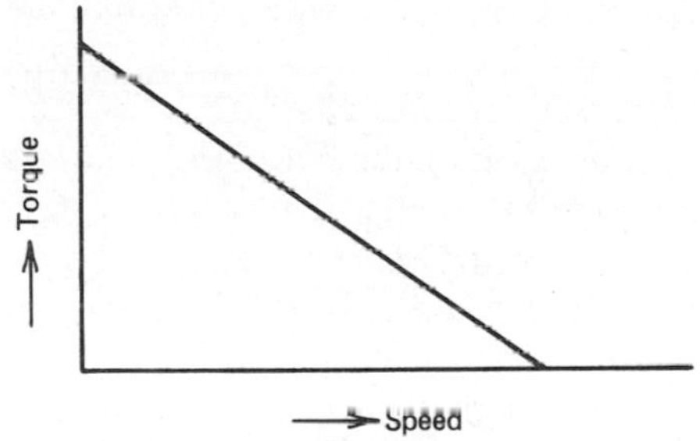

Figure 5.32. Torque-speed characteristic of a control motor.

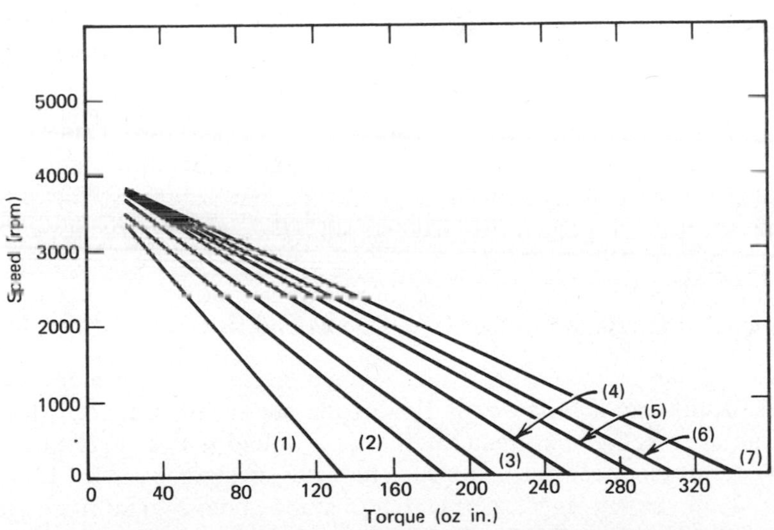

Figure 5.33. Speed-torque characteristics of a series of 12-V permanent magnet motors.

The fixed excitation is obtained either by means of permanent magnets or wound field.

An important parameter in control motors is the inertia of the rotating element, which generally should be minimized to permit fast response to step changes in the input control function. A useful parameter in evaluating control motors is the *torque/inertia ratio*, which is usually calculated using the motor stall torque (the torque at which the torque/speed characteristic meets the ordinate in Figure 5.32).

5.12.6 PERMANENT MAGNET MOTORS

The use of magnet excitation instead of electrically excited wound fields offers certain advantages in dc commutator machines and carries with it certain risks. In very small machines, permanent magnet excitation almost always results in a lower manufacturing cost. As the size of the machine increases, this cost advantage generally decreases, depending upon the particular application. Commercial permanent magnet motors have been built in sizes of 5 hp and larger.

The efficiency of a permanent magnet machine is generally higher than that of a wound-field machine, because of the elimination of the field electrical loss. In some cases, a permanent magnet machine will be smaller than a wound-field machine of equal power rating. Also, where field control is not required, the use of permanent magnet excitation simplifies the construction of the machine and eliminates two electrical terminals and their associated wiring.

The problems and risks of using permanent magnet excitation include:

1. **The field magnet** can be demagnetized by armature reaction mmf, causing the machine to become inoperative. Demagnetization can result from improper design, excessive armature current caused by a fault or transient or improper connection in the armature circuit, improper brush shift, or temperature effects.

2. **The flux densities** at which permanent magnets can operate (see Table 2.2) result in air-gap flux densities that are much lower than those in wound-field machines. This is particularly true of the ferrite (ceramic) magnets, which are the lowest-cost class of permanent magnets and the most widely used in machine application. Thus an increase in armature size (as compared to a wound field machine of the same rating) is required to offset the decreased size of the magnet in relation to the size of the wound field.

The speed versus torque characteristic of a permanent magnet motor is more flat than that of the shunt motor because the flux of the permanent magnet is less affected by armature reaction mmf than the flux of a wound pole configuration. Figure 5.33 illustrates the speed versus torque characteristics of a series of 12-V, 3-in. diameter, and varying-length permanent magnet motors. This type of motor is used with battery power supplies in applications that include automotive heater and air conditioner blowers, electric fuel pumps, marine engine starters, wheelchairs, and cordless power tools.

The design and application of a permanent magnet motor is similar to that of wound-field motors except for the consideration of armature reaction demagnetizing mmf. The design must assure that under all foreseeable conditions of operation, including fault conditions, the permanent magnet will not be demagnetized. Magnet demagnetization results from the crossmagnetizing armature reaction underneath the pole faces, which is described in Section 5.6 and illustrated in Figure 5.19. As noted in that discussion, armature reaction aids the main field flux under one-half of the pole face and opposes the main field flux underneath the other half. In a motor, the opposing mmf occurs at what is called the "trailing edge" of the pole, that is, the half of the pole face over which an armature conductor passes last as it rotates through the pole span.

It is this mmf which will tend to demagnetize the magnet. The maximum armature reaction effect occurs at the maximum level of armature current which exists in a motor at the start or stalled condition with maximum armature voltage applied. The trailing-edge armature reaction mmf is, from (5.25),

$$mmf_{te} = \left(\text{conductors under}\,\tfrac{1}{2}\,\text{pole face}\right) \times I_{st}/a \tag{5.41}$$

where I_{st} is the starting or stalled armature current with the maximum expected armature voltage applied. The magnetic field intensity resulting from this mmf is

$$H_{te} = \frac{mmf_{te}}{(l_m + l_g)} \tag{5.42}$$

where l_m and l_g are the radial lengths of the magnet (pole) and the air gap, respectively. In using the normal SI units, these lengths will be expressed in meters, and H_{te} will be in A/m. *It is still common practice to express the characteristics of permanent magnets in cgs units* (H in oersteds and B in gauss). Therefore, it may be desirable to convert H_{te} in (5.42) to cgs units

$$H(\text{Oe}) = \frac{H(\text{A/m})}{79.577} \tag{5.43}$$

An analysis of a permanent magnetic system with a variable air gap is given in Chapter 2 and illustrated in Figure 2.17. It was shown in Section 2.11 and in Figure 2.17 that increasing the air gap or the permeance external to the permanent magnet had a demagnetizing effect upon the magnet. The demagnetizing effect of armature reaction is a similar phenomenon but requires a different graphical analysis than that shown for the variable air-gap demagnetization in Chapter 2. Variations in the permeance of the external magnetic circuit change the slope of the operating line, *OA* in Figure 2.17, whereas a demagnetizing mmf appears as a component of magnetic field intensity along the abscissa of Figure 2.17.

To illustrate this analysis, the permanent magnet demagnetizing characteristic of a ceramic magnet is shown in Figure 5.34. It can be assumed that the air gap of a permanent magnet motor is constant for all rotor positions. Assume that this air gap results in the circuit permeance to give the operating line, *OA*, at a slope, α_A, in Figure 5.34. Shown also in Figure 5.34 is the *intrinsic demagnetization curve. Intrinsic induction* (magnetic flux density) is defined as[9] "The vector difference at a point in a magnetic body between the magnetic induction at that point and the magnetic induction that would exist in a vacuum under the influence of the same magnetizing force" and expressed by the equation

$$\mathbf{B}_i = \mathbf{B} - \mu_0 \mathbf{H} \tag{5.44}$$

Intrinsic induction is also defined as the contribution of the magnetic material alone to the total magnetic induction. In a linear, homogeneous material, the vector difference of (5.44) can be replaced by an algebraic difference.

In the second and third quadrants of the hysteresis loop for a magnetic material, *H* is negative and the subtraction process indicated in (5.44) becomes a

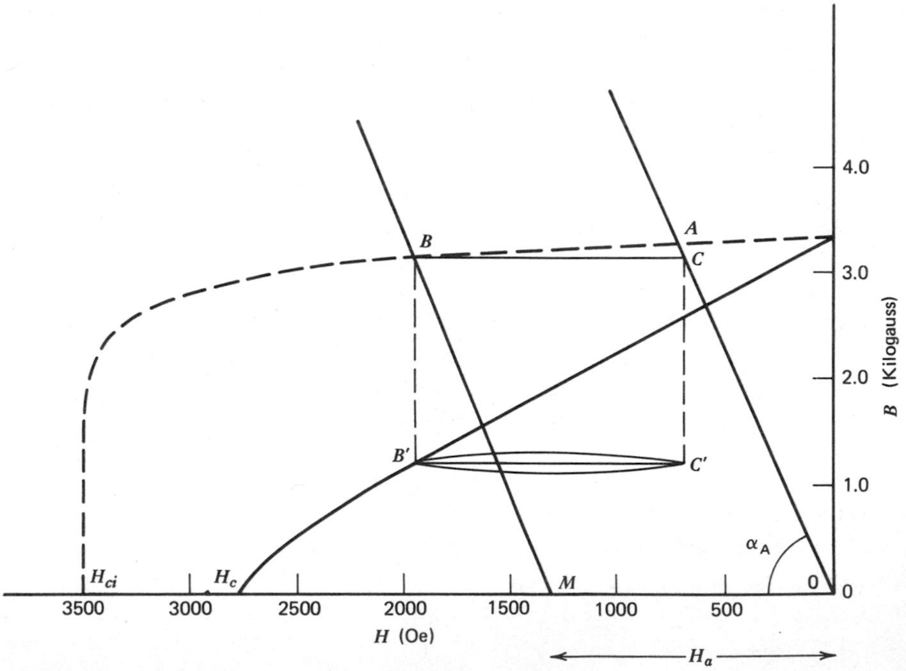

Figure 5.34. *BH*-curve of a permanent magnet.

summation. Therefore, in the second quadrant, the intrinsic demagnetization curve is above the normal demagnetization curve. This curve may be obtained easily when the normal curve is plotted in cgs units since, in this set of units, $\mu_o = 1$, and a point on the intrinsic curve can be obtained for any point, with the coordinates B_d and H_d on the normal curve from the relationship

$$B_i = B_d + H_d \tag{5.45}$$

It is seen from (5.45) that for materials with a very large coercive force, H_c (such as ferrites and rare earth cobalt alloys), the intrinsic demagnetization curve will be considerably different from the normal curve.

The graphical construction for evaluating the demagnetizing effects of armature reaction on a permanent magnet will now be explained with the aid of Figure 5.34. Application of an external magnetizing field intensity, H_a, causes the operating line, OA, to be shifted to the left parallel to OA to a new position shown by MB, where B is the point where the new operating line crosses the intrinsic curve. The vertical projection of point B to the normal curve at point B' gives the flux density that would exist in the presence of the external magnetic field. If the external field is removed, the operating point does not return along the normal curve but, rather, along a minor hysteresis loop to point C', which is the projection of point C on the original operating line.

If the external magnetic field is reapplied, the operating point will move back to point B' on the upper portion of the minor hysteresis loop. The mean slope of this minor loop is determined by the permanent magnet characteristic known as *recoil permeability*, which is defined in Section 2.11. This slope is very close to unity for ceramic permanent magnets, implying that the line, $B'C'$, in Figure 5.34 is nearly horizontal. This means that the variation of flux density in a ceramic magnet motor caused by demagnetizing armature reaction is small compared to the variation in a wound-field motor.

If the shift in the operating line, H_a, is greater than H_{ci}, the trailing-edge half of the magnet will be demagnetized. Actually, H_a should be considerably less than H_{ci} to provide a margin of safety to account for variations of the demagnetization curve with temperature, changes in other materials constants, inaccuracies in the design calculations, minor variations in the air gap with bearing wear, and so forth. Reference 8 suggests that if $H_a - H_{te}$, the field intensity resulting from the maximum trailing edge demagnetizing mmf of armature reaction [see (5.42)], the line MB in Figure 5.34 should be such that $B = 0.8 B_r$, where B_r is the residual flux density.

5.13 DYNAMICS OF DC COMMUTATOR MACHINES

As in the case of developing steady state relationships, the dynamic equations will vary somewhat depending upon the configuration of the specific machine—whether series or shunt, the type of voltage sources and loads, and so forth. Most of the following discussion is based upon the separately excited configuration illustrated schematically in Figure 5.35. In terms of setting up the dynamic equations, this configuration is the most general and can be modified for use in other configurations. Lowercase symbols are used to represent time functions or time-varying parameters.

By summing voltages in the armature circuit, we obtain

$$v_a = e + i_a R_a + L_a \frac{di_a}{dt} \tag{5.46}$$

$$e = K_t i_f \omega_m \tag{5.47}$$

where L_a is armature inductance in henries. Equation (5.46) has been written for motor operation. To describe generator operation, the sign of i_a and di_a/dt are changed. The similarity between these equations and their steady state counterparts, (5.23) and (5.15), should be apparent to the reader. By summing voltages in the field circuit, we obtain

$$v_f = i_f R_f + L_f(i_f) \frac{di_f}{dt} \tag{5.48}$$

The field circuit inductance, $L_f(i_f)$, is shown as a nonlinear function of i_f to give generality to the set of equations. This nonlinear function is related to the magnetization curve of the machine or the flux versus ampere-turn characteristic of the magnetic circuit of the machine. Summation of torques acting upon the motor shaft gives

$$T_d = T_L + D\omega_m + J \frac{d\omega_m}{dt} \tag{5.49}$$

$$T_d = K_t i_f i_a \tag{5.50}$$

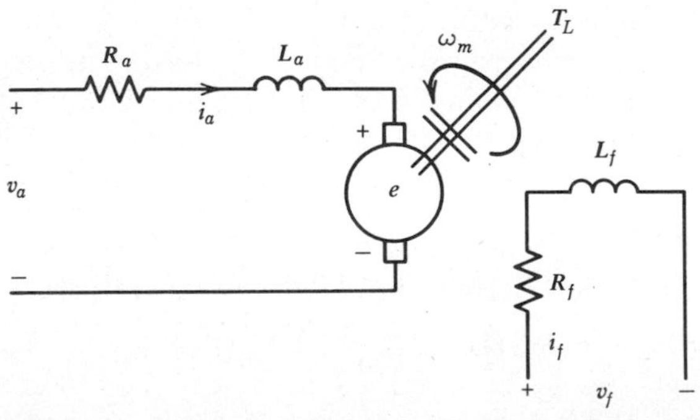

Figure 5.35. Schematic diagram for dynamics analysis.

where D is a viscous damping coefficient representing rotational loss torque in newton-meter-seconds; J is the moment of inertia of the total rotating system, including machine rotor, load, couplings, shaft, in $kg-m^2$ or $N-m-s^2$; and T_L is the load torque, in newton-meters.

Equation (5.49) has been written for motor operation. To describe generator operation, the signs of T_d and T_L are reversed. A linear damping coefficient is another linearization that is used in machine analysis. The load torque, T_L, is a function of time even though we express it in uppercase symbols to distinguish it from the symbol for time. The set of equations for the dc commutator machine is nonlinear not only because of the nonlinear coefficients, such as L_f and, possibly D, but also because of the product term in the torque and voltage expressions, (5.47) and (5.50).

Equations (5.46) to (5.50) provide a set of state equations that are useful in the analysis of a great number of machine problems. In order to apply these equations, the physical conditions of the specific problem must be introduced in an analytical manner. These conditions include numerical values for the circuit coefficients, the R's, L's, D, and J; descriptions of the three input terms, v_a, v_f, and T_L and initial conditions for the state variables. The conditions also include modifications of the equations to conform with the circuit configuration of the specific problem, for example, modification for series-field excitation. The relative significance of the nonlinearities must be determined and mathematical techniques for handling the nonlinearities must be developed. Before proceeding with some simple examples of applying these equations, let us take a brief look at some of their physical implications.

Equation (5.49) is helpful in understanding the mechanism by which a dc commutator machine is loaded. For this purpose, it is convenient to rearrange (5.49) as follows:

$$T_d - T_L = D\omega_m + J\frac{d\omega_m}{dt} \tag{5.51}$$

Let us consider motor operation first. Motor start-up is possible only if the electromagnetic torque, T_d, is larger than the load torque, T_L. At start-up, the speed is zero and the difference between these two torques determines the initial acceleration of the machine. Direct current commutator motors are capable of developing very high starting torques (as the torque at zero speed is called) and are used frequently because of this capability. As long as the left-hand side of (5.51) is larger than the right-hand side, the motor will continue to accelerate.

An equilibrium condition is reached when $T_d - T_L = D\omega_m$, at which time the acceleration is zero and the motor operates at a constant speed. This situation is shown in Figure 5.36 for two different types of load torques—constant torque, which might represent a cutting tool or lathe, and a torque varying approximately with cubic power of speed, which might represent a fan load. If the left-hand side of (5.51) becomes negative, which can occur if the load torque is

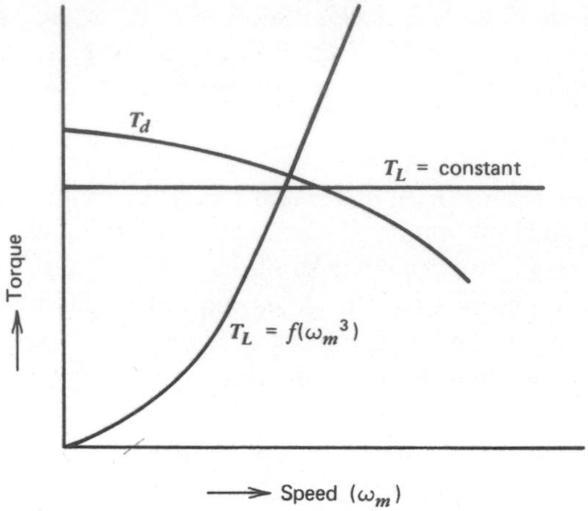

Figure 5.36. Motor and load torque characteristics.

increased or the electromagnetic torque is decreased from their values at the equilibrium speed, the right-hand side must also go negative. This situation is accomplished by means of a negative acceleration, that is, deceleration, causing the speed to decrease to a new equilibrium condition. The process of loading and speed variation is generally a stable process in dc commutator machines, except during conditions of very low field excitation or of excessive armature reaction with low values of load torque, which may result in excessive speed.

5.13.1 MOTOR START-UP

The start-up and the stalled condition of a dc commutator motor must be considered in the use of the machine. Stalling can occur if T_L in (5.51) is too large, or as a result of a variety of malfunctions in the machines or its controls. At start-up or stall, the back emf of the motor is zero and the armature current is equal to the applied voltage divided by the armature resistance. This condition results in an armature current far in excess of the rated current for the brush/commutator assembly and the armature winding.

Such a situation can be tolerated only for a matter of seconds in most machines before permanent damage is done to some armature components. In many machines using armature voltage control, this situation is avoided by means of control circuitry, which includes armature current feedback signals. In machines with simpler control schemes, an armature resistance is usually added at start-up to limit armature current to safe values. Direct current commutator motor starters consist of discrete steps of resistance and have facilities for switching in various values of resistance to control starting current and torque.

5.13.2 DYNAMIC AND REGENERATIVE BRAKING

A most useful characteristic of dc commutator machines is the capability of a smooth transition from motor to generator operation, and vice versa. This is one of the reasons why this class of machine is still so widely applied as a traction device, despite its obvious disadvantages due to the commutator/brush assembly. In most traction applications, much energy is wasted in the process of braking or decelerating the vehicle. This energy ends up as heat in the brake shoes or brake bands. There are two schemes of eliminating the wear of braking components in traction applications where dc commutator machines are used. One, is *dynamic braking*. The kinetic energy of a moving vehicle is dumped into a resistance by means of the generator action of a dc commutator machine. In *regenerative braking*, also most valuable as an energy-saving scheme, the kinetic energy of the vehicle is stored in a battery, flywheel, or other type of energy storage system by means of the generator action of the machine.

There is a third type of electrical braking, sometimes employed in industrial dc commutator motors such as overhead crane or light truck motors, which is known as *plugging*. This involves the sudden reversal of the connections of either the field or armature winding during motor operation, which causes a torque of the opposite direction to be developed at a large magnitude. In this case, the kinetic energy of the moving system is dissipated in the armature resistance. This braking scheme requires an overdesigned brush/commutator assembly and heavy armature windings. Electrical braking of any type becomes less effective as speed decreases, because the torque decreases as the speed approaches zero. Therefore, back-up mechanical or hydraulic brakes are usually required with electrical braking.

In regenerative braking, only a portion of the stored kinetic energy is available for charging a battery or flywheel, since a portion of the kinetic energy is dissipated in the windage and friction losses of the motor and vehicle and in the electrical losses of the motor during deceleration. Therefore, the effectiveness of regenerative braking as a means of conserving energy in traction applications is governed by the energy efficiency of the electrical components in the drive train. It has been shown[10] that about 35% of the energy put into an automotive vehicle during typical urban driving is theoretically recoverable by means of regenerative braking. When this energy is reduced by the amount necessary to overcome the electrical system losses and the mechanical losses of the vehicle, about 10 to 15% of the energy input to the vehicle can be recovered and stored in batteries, flywheels, or other devices. The exact value of the recoverable energy depends upon the type of driving, the terrain, the efficiency of the drive train, gear ratios in the drive/train, and so forth.

Dynamic braking, that is, the dissipation of kinetic energy in a resistance external to a traction motor, is readily described analytically. For this case, (5.46) and (5.49) become

$$e = i(R_a + R_b) + L_a \frac{di_a}{dt} \tag{5.52}$$

where r_b is the resistance added to the armature circuit to dissipate the kinetic energy of braking;

$$-T_L - T_d = D\omega_m + J\frac{d\omega_m}{dt} \tag{5.53}$$

Equations (5.47), (5.48), and (5.50) are unchanged by motor or generator operation. The solution of these equations representing dynamic braking is given as a problem at the end of this chapter.

Important parameters in the analysis of a machine dynamics are the inertia term, J, and the viscous damping term, D. A simple experimental method for obtaining these parameters is known as the *retardation test*. This test consists of accurately measuring rotor speed as a function of time while the rotor is allowed to coast to a standstill with zero armature or field excitation. Speed is best measured by obtaining a voltage proportional to speed from a small tachometer and recording this voltage on a strip recorder.

A typical retardation speed characteristic is shown in Figure 5.37. The parameters can be obtained from this test by means of (5.53). During a coastdown with no armature or field excitation and no load torque, (5.53) becomes

$$D\omega_m = J\frac{d\omega_m}{dt} \tag{5.54}$$

Multiplying both sides (5.53) by the mechanical speed, ω_m, will result in two power terms

$$D\omega_m^2 = J\omega_m\frac{d\omega_m}{dt} \tag{5.55}$$

The left-hand side of (5.55) is the mechanical loss of the machine and the right-hand side is the rate at which kinetic energy is supplied to the rotating system.

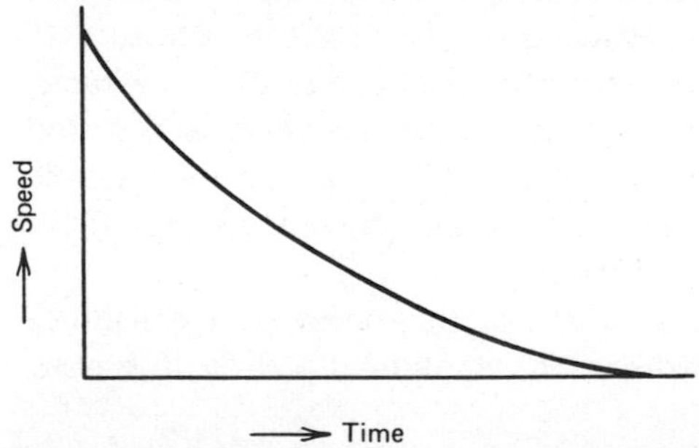

Figure 5.37. Speed versus time in a retardation test.

The inertia is found by picking a certain value of speed, determining the mechanical losses at this speed by the method outlined in Section 5.9, determining the speed-time slope at this value of speed from a retardation test, and solving for J from (5.55). Note that the retardation curve around the chosen speed only is required. A value for the linear mechanical damping coefficient can be obtained for the chosen speed by dividing the mechanical losses by the square of mechanical speed. The value of the damping coefficient will change considerably over the speed range of a machine.

5.13.3 TRANSFER FUNCTIONS FOR DC COMMUTATOR MACHINES

Let us return now to (5.46) to (5.50). There are several types of problems for which the linear versions of these equations are acceptable, and the methods of linear control theory are commonly used to study machine dynamics or to model the machine as part of a larger system. One configuration for which linear methods are applied and which is widely used in position and speed control systems is the permanent magnet or separately excited motor with constant excitation.

For this configuration, (5.48) is unnecessary. Let us also neglect the load torque, T_L, for the time being. Taking the Laplace transform of the remaining equation gives

$$V_a = E + (R_a + sL_a)I_a \tag{5.56}$$

$$E = K_s\Omega_m \tag{5.57}$$

$$T_d = (D + sJ)\Omega_m \tag{5.58}$$

$$T_d = K_s I_a \tag{5.59}$$

Uppercase letters for the variables in these equations indicate the transformed variable, $F(s)$. The new constant, K_s, is the product of $K_f I_f$ for separately excited machines and a similar term for permanent magnet machines. The constant K_s is generally obtained as the slope of the open-circuit voltage versus speed (in rad/s) curve. There are several means of representing these equations in linear control theory format. The simplest representation is a block diagram that leaves the back emf term as a feedback signal—which it actually is—and is shown in Figure 5.38. The two principal time constants of this type of machine, the electrical, τ_a, and the mechanical, τ_m, are introduced in Figure 5.38. These time constants are defined as

$$\tau_a = \frac{L_a}{R_a} \tag{5.60}$$

$$\tau_m = \frac{J}{D} \tag{5.61}$$

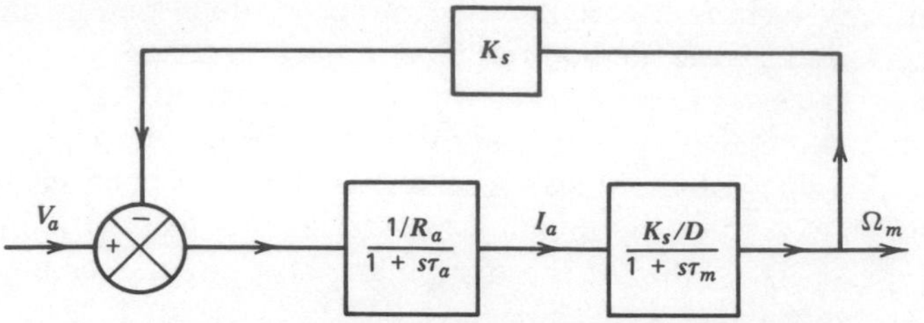

Figure 5.38. Block diagram of a motor with constant excitation.

In systems where the load torque must be considered, the block diagram of Figure 5.38 can be modified to become a multiple-input diagram with T_L as the second input. This is illustrated in Figure 5.39. It is often more convenient to describe the motor dynamics by combining (5.56) to (5.59):

$$\frac{\Omega_m}{V_a} = \frac{K_s}{K_s^2 + R_a D + s(L_a D + R_a J) + s^2 J L_a} \tag{5.62}$$

A similar expression can be obtained with T_L as the input function. The machine time constants are obtained experimentally by means of frequency-response measurements. They result from factoring the denominator of (5.62) rather than explicit component time constants given by (5.60) and (5.61). For position control applications, the angular position function, Θ, can be introduced into (5.56) to (5.59) by the relationship, $\Omega_m = s\Theta$. The overall motor transfer function then becomes

$$\frac{\Theta}{V_a} = \frac{K_s}{s\left[K_s^2 + R_a D + s(L_a D + R_a J) + s^2 J L_a\right]} \tag{5.63}$$

This same configuration is frequently used as a generator in the form of a tachometer or speed sensing device. Since the output of a tach-generator is

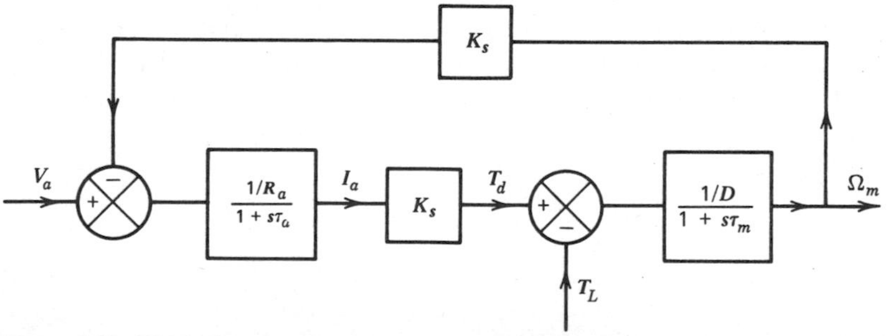

Figure 5.39. Block diagram of a motor with multiple inputs.

generally fed into a high-impedance load, such as a voltmeter, armature resistance and inductance can be neglected and the transfer function becomes from (5.57)

$$\frac{E}{\Omega_m} = K_s \tag{5.64}$$

The separately excited generator can be considered as an amplifier and was formerly used in this capacity before the advent of high-power semiconductor devices. As an amplifier, the generator is operated at a constant speed and controlled by means of the input field voltage, v_f. The amplifying characteristics can be observed by bringing back (5.48) into the discussion and dropping (5.49) and (5.50). We shall also add a load resistance, R_L, in the armature circuit and define the amplifier output as the voltage across this resistance. The transformed equations for this configuration are

$$V_L = I_a R_L \tag{5.65}$$

$$E = \left[(R_a + R_L) + sL_a\right] I_a \tag{5.66}$$

$$E = K_f I_f \tag{5.67}$$

$$V_f = (R_f + sL_f) I_f \tag{5.68}$$

We have assumed a linear field inductance, L_f, and voltage constant, K_f, in order to make use of linear control theory notation. This approach is acceptable if the range of field control is totally within the saturated or the unsaturated region of the magnetization curve, or if piecewise linear techniques are to be used. Note that K_f is the slope of the magnetization curve (Figure 5.14).

The transfer function for this system is

$$\frac{V_L}{V_f} = \frac{\dfrac{K_f}{R_f}\left(\dfrac{R_L}{R_L + R_a}\right)}{(1 + s\tau_f)(1 + s\tau_L)} \tag{5.68}$$

$$\tau_f = \frac{L_f}{R_f} \tag{5.69}$$

$$\tau_L = \frac{L_a}{(R_a + R_L)} \tag{5.70}$$

Since R_L generally is much larger than the armature resistance, R_A, it is seen that the steady state gain of this "amplifier" is K_f/R_f.

A second stage of amplification can be added to this amplifier by shorting the armature brushes and adding a second set of brushes on an axis that is 90 electrical degrees from the main brush axis. From Figures 5.17 to 5.19, it should be apparent to the reader that this second set of brushes is in the proper position to pick off from the armature winding the voltage induced by the armature reaction flux. This is an example of a machine configuration in which armature reaction serves a beneficial function. It is known as the *amplidyne*,[11] a General Electric trade name, and has found many applications as a power amplifier. The shorting of the brushes in the normal position sets up a very large armature reaction field, which causes a relatively large voltage to be induced across the second set of brushes. In effect, this adds a second K_f multiplier to the steady state gain portion of (5.69).

REFERENCES

1. C. G. Veinott, *Fractional-and Subfractional-Horsepower Electric Motors*, 3rd. ed., McGraw-Hill Book Company, New York, 1970.
2. C. S. Siskind, Direct-current Armature Windings—Theory and Practice, McGraw-Hill Book Company, New York, 1950.
3. "Airesearch 5HP Motor Development and Test Summary Report," Report No. F-4095-R, Airesearch Div. Garret Corp,. Los Angeles, Calif., February 1963.
4. J. E. Vrancik, "Prediction of Windage Power Loss in Alternators," NASA Report No. TN D-4849, Lewis Research Center, Cleveland, Ohio, July 1968.
5. A. Kusko, *Solid State DC Motor Drives*, MIT Press, Cambridge, 1968.
6. B. M. Emunson, and J. S. Ewing, "New NEMA Standards for DC Motors for Use on Rectified Power," *IEEE Transactions on Industry and General Applications*, Vol. IGA-7, No. 4, New York, July/August 1971.
7. "Test Procedure for Evaluation and Classification of Insulation Systems for DC Machines," *IEEE Standard* 304–1969, New York, 1969.
8. J. R. Ireland, *Ceramic Permanent-Magnet Motors*, McGraw-Hill Book Company, New York, 1968.
9. *IEEE Standard Dictionary of Electrical And Electronic Terms*, IEEE Standard No. 100-1972, Wiley Interscience, New York, 1972.
10. L. E. Unnewehr, et al., "Energy Saving Potential of Engine-Electric Vehicular Drives," Proceedings of the Eleventh Intersociety Energy Conversion Engineering Conference, Vol. 1, September 1976.
11. T. A. Murray, "Amplidynes—Sensitive Control Systems for Brute-Force Applications," *Machine Design*, Cleveland, Ohio, December 17, 1964.

PROBLEMS

5.1. A 10-pole motor has 123 slots and is wave-wound with two coils per slot and four turns per coil. Determine the armature factor, K_a, and the pole pitch, α, in degrees.

5.2. A four-pole, lap-wound machine has 48 slots, two coils/slot, and three turns/coil. Determine the armature factor, K_a, and the pole pitch, α, (in degrees).

5.3. The machine of Problem 5.2 is driven at 5000 rpm and excited to give a flux of 0.015 Wb/pole. Determine the armature induced voltage, E.

5.4. By completing the winding layout diagrams shown in Figures 5.10 and 5.11, determine the number of parallel electrical circuits between the armature terminals in each case. Does your answer correlate with the general rules given in Section 5.2?

5.5. The motor of Problem 5.1 is operated with an air-gap flux of 0.012 Wb/pole and an armature current of 100 A. Determine the electromagnetic developed torque.

5.6. A separately excited motor has a nameplate rating of 50 hp, 440 V, and 3000 rpm.
   (a) Determine the rated torque in lb-ft and in N-m.
   (b) Assuming that the machine is 85% efficient at its rated output, what is the armature current at rated output with rated voltage applied?

5.7. Show that the constants, K_t, in (5.15) and (5.22) are dimensionally equivalent.

5.8. Repeat Problem 5.7 for the constants, K_s, in (5.57) and (5.59). What is the SI (MKS) unit of K_s?

5.9. In practical usage and on many motor nameplates, the units of K_s in (5.57) are given as volt-s/rad or just volt-s and the units of K_s in (5.59) as lb-ft/A. Show that these two units are dimensionally equivalent.

5.10. The motor characterized by Curve (4) of Figure 5.33 has a stall (zero rpm) current of 42 A. Determine the speed constant, K_s. What is the no-load (zero current or zero torque) induced voltage of this motor?

5.11. A separately excited machine has the magnetization characteristic as shown in Figure 5.14. The machine has an armature resistance of 0.1 ohm and a field circuit resistance of 40 ohm; there are 500 turns/pole in the field winding. With 120 V applied to the field circuit,
   (a) What is the armature no-load or open-circuit voltage at 5000 rpm?
   (b) If the machine is operated as a generator with an armature current of 100 A, what is the armature terminal voltage? (Neglect the armature reaction.)
   (c) If the armature reaction at 100 A results in a demagnetizing mmf of 220 At/pole, determine the armature terminal voltage at 100 A.

5.12 The motor of Problem 5.10 has an armature resistance of 0.4 ohm and is operated from a 12 V source. When the motor draws 10 A, at what value of speed and torque is the motor operating?

5.13 Test data taken on a shunt machine at 2400 rpm gives the following points for the magnetization curve:

Field current (A)	0	1.0	2.0	3.0	4.0	5.0
Armature voltage (V)	4	75	133	160	172	182

(a) Express this result graphically.
(b) If the machine is to be operated as a self-excited shunt generator, determine the shunt-field resistance necessary to obtain a no-load voltage of 150 V at 2400 rpm.
(c) Determine the "critical" shunt-field resistance for self-excited voltage buildup.

5.14. A separately excited 120-volt motor develops a torque of 20 lb-ft when the armature current is 8 A.
(a) Determine the armature current when the developed torque is 60 lb-ft, assuming no change in field current and neglecting armature reaction.
(b) If the speed is 6% higher with the 20 lb-ft torque than with the 60 lb-ft torque, determine the armature resistance.

5.15. A long-shunt compound generator has the magnetization curve given Figure 5.14. The shunt field has 700 turns/pole, the series field 5 turns/pole. Armature circuit resistance (including brushes) is 0.08 ohm and series field resistance is 0.005 ohm. At 5000 rpm, determine the terminal voltage when the shunt field current is 2.0 A and the armature current is 150 A. Neglect the armature reaction.

5.16. The permanent magnet motor whose speed-torque characteristic is given here as curve (7) of Figure 5.33 is frequently used to drive air blowers for air conditioning purposes. Assume a blower speed-torque characteristic described by the equation, $T = 0.15 \times 10^{-6} \omega_m^3$, where T is in N-m and ω_m is in rad/s. Determine the steady state speed at which this motor will drive the blower. *Note*: This problem can be solved either by graphical or analytical methods. If graphical techniques are used, it will be more convenient to replot curve 7 of Figure 5.33 with speed as the abscissa. Also keep in mind the difference in units between the data of curve 7 and that of the above equation.

5.17. A series motor has the magnetic characteristic shown in III-1 of Appendix III. The series field consists of six turns/pole and has a resistance of 0.008 ohm. The armature circuit (including brushes) has a resistance of 0.017 ohm. The armature constant, K_a, is 32. At a certain load, the current is 400 A with 24 V applied.
(a) Determine the speed and torque at this value of armature current.
(b) Repeat "a" for currents of 200, 100, and 50 A at 24 V.
(c) Plot the resulting speed-torque characteristic.

5.18. The mechanical and magnetic losses for the motor described in Problem 5.17 have been determined by test and are shown in the curves of Figure 5.40. Determine the efficiency of this series motor at the four values of current used in Problem 5.17. Neglect stray-load losses.

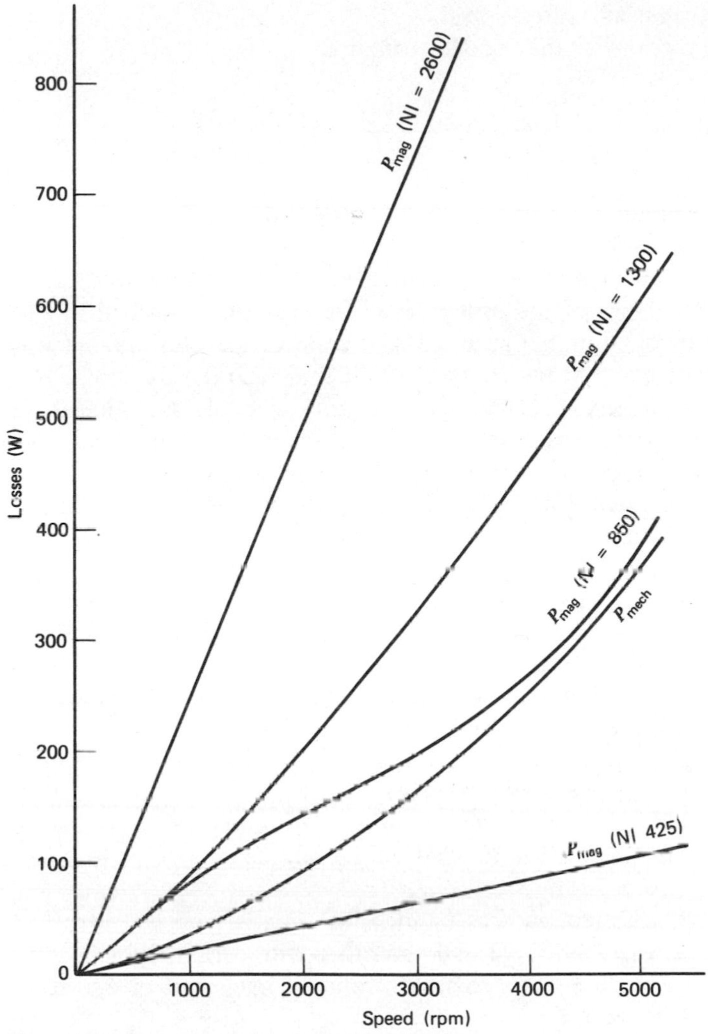

Figure 5.40. Series motor losses (Problem 5.11).

5.19. Considering the machine described in Problem 5.11

(a) Determine the initial ($t=0$) armature current if the armature is directly connected across a 120-V supply.

(b) Determine the resistance to be added in series with the armature to limit the armature current to 200 A.

5.20. For the motor of Example 5.5 calculate the required armature terminal voltage, input power, and efficiency when

(a) Operated as a motor to supply one-half rated torque at one-half rated speed.

(b) Operated as a generator to supply rated terminal voltage at rated armature current and rated speed.

5.21. Develop an expression for the transfer function

$$G_T(s) = \frac{\Omega_m(s)}{T_L(s)}$$

for analysis of a motor when considering the load torque, T_L, as input and shaft speed, Ω_m as output.

5.22. The permanent magnet material used in a motor has a profound effect upon the characteristics of the motor. For the example described in this Chapter (Section 5.13.6 and Figure 5.34), a ceramic (ferrite) magnet was used. This type of material results in a motor that operates at a relatively low air-gap flux density but that can sustain relatively high levels of reaction without demagnetization. Compare the ceramic motor with one using Alnico VI (Figure 5.13 and Table 2.2) in the following steps:

(a) Plot the intrinsic demagnetization curve for Alnico VI. What is H_{ci} for Alnico VI? The normal demagnetization curve will also be necessary.

(b) It is seen that Alnico VI must be operated at a much higher permeance ratio (slope of line OA in Figure 5.34) than the ceramic material. Since the permeance ratio is a function of the magnetic permeance of the external magnetic circuit, mainly the air-gap length, what does this imply concerning the length of practical air gaps usable in an Alnico VI motor as compared to those in ceramic motors?

(c) Assume a permeance ratio (B_d/H_d) of 50 for Alnico VI. What permeance ratio was used in Figure 5.34?

(d) What armature reaction (in terms of a field intensity, H_a) can be tolerated in the Alnico VI motor before the flux density drops to 0.8 B_r (point B in Figure 5.34)?

(e) Assuming two different motors of identical dimensions and identical windings, one with Alnico VI and the other with ceramic poles (as in Figure 5.33), what is the approximate ratio of back emf or armature voltage levels between the two motors.

(f) Try to estimate the power capability of the two motors (related to the product of E and I_a), and discuss the relative merits of both types of magnetic materials for motor applications.

5.23. Assuming linear damping coefficient, D, as in (5.53) and letting $T_L = 0$, solve for motor speed as a function of time during a dynamic braking operation. See (5.47), (5.48), (5.50), and (5.52). Assume that a braking operation is initiated at $t=0$ when $i_a=I_0$ and $\omega_m=\Omega_0$. Also assume that V_f and i_f are maintained constant during the brake operation.

5.24. A control motor (permanent magnet excited) drives an inertial load through a gear train of overall gear ratio, R. If the load inertial is designated J_L and load torque, T_L, is zero, derive the transfer function for speed control,

$$G_s(s)=\frac{\Omega_m(s)}{V_a(s)}$$

How can the gear ratio be used to alter the mechanical time constant, τ_m, of the system?

5.25. For an armature-controlled, separately excited motor described by (5.56) to (5.59) and Figure 5.38, assume T_L and D are zero. The motor is initially at standstill and excited at rated field current. Rated armature voltage is suddenly applied at $t=0$. Find the total energy dissipated in the armature resistance, R_a, during the time required for the motor to reach its final steady-state speed. Compare this energy with the kinetic energy stored in the rotating inertia at steady state speed. Repeat if D is nonzero.

Chapter 6
Introduction Machines

The induction motor is the most common of all motors. Like the dc machine discussed in the last chapter, an induction machine consists of a stator and a rotor mounted on bearings and separated from the stator by an air gap. Electromagnetically, the stator consists of a core made up of punchings (or laminations) carrying slot-embedded conductors. These conductors are interconnected in a predetermined fashion and constitute the armature windings. We shall consider some details of armature windings in the next section.

Alternating current is supplied to the stator windings and the currents in the rotor windings are induced by the stator currents. The rotor of the induction machine is cylindrical and carries either (1) conducting bars short-circuited at both ends, as in a *cage-type* machine (Figure 6.1), or (2) a polyphase winding with terminals brought out to slip rings for external connections, as in a wound-rotor machine (Figure 6.2). A wound-rotor winding is similar to that of the stator. Sometimes the cage-type machine is also called a *brushless* machine and the wound-rotor machine termed a *slip-ring* machine. The stator and the rotor, in its three different stages of production, are shown in Figure 6.3. The motor is rated at 2500 kW, 3 kV, 575 A, two-pole, and 400 Hz. A finished cage-type rotor of a 3400-kW 6-kV motor is shown in Figure 6.4, and Figure 6.5 shows the wound rotor of a three-phase slip ring 16,200-kW four-pole induction motor. A cutaway view of a completely assembled motor, having a cage-type rotor, is shown in Figure 6.6.

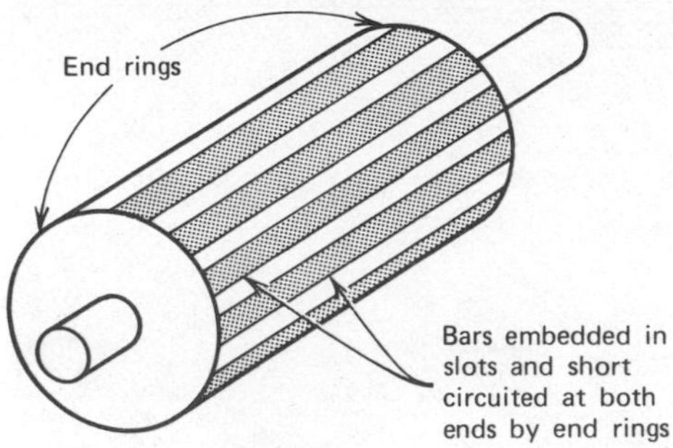

Figure 6.1. A cage-type rotor.

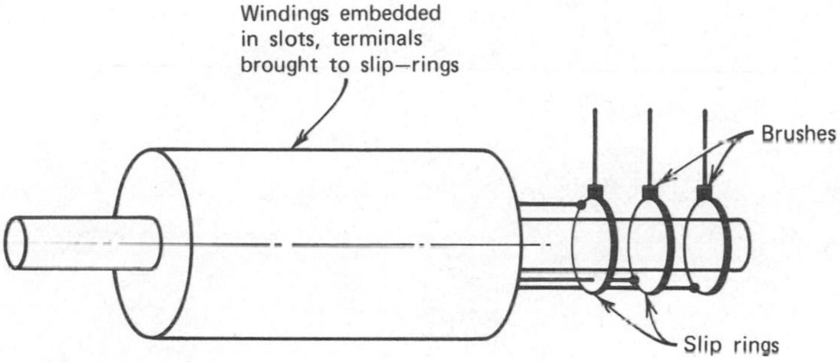

Figure 6.2. A wound rotor.

Figure 6.3. Rotor for a 2500-kW 3-kV two-pole 400-Hz motor in different stages of production. (Courtesy Brown Boveri Company.)

Figure 6.4. Complete rotor of a 3400-kW 6-kV 990-rpm motor. (Courtesy Brown Boveri Company.)

Figure 6.5. Rotor of a three-phase 15,200-kW 2.4-kV slip-ring induction motor. (Courtesy Brown Boveri Company.)

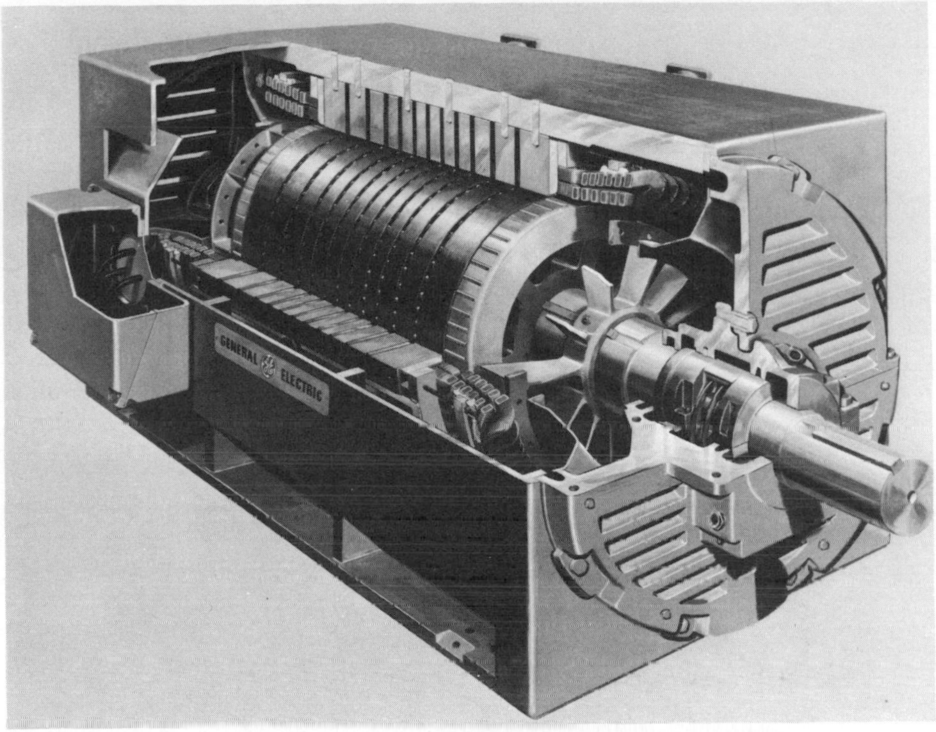

Figure 6.6. Cutaway of an induction motor. (Courtesy General Electric Company.)

An induction machine operates on the basis of interaction of induced rotor currents and the air-gap fields. If the rotor is allowed to run under the torque developed by this interaction, the machine will operate as a motor. On the other hand, the rotor may be driven by an external source beyond a speed such that the machine begins to deliver electrical power and operates as an induction generator (rather than an induction motor, which absorbs electrical power). Thus, we see that the induction machine is capable of functioning as a motor as well as a generator. However, almost invariably it is used as a motor and seldom as a generator.

Before we consider the induction motor in detail, it is worthwhile to study its stator construction and the magnetic field produced by the stator (or armature) windings as follows.

6.1 MMF'S OF ARMATURE WINDINGS

We recall from the last chapter that the armature of a dc machine has a winding that is distributed around the periphery of the armature. Thus, slot-embedded conductors, covering the entire surface of the armature and interconnected in a predetermined manner, constitute the armature winding of a dc machine.

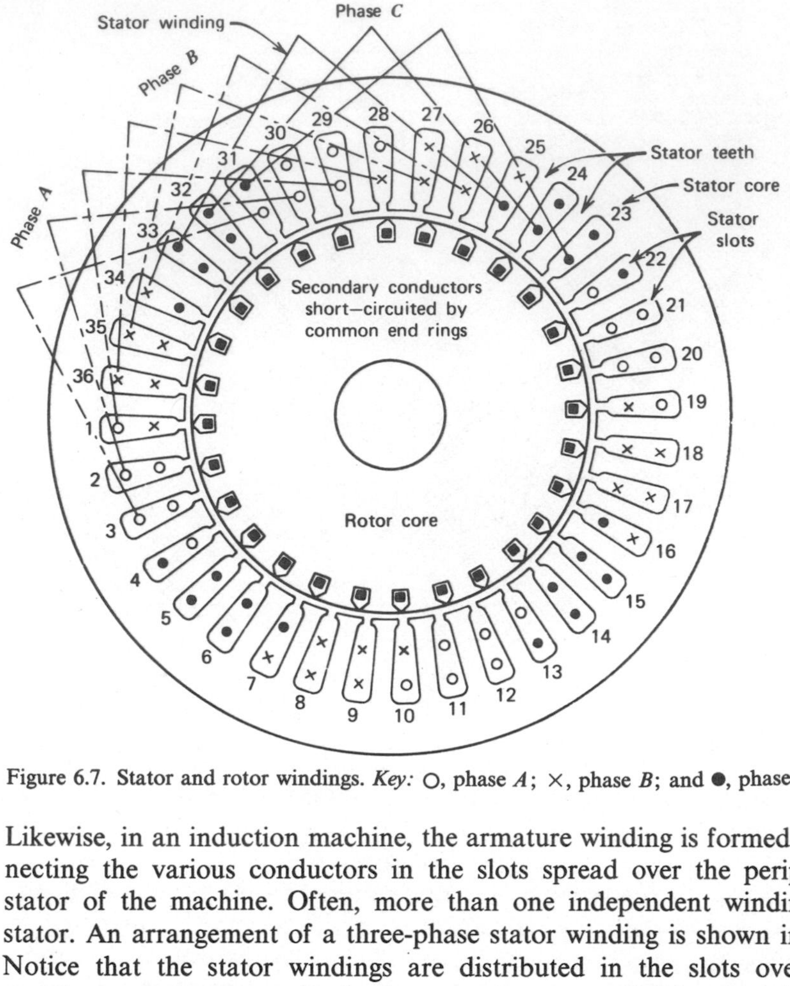

Figure 6.7. Stator and rotor windings. *Key:* ○, phase A; ×, phase B; and ●, phase C.

Likewise, in an induction machine, the armature winding is formed by interconnecting the various conductors in the slots spread over the periphery of the stator of the machine. Often, more than one independent winding is on the stator. An arrangement of a three-phase stator winding is shown in Figure 6.7. Notice that the stator windings are distributed in the slots over the entire periphery of the stator. Each slot contains two coil-sides. For instance, slot Number 1 has coil-sides of phases A and B, whereas slot Number 2 contains two layers (or two coil-sides) of phase A only. Such a winding is known as a *double-layer* winding. Furthermore, it is a four-pole winding laid in 36 slots (Figure 6.7) and we thus have three slots per pole per phase.

In order to produce the four-pole flux, each coil should have a span (or *pitch*) of one quarter of the periphery. In practice, the pitch is made a little less and, as shown in Figure 6.7, each coil embraces eight teeth. The coil-pitch is about 89% of the pole-pitch, and the winding is, therefore, a *fractional-pitch* (or *chorded*) *winding*. Further details on armature windings are available in Reference 1.

For the present we shall consider the mmf's produced by the armature windings. First of all, assume a single full-pitch coil having N-turns as shown in Figure 6.8, where the slot opening is negligible. Clearly, the machine has two

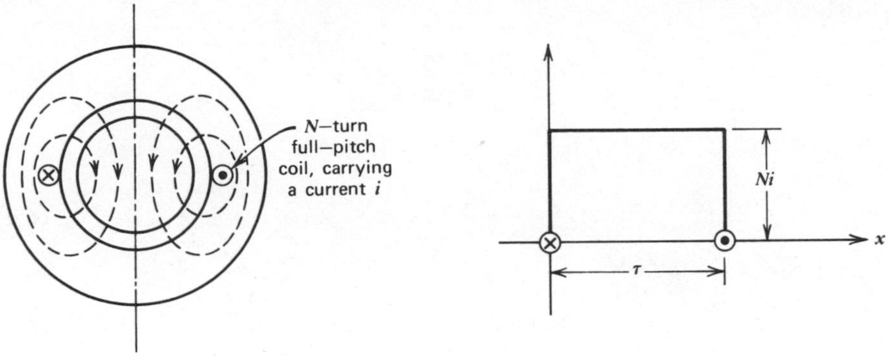

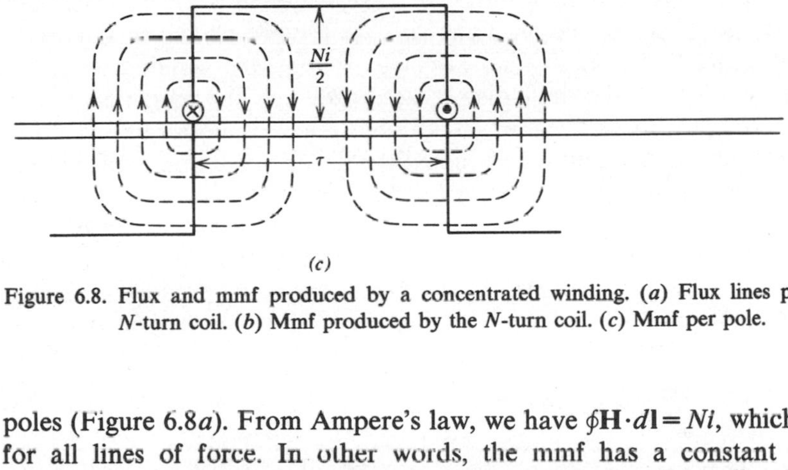

Figure 6.8. Flux and mmf produced by a concentrated winding. (*a*) Flux lines produced by an *N*-turn coil. (*b*) Mmf produced by the *N*-turn coil. (*c*) Mmf per pole.

poles (Figure 6.8*a*). From Ampere's law, we have $\oint \mathbf{H}\cdot d\mathbf{l} = Ni$, which is the same for all lines of force. In other words, the mmf has a constant value of Ni between the coil sides, as shown in Figure 6.8*b*. Traditionally, the magnetic effects of a winding in an electric machine are considered on a per pole basis. Thus, if i is the current in the coil, the mmf per pole is $Ni/2$, which is plotted in Figure 6.8*c*. The reason for such a representation is that Figure 6.8*c* also represents a flux density distribution, but to a different scale. Obviously, the flux density over one pole (say the north pole) must be opposite to that over the other (south) pole, and thus keeping the flux entering the rotor equal to that leaving the rotor surface. Comparing Figures 6.8*b* and 6.8*c*, we notice that the representation of the mmf curve with positive and negative areas (Figure 6.8*c*) has the advantage that it gives the flux density distribution, which must contain positive and negative areas. The mmf distribution shown in Figure 6.8*c* may be resolved into its harmonic components by Fourier analysis. The period of the fundamental component is the same (2τ) as that of the rectangular mmf wave. The amplitude of the fundamental wave is $4/\pi$-times the amplitude of the rectangular wave. Therefore, the fundamental component of the mmf distribution is given by

$$F(x,t)=\frac{4}{\pi}\frac{Ni}{2}\cos\frac{\pi}{\tau}x \tag{6.1}$$

If i is sinusoidal ac, so that $i=I_m \sin\omega t$, (6.1) becomes

$$F(x,t)=\frac{\sqrt{2}}{2}\frac{4}{\pi}NI\cos\frac{\pi x}{\tau}\sin\omega t \tag{6.2}$$

which simplifies to

$$F(x,t)=0.9NI\cos\frac{\pi x}{\tau}\sin\omega t \tag{6.3}$$

where I is the rms value of the current. Notice that the time variation of the flux resulting from the mmf is alternating and the flux is stationary in space.

In an electric machine we seldom have a single N-turn coil (Figure 6.8*a*) as a source of the armature mmf. Rather, we have windings that are distributed over the entire periphery of the machine, such as the one shown in Figure 6.7. Besides utilizing all the space available, by distributing the winding we reduce the harmonic content in the mmf distribution, although the magnitude of the fundamental component will be less than that for a concentrated winding. Ideally, we attempt to distribute the winding so that the resulting mmf distribution is purely sinusoidal.

In practice, as a first approximation we assume a sinusoidal mmf distribution. For the study of the induction motor, we shall assume such a distribution. Thus, we let the mmf (or flux density) space distribution produced by three identical coils that are displaced from each other by 120° (Figure 6.9) be given by

$$\begin{aligned} F_a &= F_m \sin\omega t\cos\frac{\pi x}{\tau} \\ F_b &= F_m \sin(\omega t-120°)\cos\left(\frac{\pi x}{\tau}-120°\right) \\ F_c &= F_m \sin(\omega t+120°)\cos\left(\frac{\pi x}{\tau}+120°\right) \end{aligned} \tag{6.4}$$

Notice that the three coils are excited by a three-phase source. Because we have shown the three-phase windings to consist of three (independent) N-turn coils, $F_m=0.9NI$, from (6.3) and (6.4). The space and time variations of the resultant mmf is then the sum of the three mmfs of (6.4). Observing that $\sin A\cos B = \frac{1}{2}\sin(A-B)+\frac{1}{2}\sin(A+B)$ and adding F_a, F_b and F_c we obtain the resultant mmf as

$$F(x,t)=1.5F_m\sin\left(\omega t-\frac{\pi x}{\tau}\right) \tag{6.5}$$

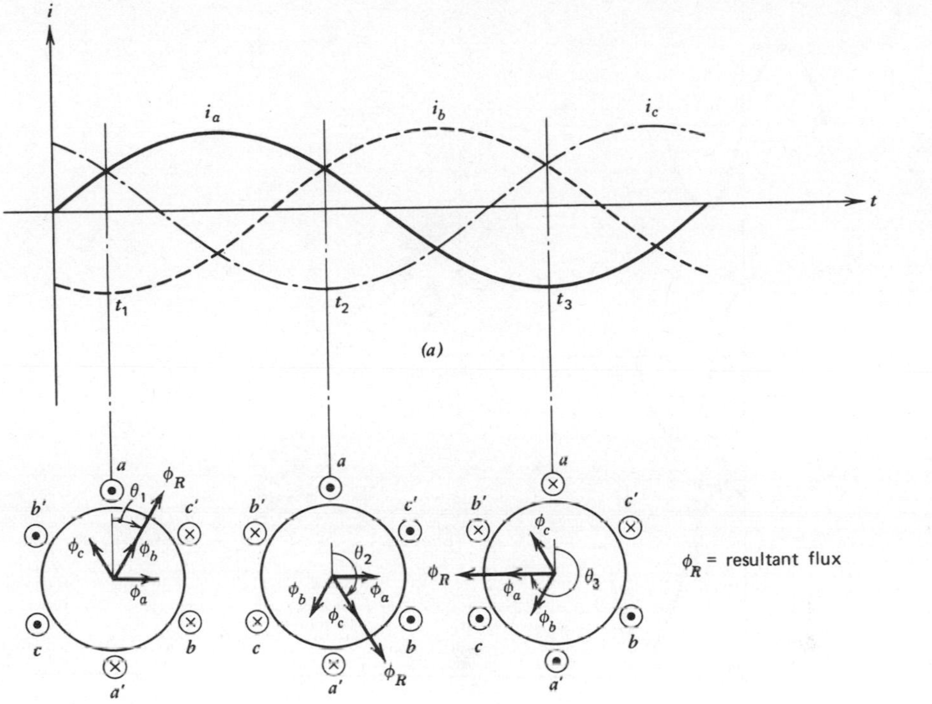

Figure 6.9. Production of a rotating magnetic field by a three-phase excitation: (*a*) time diagram; (*b*) space diagram.

Figure 6.10 shows the position of the resultant mmf at three different instants $t_1<t_2<t_3$. Notice that as time elapses a fixed point P moves to the right, implying that the resultant mmf is a traveling wave of a constant amplitude. The magnetic field produced by this mmf in an electric machine is then known as a *rotating magnetic field*. We may arrive at the same conclusion by considering the resultant mmf at various instants as shown in Figure 6.10. From these diagrams it is clear that, as we progress in time from t_1 to t_3, the resultant mmf rotates in space from θ_1 to θ_3. The existence of the rotating magnetic field is essential to the operation of an induction motor.

To determine the velocity of the traveling field given by (6.5), imagine an observer traveling with the mmf wave from a point P. To this observer, the magnitude of the mmf wave will remain constant (independent of time), implying that the right-hand side of (6.5) would appear constant. Expressed mathematically this would be

$$\sin\left(\omega t-\frac{\pi x}{\tau}\right)=\text{constant}$$

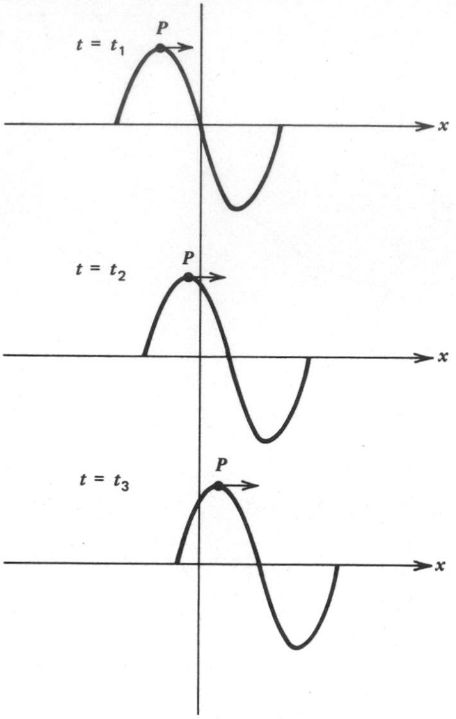

Figure 6.10. The function $\sin[\omega t-(\pi x/\tau)]$ at different time intervals $t_1<t_2<t_3$.

Or

$$\omega t - \frac{\pi x}{\tau} = \text{constant} \tag{6.6}$$

Differentiating both sides of (6.6) with respect to t, we obtain

$$\omega - \frac{\pi}{\tau}\dot{x} = 0$$

Or

$$\dot{x} = \frac{\omega\tau}{\pi} = 2f\tau = \frac{2\tau}{T} = \frac{\lambda}{T}\, m/s \tag{6.7}$$

where τ = pole-pitch, $\omega = 2\pi f$, f is the frequency of input currents and T is the corresponding period (i.e., $f = 1/T$). From (6.7) we conclude that the mmf (or flux) wave travels, during one cycle of the current, a distance twice the pole pitch (or wavelength, λ). Therefore, for a given pole-pitch and frequency, the velocity of the traveling field is constant and is known as the *synchronous velocity*.

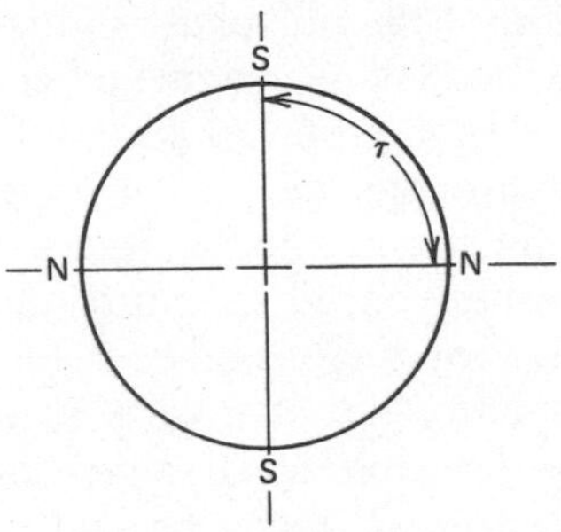

Figure 6.11. Definition of pole pitch, τ.

We can now relate the synchronous velocity (m/s) to speed in revolutions per minute (rpm) by observing that one revolution around the air gap of the machine corresponds to a linear distance, $p\tau$, where p is the number of poles (Figure 6.11). The distance traveled by the wave in one minute is $60(2f\tau)$. Thus, the speed of the traveling field, which now corresponds to a rotating field in rpm, n, is given by

$$n = \frac{60(2f\tau)}{p\tau} = \frac{120f}{p}\text{ rpm} \tag{6.8}$$

Again, this is the fixed speed, called synchronous speed, and is usually designated by n_s. So we rewrite (6.8) as

$$n_s = \frac{120f}{p}\text{ rpm} \tag{6.9}$$

6.2 ACTION OF A POLYPHASE INDUCTION MOTOR

We recall from the last section that a three phase stator excitation produces a rotating magnetic field in the air gap of an induction motor, and the field rotates at a synchronous speed given by (6.9). As the magnetic field rotates, it "cuts" the rotor conductors. By this process, voltages are induced in the conductors. The induced voltages give rise to rotor currents, which interact with the air-gap field to produce a torque. The torque is maintained as long as the rotating magnetic field and the induced rotor currents exist. Consequently, the rotor starts rotating in the direction of the rotating field. (*Note*: It may be readily verified, by applying the principle of conservation of energy, that the rotor will not run on its own torque in a direction opposite to that of the rotating field.) The rotor will achieve a steady-state speed, n, such that $n < n_s$. Clearly, when $n = n_s$ there will be no induced currents and hence no torque. The condition $n > n_s$ corresponds to the generator mode.

An alternate approach to explaining the operation of the polyphase induction motor is by considering the interaction of the (excited) stator magnetic field with the (induced) rotor magnetic field. The stator excitation produces a rotating magnetic field, which rotates in the air gap at a synchronous speed. This field induces polyphase currents in the rotor, thereby giving rise to another rotating magnetic field, which also rotates at the same synchronous speed as that of the stator and with respect to the stator. Thus, we have two rotating magnetic fields, rotating at a synchronous speed with respect to the stator but stationary with respect to each other. Consequently, according to the principle of alignment of magnetic fields (Chapter 4), the rotor experiences a torque. (It might be said to be dragged along by the stator magnetic field.) The rotor rotates in the direction of the rotating field of the stator.

6.3 SLIP AND FREQUENCY OF ROTOR CURRENTS

The acutal speed, n, of the rotor is often expressed as a fraction of the synchronous speed, n_s, as related by *slip*, s, defined as

$$s=\frac{n_s-n}{n_s} \tag{6.10}$$

The slip may also be expressed as percent slip as follows

$$\text{percent slip}=\frac{n_s-n}{n_s}\times 100 \tag{6.11}$$

At standstill, the rotating magnetic field produced by the stator has the same relative speed with respect to the rotor windings as with respect to the stator windings. Thus, the frequency of the rotor currents, f_2, is the same as the frequency of stator currents, f_1. At synchronous speed, there is no relative motion between the rotating field and the rotor, and the frequency of rotor current is zero. At other speeds the rotor frequency is proportional to the slip, s. This may be demonstrated as follows.

First, we rewrite (6.10) as

$$s=\frac{\omega_s-\omega_m}{\omega_s} \tag{6.10a}$$

where ω_m = actual speed of the rotor in rad/s. From (6.9) we observe that

$$\omega_s=\frac{\omega}{p/2} \tag{6.12}$$

Because the relative speed between the rotor and the rotating field produced by the stator is $(\omega_s-\omega_m)$, the angular frequency of the rotor induced currents, ω_r, is

$$\omega_r=\frac{p}{2}(\omega_s-\omega_m) \tag{6.13}$$

Substituting (6.10a) in (6.13) yields, in conjunction with (6.12),

$$\omega_r=s\omega_s\frac{p}{2}=s\omega \tag{6.14}$$

Or

$$f_r=sf \tag{6.15}$$

which is known as slip frequency. In (6.15) f_r = frequency of the rotor currents and f = frequency of the stator (input) currents (or voltages).

We now summarize the preceding discussions.

1. The stator rotating magnetic field rotates at the synchronous speed, ω_s (with respect to a stationary observer).
2. The rotor mmf produces a rotating magnetic field which also rotates at the synchronous speed and in the same direction as the field produced by the stator mmf. Thus, the rotating fields produced by the stator and rotor are stationary with respect to each other.
3. The rotating field produced by the rotor rotates at a speed $(\omega_s-\omega_m)$ with respect to the rotor, where ω_m is the actual mechanical speed of the rotor.
4. Currents (and voltages) induced in the rotor are of slip frequency.

6.4 THE ROTOR EQUIVALENT CIRCUIT

Recognizing the fact that the frequency of rotor currents is the slip frequency, we may express the per phase rotor leakage reactance X_2', at a slip s, in terms of the standstill per phase reactance X_2 as

$$X_2'=sX_2 \tag{6.16}$$

Next, we observe that the magnitude of the voltage induced in the rotor circuit is also proportional to the slip.

A justification of this statement follows from transformer theory (Chapter 3), because we may view the induction motor at standstill as a transformer with an air gap. For the transformer we know that the induced voltage, say E_2, is given by

$$E_2=4.44fN\phi_m \tag{6.17a}$$

But at a slip, s, the frequency becomes sf, according to (6.15); substituting this

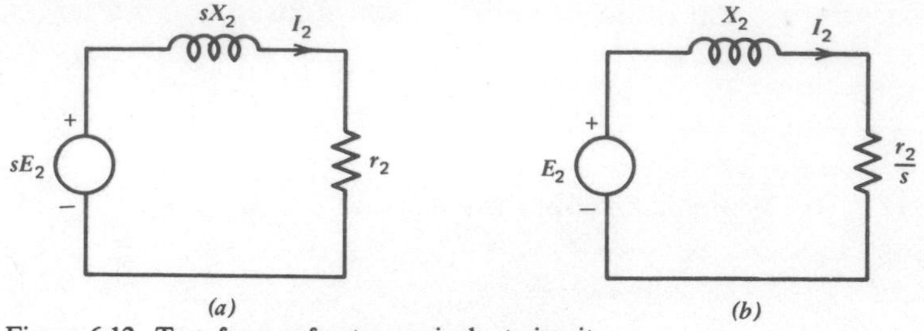

Figure 6.12. Two forms of rotor equivalent circuit.

value of frequency in (6.17*a*) yields the voltage E_2' at a slip s as

$$E_2' = 4.44 sfN\phi_m = sE_2 \tag{6.17b}$$

We conclude, therefore, that if E_2 is the per-phase voltage induced in the rotor at standstill, the voltage E_2' at a slip, s, is given by

$$E_2' = sE_2 \tag{6.17c}$$

Using (6.16) and (6.17*c*), we obtain the rotor equivalent circuit shown in Figure 6.12*a*. The rotor current I_2 is given by

$$I_2 = \frac{sE_2}{\sqrt{r_2^2 + (sX_2)^2}} \tag{6.18}$$

which may be rewritten as

$$I_2 = \frac{E_2}{\sqrt{\left(\frac{r_2}{s}\right)^2 + X_2^2}} \tag{6.19}$$

resulting in the alternate form of the equivalent circuit shown in Figure 6.12*b*. Notice that the circuits shown in Figure 6.12 are drawn on a per-phase basis. To this circuit we may now add the per-phase stator equivalent circuit to obtain the complete equivalent circuit of the induction motor, which will be discussed in the next section.

6.5 DEVELOPMENT OF THE COMPLETE EQUIVALENT CIRCUIT

We recall that in an induction motor, only the stator is connected to the ac source. The rotor is not generally connected to an external source, and rotor voltage and current are produced by induction. In this regard, the induction motor may be viewed as a transformer with an air gap, having a variable resistance in the secondary. Thus, we may consider that the primary of the transformer corresponds to the stator of the induction motor, whereas the secondary corresponds to the rotor on a per-phase basis. Because of the air gap, however, the value of the magnetizing reactance, X_m, tends to be relatively low compared with that of a transformer. As in a transformer (discussed in Chapter 3), we have a mutual flux linking both the stator and rotor, represented by the magnetizing reactance and various leakage fluxes. For instance, the total rotor leakage flux is denoted by X_2 in Figure 6.12. Although the leakage fluxes are subdivided into various components, such as end-connection leakage flux, slot leakage flux, tooth-top leakage flux, and so forth, they will not be considered here. We note, however, that an appropriate leakage reactance component is assigned to each leakage flux component and such components do not exist in a transformer.

Returning to the analogy of a transformer and considering that the rotor is coupled to the stator as the secondary of a transformer is coupled to its primary, we may draw the circuit shown in Figure 6.13. To develop this circuit further, we need to express the rotor quantities as referred to the stator. For this purpose we must know the transformation ratio, as in a transformer. (See Chapter 3.).

Care must be exercised in defining the transformation ratio. The voltage transformation ratio in the induction motor must include the effect of the stator and rotor winding distributions. Therefore, the ratio of the rotor to stator voltage becomes

$$\frac{E_2}{E_1}=\frac{k_{w2}N_2}{k_{w1}N_1} \tag{6.20}$$

where k_{w1} is the winding factor of the stator having N_1 series-connected turns-per-phase. For a cage-type rotor, the number of turns-per-phase per-pole-pair$=1/2$. Thus, the number of turns-per-phase is $N_2=(1/2)(p/2)$, where $p=$number of poles. Also, for the cage rotor, $k_{w2}=1$. For a wound rotor these quantities are found in a manner similar to that for the stator. The winding factor, k_w, is defined in Examples 6.1 and 6.2. Next, considering the stator and rotor mmf's F_1 and F_2, from (6.3), we may write

$$F_1=0.9m_1k_{w1}\frac{N_1I_1}{p} \tag{6.21a}$$

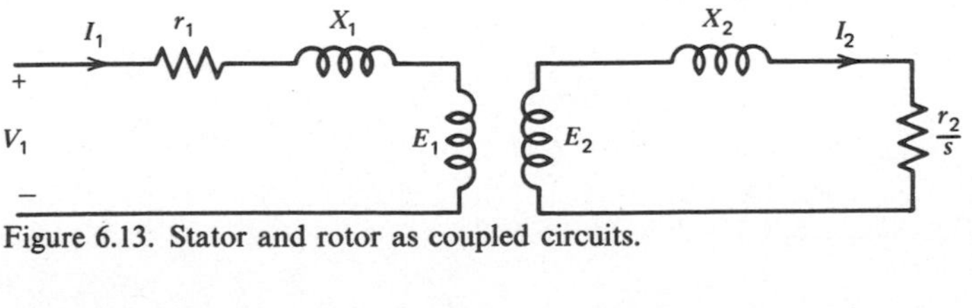

Figure 6.13. Stator and rotor as coupled circuits.

and

$$F_2 = 0.9 m_2 k_{w2} \frac{N_2 I_2}{p} \tag{6.21b}$$

where m_1 and m_2 are the number of phases on the stator and rotor, respectively, where $m_2 = m_1$ in a wound rotor, and other symbols are as defined earlier. For a cage-type rotor, the number of phases m_2 = number of bars per-pole-pair = $Q_2/(p/2)$, Q_2 being the number of rotor bars. Referring the rotor quantities to the stator implies that the rotor has in effect the same mmf as the stator. That is, a rotor current I_2', referred to the stator and flowing in N_1 turns having m_1 phases, produces the same mmf as did the original F_2. From (6.21*b*), therefore, we have

$$0.9 m_2 k_{w2} \frac{N_2 I_2}{p} = 0.9 m_1 k_{w1} \frac{N_1 I_2'}{p}$$

or

$$I_2' = \frac{m_2 k_{w2} N_2}{m_1 k_{w1} N_1} I_2 \tag{6.22}$$

Furthermore, rotor volt-ampere per-phase referred to the stator must be the same as the original rotor volt-ampere. Thus

$$m_1 E_2' I_2' = m_2 E_2 I_2 \tag{6.23}$$

We substitute (6.20) and (6.22) in (6.23) to obtain the rotor voltage referred to the stator, E_2', as

$$E_2' = \frac{k_{w1} N_1}{k_{w2} N_2} E_2 = E_1 \tag{6.24}$$

The next condition that must be fulfilled is that the rotor I^2r – losses be invariant. Expressed mathematically, this means that

$$m_1 (I_2')^2 r_2' = m_2 I_2^2 r_2 \tag{6.25}$$

where r_2' is the rotor resistance per-phase referred to the stator, and r_2 is the resistance of one bar. Equations (6.22) and (6.25) yield

$$r_2' = \frac{m_1}{m_2} \left(\frac{k_{w1} N_1}{k_{w2} N_2} \right)^2 r_2 \tag{6.26}$$

Finally, we require that the magnetic energy stored in the standstill rotor leakage reactance also remain unchanged; that is,

$$\frac{1}{2}m_1L_2'(I_2')^2=\frac{1}{2}m_2L_2I_2^2 \tag{6.27}$$

Multiplying both sides of (6.27) by the stator angular frequency ω and substituting (6.22), we get

$$X_2'=\frac{m_1}{m_2}\left(\frac{k_{w1}N_1}{k_{w2}N_2}\right)^2X_2 \tag{6.28}$$

which is the rotor leakage reactance referred to the stator.

To summarize, the rotor current, voltage, resistance, and reactance, when referred to the stator, must be multiplied by the factors contained in (6.22), (6.24), (6.26), and (6.28), respectively.

Having demonstrated the similarity between an induction motor and a transformer, and recognizing the essential differences, we can now refer the rotor quantities to the stator. Thus, we obtain the exact equivalent circuit (per phase) shown in Figure 6.14 *a* from the circuit given in Figure 6.13. For reasons that will become immediately clear, we split r_2'/s as

$$\frac{r_2'}{s}=r_2'+\frac{r_2'}{s}(1-s)$$

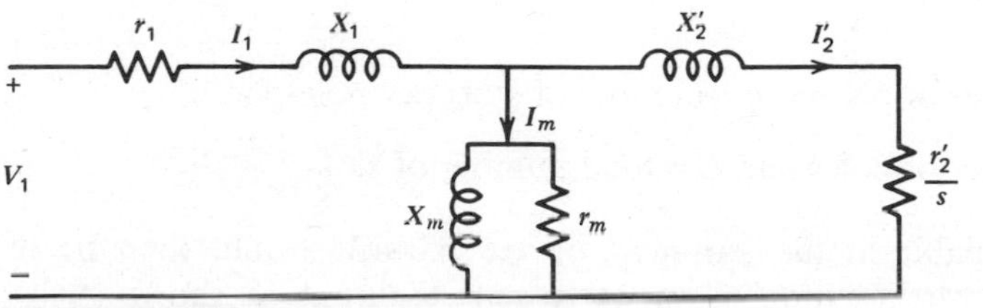

(a)

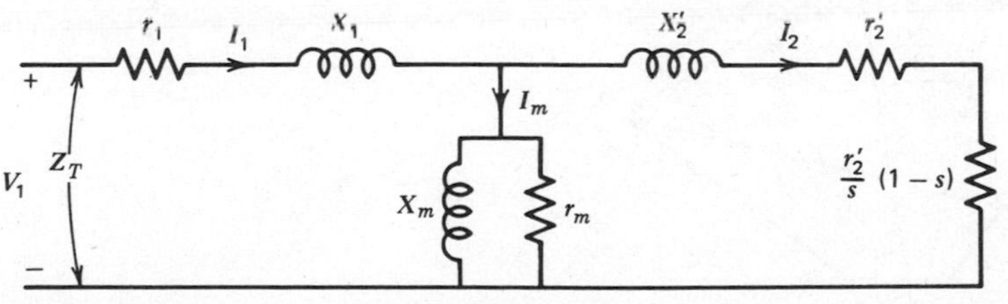

(b)

Figure 6.14. Two forms of equivalent circuits of an induction motor.

to obtain the circuit shown in Figure 6.14 *b*. Here, r_2' is simply the per-phase standstill rotor resistance referred to the stator and $r_2'(1\text{-}s)/s$ is a dynamic resistance which depends on the rotor speed and corresponds to the load on the motor. Notice that all the parameters shown in Figure 6.14 *a* and *b* are standstill values and the circuit is the per-phase exact equivalent circuit referred to the stator. We shall show the usefulness of the circuit later, but first we consider an example to show the calculations of the factors used in referring rotor quantities to the stator.

Example 6.1

Derive a general expression for the distribution factor for an ac armature winding. Indicate how the voltage equation is modified by the distribution factor.

We recall from earlier chapters (on transformers and dc machines) that the voltage, E, induced in an N-turn coil (all turns) linking a flux, ϕ, alternating at a frequency, f, is given by

$$E = 4.44 f\phi N \tag{6.29}$$

If these N-turns are distributed in a number of slots, such as those shown in Figure 6.7, the voltages induced in the coils would be displaced from each other in phase by the slot angle α, defined by

$$\alpha = \frac{180p}{Q} = \frac{180}{mq} \tag{6.30}$$

where

m = number of phases, q = number of slots per-pole-phase,

p = number of poles and Q = total number of slots.

The net voltage available at the terminals of the N-turns would then be the phasor sum of the voltages induced in each coil. Figure 6.15 shows such a

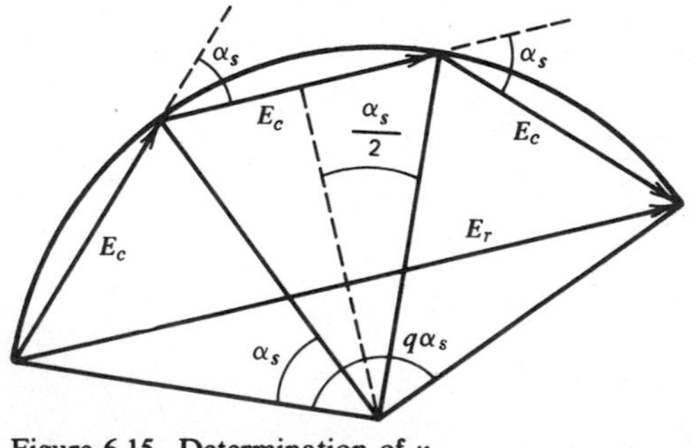

Figure 6.15. Determination of κ_d.

phasor addition, from which the ratio

$$k_d = \frac{\text{resultant voltage}}{\text{sum of individual coil voltage}} = \frac{E_r}{qE_e} \tag{6.31}$$

and k_d is known as the *distribution factor*. From Figure 6.15 we obtain

$$k_d = \frac{E_r}{qE_e} = \frac{2\,a \sin q(\alpha/2)}{2\,a\,q \sin(\alpha/2)} = \frac{\sin q\,\alpha/2}{q \sin \alpha/2} \tag{6.32}$$

The voltage equation (6.29) is modified by (6.32) as follows

$$E = 4.44\, k_d\, f\phi N \tag{6.33}$$

The distribution factors for a few three-phase windings are given in Table 6.1

***Table 6.1.* Distribution Factors for Three-Phase Winding**

Slot/pole/phase	2	3	4	5	6	8	∞
k_d	0.966	0.960	0.958	0.957	0.957	0.956	0.955

Example 6.2
Recall from Figure 6.7 that the coil pitch is not equal to the pole pitch. Such a winding was termed as a fractional-pitch winding. The voltage induced in a fractional-pitch coil is reduced by a factor known as *pitch factor*, compared to the voltage induced in a full-pitch coil. Derive an expression for the pitch factor.

In a sinusoidally distributed flux density we show a full-pitch and a fractional-pitch coil in Figure 6.16. The coil span of the full-pitch coil = pole-pitch = τ. Let the coil span of the fractional-pitch coil be β, as shown. The flux

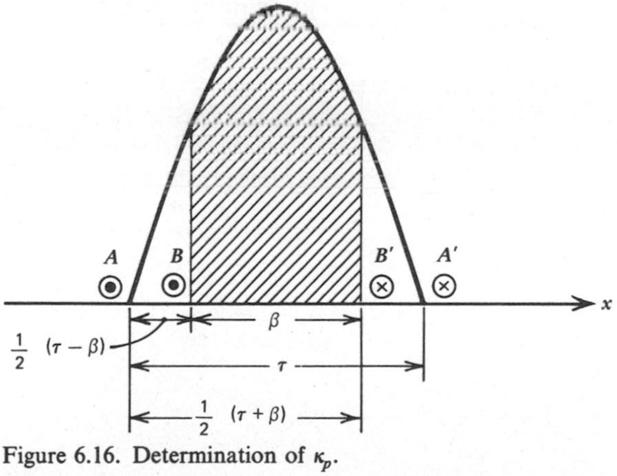

Figure 6.16. Determination of κ_p.

linking the fractional-pitch coil will be proportional to the shaded area (Figure 6.16), as compared to the flux linking the full-pitch coil (i.e., proportional to the entire area under the curve). The ratio of the shaded area to the total area is, therefore, the *pitch factor*, k_p. Thus

$$k_p = \int_{\frac{(\tau-\beta)}{2}}^{\frac{(\tau+\beta)}{2}} \sin\frac{\pi x}{\tau}\,dx \Big/ \int_{x=0}^{\tau} \sin\frac{\pi x}{\tau}\,dx = \sin\frac{\pi\beta}{2\tau} \tag{6.34}$$

Accounting for the pitch factor, the emf equation (6.33) further modifies to

$$E = 4.44 k_d k_p f\phi N = 4.44 k_w f\phi N \tag{6.35}$$

where $k_w = k_d k_p$, and was termed the *winding factor* in the preceding section; for instance in (6.20) to (6.28).

Example 6.3
The stator winding of a cage-type induction motor shown in Figure 6.7 has 24 turns-per-phase. Calculate the factor by which the rotor standstill resistance must be multiplied to refer it to the stator.

We notice from Figure 6.7 that the stator has three phases, or, $m_1 = 3$. Also, it has four poles, or, $p = 4$; the number of slots per-pole per-phase, $q = 36/4\times3 = 3$, and the slot angle $\alpha = 180/3\times3 = 20$, from (6.30). Therefore, from (6.32) or Table 6.1

$$k_{d1} = \frac{\sin(3\times20/2)}{3\sin(20/2)} = 0.96$$

Again, from Figure 6.7, $\tau = 9$ slots and $\beta = 8$ slots. Thus, from (6.34), we get

$$k_{p1} = \sin\frac{8\pi}{18} = \sin 80° = 0.985$$

The winding factor for the stator

$$k_{w1} = k_{d1} k_{p1} = 0.945.$$

For the rotor we have

$$k_{w2} = 1, \qquad N_2 = \frac{p}{4} = 1, \text{ and } m_2 = \frac{2Q_2}{p} = 2\times\frac{28}{4} = 14.$$

Substituting these values in (6.26), we obtain

$$r_2' = \frac{3}{14}\left(\frac{0.945\times 24}{1\times 1}\right)^2 r_2 = 110\, r_2$$

or the required factor = 110.

6.6 PERFORMANCE CALCULATIONS FROM EQUIVALENT CIRCUITS

The major usefulness of the equivalent circuit of an induction motor is in its performance calculations. Of course, here we assume that all the circuit parameters are known and either the input or output conditions or a combination of the two are specified. For instance, the input voltage and operating slip may be given, and we may be required to determine input current, power factor, efficiency, and so forth. We emphasize that all calculations are made on a *per-phase basis*, assuming a balanced operation of the machine. The total quantities are obtained by using appropriate multiplying factor as shown in Example 6.4.

For the sake of illustration we refer to Figure 6.14. We redraw this circuit in Figure 6.17, where we also show approximately the power flow and various power losses in one phase of the machine. Notice that we have neglected the core losses, most of which losses are in the stator. We shall include core losses only in efficiency calculations. Therefore, the power crossing the air gap, P_g, is the difference between the input power, P_i, and the stator $I_1^2 r_1$ – loss; that is

$$P_g = P_i - I_1^2 r_1 \qquad \text{watt/phase} \tag{6.36}$$

Clearly, this power is dissipated in the resistance r_2'/s, Figure 6.14. Therefore

$$P_g = I_2^2 \frac{r_2'}{s} \tag{6.37}$$

If we subtract the rotor $I_2^2 r_2'$ – loss from P_g, we obtain the developed electromagnetic power, P_d, so that

$$P_d = P_g - I_2^2 r_2' = (1-s)P_g \tag{6.38}$$

This is the power that appears across a resistance having an ohmic value $r_2'[(1-s)/s]$, which corresponds to the load. The rotational power, P_r, may be subtracted from P_d to obtain the shaft output power, P_0. Thus

$$P_0 = P_d - P_r \tag{6.39}$$

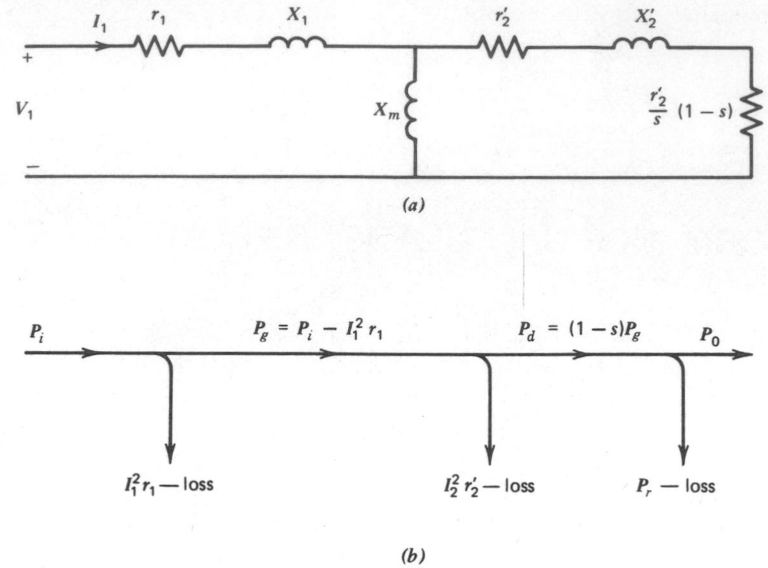

Figure 6.17. Power flow in an induction motor.

Also

$$P_i = V_1 I_1 \cos\phi_1 \tag{6.40}$$

and the efficiency, η, is the ratio P_0/P_i.

We now illustrate the above procedure by the following example.

Example 6.4

The parameters of the equivalent circuit, Figure 6.17*a*, for a 220 V three-phase 60 Hz induction motor are

$$r_1 = 0.2 \text{ ohm} \quad r_2' = 0.1 \text{ ohm}$$

$$X_1 = 0.5 \text{ ohm} \quad X_2' = 0.2 \text{ ohm}$$

$$X_m = 20.0 \text{ ohm}$$

The total iron and mechanical losses are 350 W. For a slip of 2.5%, calculate (a) input current, (b) output power, (c) output torque, and (d) efficiency.

Because the iron losses are known (350 W), we make an approximation by neglecting the resistance r_m. Thus, from Figure 6.17a, the total impedance is

$$Z_t = r_1 + jX_1 + \frac{jX_m\left(\frac{r_2'}{s} + jX_2'\right)}{\frac{r_2'}{s} + j(X_m + X_2')}$$

$$= 0.2 + j0.5 + \frac{j20(4 + j0.2)}{4 + j(20 + 0.2)}$$

$$= (0.2 + j0.5) + (3.8 + j0.81) = 4.21 \angle 18°$$

$$\text{Phase voltage} = 220/\sqrt{3} = 127 \text{ V}$$

$$\text{Input current} = 127/4.21 = 30 \text{ A}$$

$$\text{Power factor} = \cos 18° = 0.95$$

$$\text{Total input power} = \sqrt{3} \times 220 \times 30 \times 0.95$$

$$= 10.85 \text{ kW}$$

$$\text{Total power across the air gap} = 3 \times 30^2 \times 3.8$$

$$= 10.25 \text{ kW}$$

$$\text{Total power developed} = 0.975 \times 10.25 = 10.0 \text{ kW}$$

$$\text{Total output power} = 10 - 0.35 = 9.65 \text{ kW}$$

$$\text{Total output torque} = \text{output power}/\omega_m$$

$$= (9.65/184) \times 1000 = 52.4 \text{ N m}$$

$$\text{where } \omega_m = 0.975 \times 60 \times \pi = 184 \text{ rad/s}$$

$$\textit{Efficiency} = \frac{9.65}{10.85} = 89\%$$

Using a procedure similar to that given in the example above, we can calculate the performance of the motor at other values of the slip, ranging from 0 to 1. The characteristics thus calculated are shown in Figure 6.18.

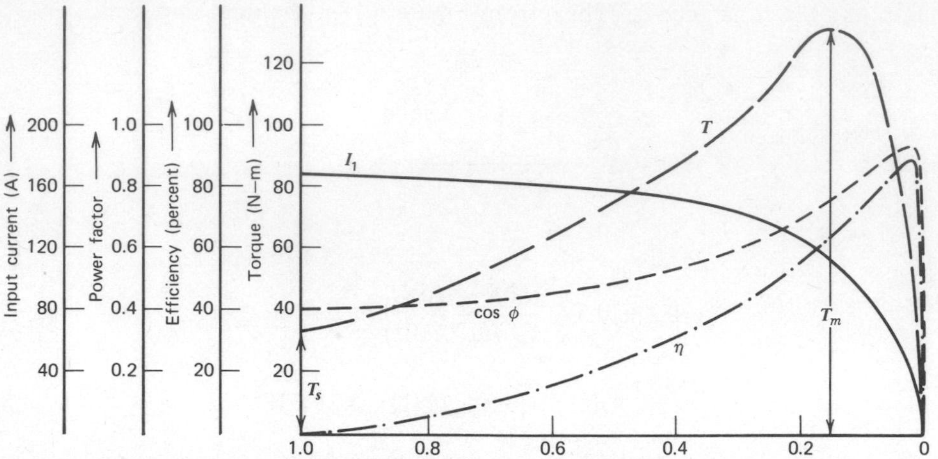

Figure 6.18. Characteristics of an induction motor. T_m = maximum torque; T_s = starting torque. *Key:* —, input current, A; — — —, power factor; — · — · —, efficiency, percent; and — — —, torque, N-m.

6.7 THE EQUIVALENT CIRCUIT FROM TEST DATA

The previous example illustrates the usefulness of an equivalent circuit. However, we did not actually use the exact circuit shown in Figure 6.10. Rather, to simplify the calculations we neglected the shunt-branch resistance r_m. In many calculations, for practical purposes the induction machine is represented by the approximate equivalent circuit shown in Figure 6.19. In order to calculate the performance of the machine, its parameters must be known. The parameters of the circuit shown in Figure 6.19 can be obtained from the following two tests.

1. **No-load test:** In this test, rated voltage is applied to the machine and it is allowed to run on no-load. Input power, voltage, and current are measured. These are reduced to per-phase values and denoted by P_0, V_0, and I_0

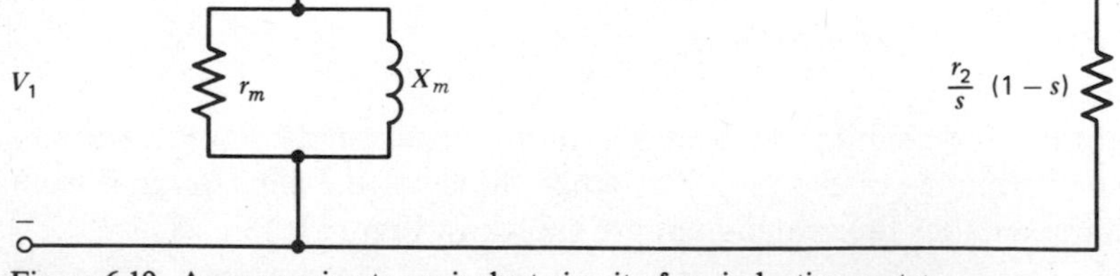

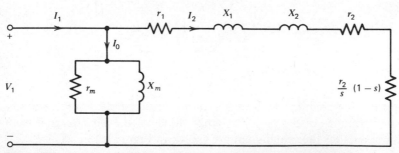

Figure 6.19. An approximate equivalent circuit of an induction motor.

respectively. When the machine runs on no-load, the slip is close to zero and the circuit to the right of the shunt branch is taken to be an open circuit. Thus, the parameters r_m and X_m are found from the following equations:

$$r_m = \frac{V_0^2}{P_0} \tag{6.41}$$

$$X_m = \frac{V_0}{I_0 \sin\phi_0} \tag{6.42}$$

where

$$\phi_0 = \cos^{-1}\frac{P_0}{V_0 I_0} \tag{6.43}$$

2. **Blocked-rotor test:** In this test, the rotor of the machine is blocked ($s=1$) and a reduced voltage is applied to the machine so that the rated current flows through the stator windings. The input power, voltage, and current are recorded and reduced to per-phase values. These are denoted, respectively, by P_s, V_s, and I_s. In this test, the iron losses are assumed to be negligible and the shunt branch of the circuit shown in Figure 6.19 is considered to be absent. The parameters are thus found from

$$r_e = r_1 + a^2 r_2 = \frac{P_s}{I_s^2} \tag{6.44}$$

$$X_e = X_1 + a^2 X_2 = \frac{V_2 \sin\phi}{I_s} \tag{6.45}$$

where

$$\phi_s = \cos^{-1}\frac{P_s}{V_s I_s} \tag{6.46}$$

In (6.44) and (6.45) a is a constant and is analogous to the transformation ratio of a transformer. It takes into account the effect of rotor resistance and reactance as referred to the stator, discussed in Section 6.5. The tests described here are approximate. Refinements can, however, be made and details are available in References 1 and 4. The stator resistance per phase, r_1, can be directly measured, and knowing r_e from (6.44) we can determine $r_2' = a^2 r_2$, the rotor resistance referred to the stator. There is no simple method of determining the leakage reactances X_1 and X_2 separately. The total value of the leakage reactance is given by (6.45) and, approximately, we may assume $X_1 = X_2$. Consider now an example to illustrate the calculations involved in the determination of the machine constants from test data.

Example 6.5
The results of the no-load and blocked-rotor tests on a three phase, Y-connected induction motor are as follows:

No-load test:	line-to-line voltage	=	220 V
	total input power	=	1000 W
	line current	=	20 A
	friction and windage loss	=	400 W
Blocked-rotor test:	line-to-line voltage	=	30 V
	total input power	=	1500 W
	line current	=	50 A

Calculate the parameters of the approximate equivalent circuit shown in Figure 6.19.

$$V_0 = \frac{220}{\sqrt{3}} = 127 \text{ V}$$

$$I_0 = 20 \text{ A}$$

$$P_0 = \frac{1}{3}(1000 - 400) = 200 \text{ W}$$

Thus from (6.41) to (6.43)

$$r_m = \frac{127^2}{200} = 80.5 \ \Omega$$

$$\phi_0 = \cos^{-1}\frac{200}{20 \times 127} = 86°$$

$$X_m = \frac{127}{20 \times 0.99} = 6.4 \ \Omega$$

Now

$$V_s = \frac{30}{\sqrt{3}} = 17.32 \text{ V}$$

$$I_s = 50 \text{ A}$$

$$P_s = \frac{1500}{3} = 500 \text{ W}$$

Thus from (6.44) to (6.46)

$$r_e = \frac{500}{50^2} = 0.2\ \Omega$$

$$\phi_s = \cos^{-1}\frac{500}{17.32 \times 50} = 54°$$

$$X_e = 17.32 \times \frac{0.8}{50} = 0.277\ \Omega$$

Knowing the circuit constants, we can calculate the machine performance, as in Example 6.4.

6.8 PERFORMANCE CRITERIA OF INDUCTION MOTORS

The preceding two examples show the usefulness of the equivalent circuit and the method of determining its parameters from test data in order to calculate the performance of the motor.The performance of an induction motor may be characterized by the following factors.

1. Efficiency.
2. Power factor.
3. Starting torque.
4. Starting current.
5. Pull-out (or maximum) torque.

Notice that these characteristics are shown in Figure 6.18. In design considerations, heating because of I^2r-losses and core losses and means of heat dissipation must be included. It is not within the scope of the book to present a detailed discussion of the effects of design changes, and consequently parameter variations, on each performance characteristic. Here we summarize the results as trends. For example, the efficiency is approximately proportional to $(1-s)$. Thus, the motor would be most compatible with a load running at the highest possible speed. Because the efficiency is clearly dependent on I^2r-losses, r_2' and r_1 must be small for a given load. To reduce core losses, the working flux density (B) must be small. But this imposes a conflicting requirement on the load current (I_2') because the torque, which is determined by the load, is dependent on the product of B and I_2'. In other words, an attempt to decrease the core losses beyond a limit would result in an increase in the I^2r-losses for a given load.

It may be seen from the equivalent circuits (developed in Section 6.5) that the power factor can be improved by decreasing the leakage reactances and increasing the magnetizing reactance. However, it is not wise to reduce the leakage reactances to a minimum, since the starting current of the motor is essentially limited by these reactances. Again, we notice the conflicting conditions for a high power factor and a low starting current. Also, the pull-out torque would be higher for lower leakage reactances.

A high starting torque is produced by a high r_2'; that is, the higher the rotor resistance, the higher would be the starting torque. A high r_2' is in conflict with a high efficiency requirement.

We may arrive at some of the above conclusions by considering the rotor circuit only as shown by the following example.

Example 6.6

From the rotor equivalent circuit shown in Figure 6.12, (a) find r_2 for which the developed torque would be a maximum. (b) What is the slip at this maximum torque? (c) Determine r_2 for a maximum starting torque. (d) What is the effect of X_2 on the torque?

From Figure 6.12, the developed power P_d per-phase is given by

$$P_d = I_2^2 \frac{r_2}{s}(1-s) = T_e \omega_m$$

But the mechanical speed ω_m is related to the synchronous speed by

$$\omega_m = (1-s)\omega_s$$

The above two equations yield the expression for the electromagnetic torque, T_e, as

$$T_e = \frac{I_2^2 r_2}{s\omega_s} \tag{6.47}$$

But the rotor current I_2 is given by

$$I_2 = \frac{sE_2}{\sqrt{r_2^2 + (sX_2)^2}} \tag{6.48}$$

From (6.47) and (6.48) we have

$$T_e = \frac{E_2^2}{\omega_s}\left(\frac{sr_2}{r_2^2 + s^2X_2^2}\right) \tag{6.49}$$

For a maximum T_e we must have $\partial T_e/\partial r_2=0$, which, together with (6.49), gives

$$\frac{\partial T_e}{\partial r_2}=\frac{sE_2^2}{\omega_s}\left[\frac{(r_2^2+s^2X_2^2)-r_2(2r_2)}{(r_2^2+s^2X_2^2)^2}\right]=0$$

or

$$r_2^2+s^2X_2^2-2r_2^2=0$$

(a)

$$r_2=sX_2$$

and
(b)

$$s=\frac{r_2}{X_2}$$

At starting $s=1$,
(c)

$$r_2=X_2$$

(d)For a given rotor resistance, the starting torque would be maximum if $X_2=0$, from (6.49).

Clearly, the above analysis is only approximate. However, we can arrive at similar conclusions by using the exact equivalent circuit as shown in Problem 6.5.

6.9 SPEED CONTROL OF INDUCTION MOTORS

Because of its simplicity and ruggedness, the induction motor finds numerous applications. However, it suffers from the drawback that, in contrast to dc motors, its speed cannot be easily and efficiently varied continuously over a wide range of operating conditions. We shall briefly review the various possible methods by which the speed of the induction motor can be varied either continuously or in discrete steps. It is beyond the scope of this book to consider all these methods in detail and the interested reader should consult the references given at the end of this chapter.

The speed of the induction motor can be varied by either (1) varying the synchronous speed of the traveling field or (2) varying the slip. Because the efficiency of the induction motor is approximately proportional to $(1-s)$, any method of speed control which depends on the variation of slip is inherently inefficient. On the other hand, if the supply frequency is constant, varying the speed by changing the synchronous speed results only in discrete changes in the speed of the motor. We shall now consider these methods of speed control in some detail.

6.9.1 SPEED CONTROL BY CHANGING THE SYNCHRONOUS SPEED

Recall that the synchronous speed n_s of the traveling field in a rotating induction machine is given by

$$n_s = 120\frac{f}{p}$$

where p = number of poles and f = supply frequency, which indicates that n_s can be varied by either (1) changing the number of poles p, or (2) changing the frequency f. Both of these methods have found applications and we consider here the pertinent qualitative details.

1. **The pole-changing method:** In this method, the stator winding of the motor is so designed that by changing the connections of the various coils (the terminals of which are brought out), the number of poles of the winding can be changed in the ratio of 2 to 1. Accordingly, two synchronous speeds result. We observe that only two speeds of operation are possible. If more independent windings (e.g., two) are provided—each arranged for pole-changing—more synchronous speeds (e.g., four) can be obtained. However, the fact remains that only discrete changes in the speed of the motor can be obtained by this technique. The method has the advantage of being efficient and reliable because the motor has a squirrel-cage rotor and no brushes.

   Another method of pole-changing is by means of pole-amplitude modulation. Single-winding squirrel-cage motors are reported to have been developed which yield three operating speeds. Another method, based on pole-changing, which produced three or five speeds, has been termed "phase-modulated pole-changing." Like the simplest pole-changing method, the pole-amplitude modulation and the phase-modulated pole-changing method give discrete variation in the synchronous speed of the motor.

2. **The variable-frequency method:** We recall that the synchronous speed is directly proportional to the frequency. If it is practicable to vary the supply frequency, then the synchronous speed of the motor can also be varied. The variation in speed is continuous or discrete according to continuous or discrete variation of the supply frequency. However, the maximum torque developed by the motor is inversely proportional to the synchronous speed. If

we desire a constant maximum torque, the supply voltage and supply frequency should be increased if we wish to increase the synchronous speed of the motor. The inherent difficulty in the application of this method is that the supply frequency, which is commonly available, is fixed. Thus, the method is applicable only if a variable-frequency supply is available. Various schemes have been proposed to obtain a variable-frequency supply. With the advent of solid-state devices with comparatively large power ratings, it is now possible to use static inverters to drive the induction motor. Solid-state control of induction motors is discussed in some detail in Chapter 8.

6.9.2 SPEED CONTROL BY CHANGING THE SLIP

The method of controlling the speed of an induction motor by changing its slip is best understood by reference to Figure 6.20. The dotted curve shows the speed-torque characteristic of the load. The curves with solid lines are the speed-torque characteristics of the induction motor under various conditions (such as different rotor resistances—r_2', r_2'', r_2''' or different stator voltages—V_1, V_2). We have four different torque-speed curves and, therefore, the motor can run at any one of four speeds—N_1, N_2, N_3, and N_4—for the given load. Note that to the right of the peak torque is the stable operating region of the motor. In practice, the slip of the motor can be changed by one of the following methods.

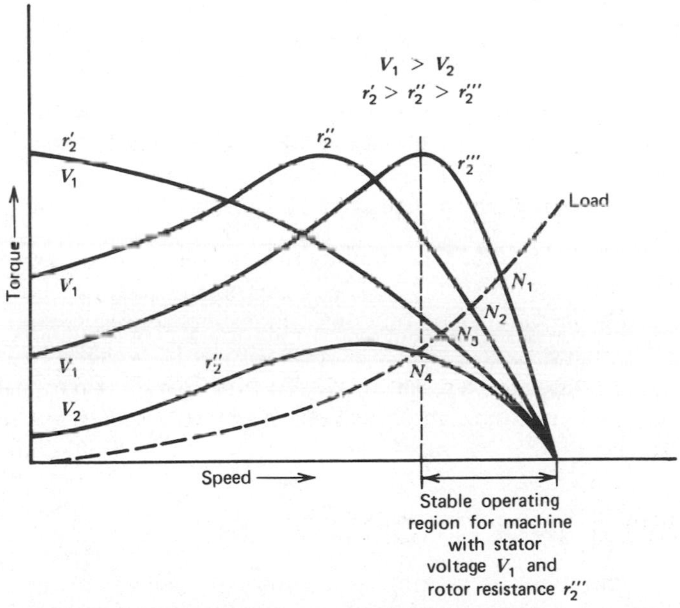

Figure 6.20. Speed control by changing the slip.

1. **Variable stator voltage method:** Since the electromagnetic torque developed by the machine is proportional to the square of the applied voltage, we obtain different torque-speed curves for different voltages applied to the motor. For a given rotor resistance, r_2, two such curves are shown in Figure 6.20 for two applied voltages V_1 and V_2. Thus, the motor can run at speeds N_2 or N_4. If the voltage can be varied continuously from V_1 to V_2, the speed of the motor can also be varied continuously between N_2 and N_4 for the given load. This method is applicable to the cage-type as well as the wound-rotor-type induction motors.
2. **Variable rotor-resistance method:** This method is applicable only to the wound-rotor motor. The effect on the speed-torque curves of inserting external resistances in the rotor circuit is shown in Figure 6.20 for three different rotor resistances r_2', r_2'', and r_2'''. For the given load, three speeds of operation are possible. Of course, by continuous variation of the rotor resistance, continuous variation of the speed is possible.
3. **Control by solid-state switching:** Other than the inverter-driven motor, the speed of the wound-rotor motor can be controlled by inserting the inverter in the rotor circuit, or by controlling the stator voltage by means of solid-state switching devices such as silicon-controlled rectifiers (SCRs or thyristors). The output from the SCR feeding the motor is controlled by adjusting its firing angle. The method of doing this is similar to the variable-voltage method outlined earlier. However, it has been found that control by an SCR gives a wider range of operation and is more efficient than other slip-control methods. (See Chapter 8.)
4. **Speed control by auxiliary machines:** There are numerous other schemes available for controlling the speed of the induction motor. These include concatenation, the Schrage motor, the Kramer control, the Scherbius control, and so forth, described in some detail in References 1 and 4.

In summary, it should be pointed out that the method of controlling the speed of an induction motor by controlling its slip is basically inefficient and may have other disadvantages also. The method based on controlling synchronous speed is efficient, but difficult to achieve in practice, especially with brushless (or cage-type) machines for continuous-speed variation. A satisfactory, economical and efficient method of speed control of the brushless motor is, it seems, yet to be developed.

6.10 STARTING OF INDUCTION MOTORS

We have discussed the conditions for starting a dc motor in the last chapter. Somewhat similar conditions apply to the starting of an induction also. To accelerate the motor from rest to a steady-state operating speed, the input energy must be greater than the kinetic energy of the motor and the load by at least the amount of the total losses. The input current must not be excessive in accomplishing this, for example, no more than six times the full-load current. The starting torque should be about 1.5 times the full-load torque.

Consider the current limitation first. Some of the common methods of limiting the stator current while starting are:

1. **Reduced voltage starting.** At the time of starting, a reduced voltage is applied to the stator and the voltage is increased to the rated value when the motor is within 25% of its final speed. This method has the obvious limitation that a variable-voltage source is needed and the starting torque drops substantially. The so-called wye-delta method of starting is a reduced-voltage starting method. If the stator is normally connected in delta, reconnection to wye reduces the phase voltage resulting in less current at starting. For example, at starting, if the line current is about five times the full-load current in a delta-connected stator, the current in the wye-connection will be less than twice the full-load value. But, at the same time, the starting torque for a wye-connection would be about one-third its value for a delta-connection. The advantage of wye-delta starting is that it is inexpensive and requires only a three-pole (or three single-pole) double-throw switch or switches, as shown in Figure 6.21.
2. **Current limiting by series resistance** Series resistances inserted in the three lines sometimes are used to limit the starting current. These resistances are shorted out when the motor has gained speed. This method has the obvious disadvantage of being inefficient because of the extra losses in the external resistances.

Turning now to the starting torque, we recall from the last section that the starting torque is dependent on the rotor resistance. Thus, a high rotor resistance results in a high starting torque. Therefore, in a wound rotor machine (See

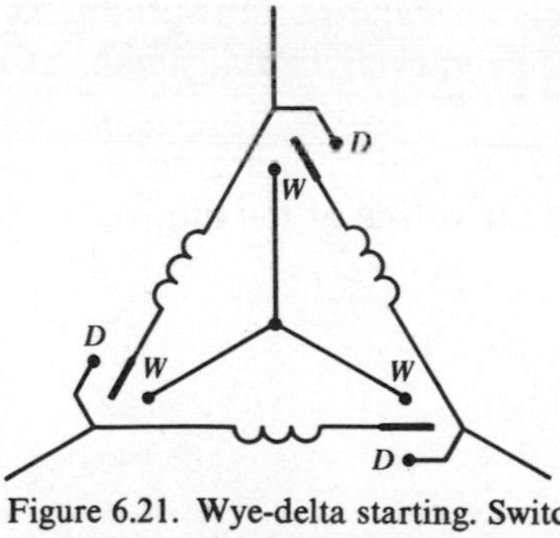

Figure 6.21. Wye-delta starting. Switches on W correspond to wye and switches on D correspond to the delta connection.

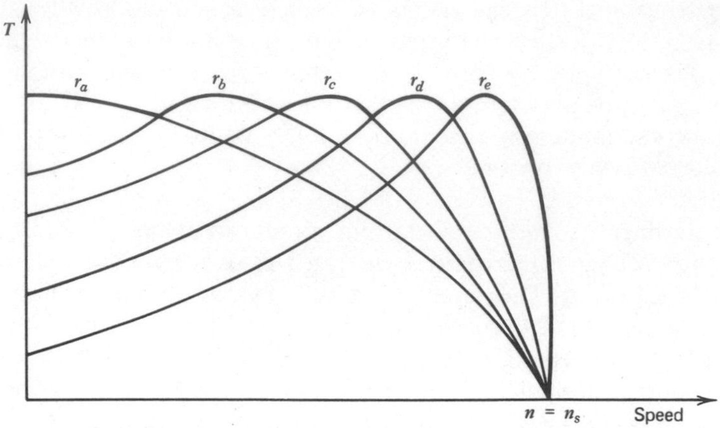

Figure 6.22. Effect of changing rotor resistance on the starting of a wound-rotor motor.

Figure 6.22.), external resistance in the rotor circuit may be conveniently used with a *deep-bar rotor* where the slot depth is two or three times greater than the slot width. (See Fig 6.23.) Rotor bars embedded in deep slots provide a high effective resistance and a large torque at starting. Under normal running conditions with low slips, however, the rotor resistance becomes lower and efficiency high. This characteristic of rotor bar resistance is a consequence of skin-effect. Because of skin-effect, the current will have a tendency to concentrate at the top of the bars at starting, when the frequency of rotor currents is high. At this point, the frequency of rotor currents will be the same as the stator input frequency (e.g., 60 Hz). While running, the frequency of rotor currents (=slip frequency=3 Hz at 5% slip and 60 Hz) is much lower. At this level of operation, skin-effect is negligible and the current almost uniformly distributes throughout the entire bar crosssection.

Skin-effect is used in an alternate form in a *double-cage* rotor (Figure 6.24), where the inner cage is deeply embedded in iron and has low resistance bars. The outer cage has relatively high resistance bars close to the stator. At starting, because of skin-effect, the influence of the outer cage dominates, thus producing a high starting torque. While running, the current penetrates to full depth into the lower cage—because of insignificant skin-effect—which results in an efficient steady-state operation. Notice that under normal running conditions both cages carry current, thus somewhat increasing the rating of the motor.

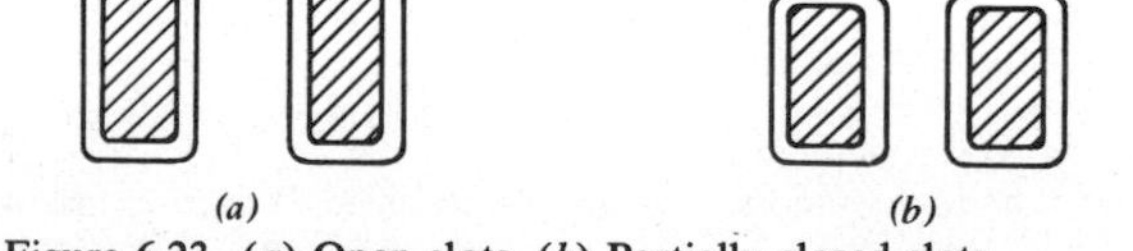

Figure 6.23. (*a*) Open slots. (*b*) Partially closed slots.

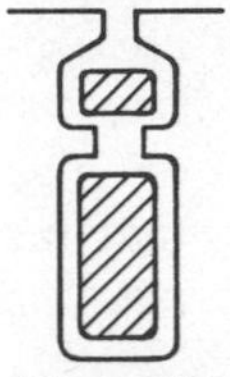

Figure 6.24. Form of a slot for a double-cage rotor.

6.11 SINGLE-PHASE INDUCTION MOTORS

In the preceding sections we have considered the polyphase—rather three-phase—induction motor operating under balanced conditions. Now, let us consider a three-phase induction motor running at light load. If one of the supply lines is disconnected, the motor will continue to run, although at a different speed. Such an operation of a three-phase induction motor may be considered as the operation of a single-phase motor.

Let us now consider the three-phase motor at rest and fed by a single-phase source. Obviously, the motor will not start because we have a pulsating magnetic field in the air gap, rather than a rotating magnetic field which is required for torque production, as discussed earlier. Thus, we conclude that a single-phase induction motor is not self-starting but will continue to run if started by some means. This implies that to make it self-starting, the motor must be provided with an auxiliary means of starting. In a later section, we shall examine the various means of starting the single-phase induction motor.

Not considering the starting mechanism, the essential difference between the three-phase and single-phase induction motor is that the single-phase induction motor has a single stator winding which produces an air-gap field that is stationary in space but alternating in time. The three-phase induction motor has a three-phase winding that produces a time-invariant rotating magnetic field in the air gap. The rotor of the single-phase induction motor is almost always a cage-type rotor and is similar to that of a polyphase induction motor. The rating of a single-phase motor of the same size as a three-phase motor would be smaller, as expected, and single-phase induction motors are rated most often as fractional horsepower motors. These are the most widely used motors in household appliances, fans, and so forth.

6.11.1 ANALYSIS OF THE OPERATION OF SINGLE-PHASE INDUCTION MOTORS

We recall from above that the magnetic field produced by the stator of a single-phase motor alternates through time. The field induces a current—and consequently,—an mmf in the rotor circuit and rotates with the rotor. A single-phase induction motor may be analyzed by considering the mmf's, fluxes, induced voltages (both rotational and transformer), and currents that are separately produced by the stator and by the rotor. Such an approach leads to the crossfield theory. However, we can also analyze the single-phase motor in a manner similar to that for the polyphase induction motor. We recall that the polyphase induction motor operates on the basis of the existence of a rotating magnetic field. This approach is based on the concept that an alternating magnetic field is equivalent to two rotating magnetic fields rotating in opposite directions. When this concept is expressed mathematically, the alternating field of the form is

$$B(\theta, t) = B_m \cos\theta \sin\omega t \tag{6.50}$$

Then (6.50) may be rewritten as

$$B_m \cos\theta \sin\omega t = \frac{1}{2} B_m \sin(\omega t - \theta) + \frac{1}{2} B_m \sin(\omega t + \theta) \tag{6.51}$$

In (6.51), the first term on the right-hand side denotes a forward rotating field, whereas the second term corresponds to a backward rotating field. The theory based on such a resolution of an alternating field into two counter-rotating fields is known as the double-revolving field theory. The direction of rotation of the forward rotating field is assumed to be the same as the direction of the rotation of the rotor. Thus, if the rotor runs at n rpm and n_s is the synchronous speed in rpm, the slip, s_f, of the rotor with respect to the forward rotating field is the same as s, defined by (6.10); or

$$s_f = s = \frac{n_s - n}{n_s} = 1 - \frac{n}{n_s} \tag{6.52}$$

But the slip, s_b, of the rotor with respect to the backward rotating flux is given by

$$s_b = \frac{n_s - (-n)}{n_s} = 1 + \frac{n}{n_s} = 2 - s \tag{6.53}$$

We know from the operation of polyphase motors that, for $n < n_s$, (6.52) corresponds to a motor operation and (6.53) denotes the braking region. Thus, the two resulting torques have an opposite influence upon the rotor.

The torque relationship for the polyphase induction motor is applicable to each of the two rotating fields of the single-phase motor. Thus, the resultant torque of a single-phase induction motor can be written as

$$T_e = \frac{I_2^2}{\omega_m}\frac{(1-s)}{s}r_2 - \frac{I_2^2}{\omega_m}\frac{(1-s)}{(2-s)}r_2 \tag{6.54}$$

Now, we notice from (6.51) that the amplitude of the rotating fields is one-half of the alternating flux. Thus, the total magnetizing and leakage reactances of the motor can be divided equally so as to correspond to the forward and backward rotating fields. The approximate equivalent circuit of a single-phase induction motor, based on the double-revolving field theory, becomes as shown in Figure 6.25*a*. The torque-speed characteristics are qualitatively shown in Figure 6.25*b*. The following example illustrates the usefulness of the circuit.

Example 6.7
With reference to Figure 6.25*a*, the constants of a 1/4 hp, 230 V, four-pole, 60 Hz single-phase induction motor are: $r_1 = 10.0\,\Omega, r_2 = 11.6\,\Omega$, $x_1 = 12.8\,\Omega = x_2$, and $x_m = 258.0\,\Omega$. For an applied voltage of 210 V, at a 3% slip, calculate (a) input current, (b) power factor, (c) developed power, (*d*) shaft power (if mechanical losses are 7 W), and (*e*) efficiency (if iron losses at 210 V are 35.5 W).

For the given circuit and from the given data we have

$$\frac{0.5r_2}{s} = \frac{11.65}{2\times 0.03} = 194.16\,\Omega$$

$$\frac{0.5r_2}{2-s} = \frac{11.65}{2(2-0.03)} = 2.96\,\Omega$$

$$j0.5x_m = j129\,\Omega$$

and

$$j0.5x_2 = j0.5x_1 = j6.4\,\Omega$$

For the forward-field circuit

$$Z_f = \frac{194.16\times j129}{194.16 + j129} = 59.2 + j86$$

and for the backward-field circuit

$$Z_b = \frac{2.96\times j129}{2.96 + j129} \cong 2.96$$

The total series impedance, Z_e, is

$$Z_e = Z_1 + Z_f + Z_b = (10 + j12.8) + (59.2 + j89) + 2.96 = 124\angle 55°$$

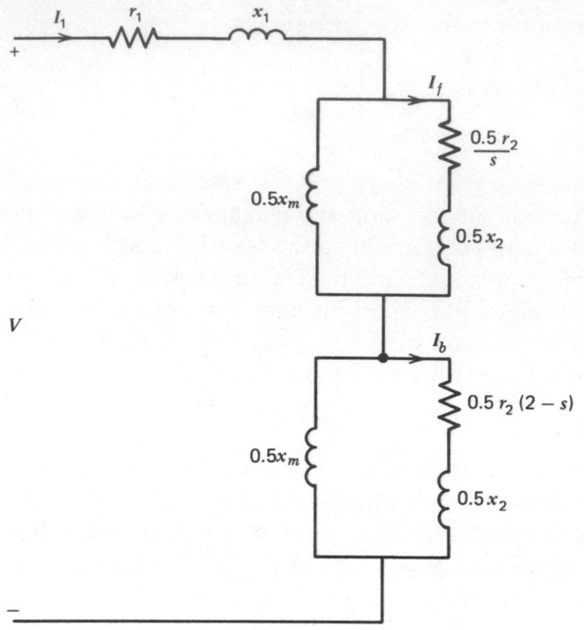

(a)

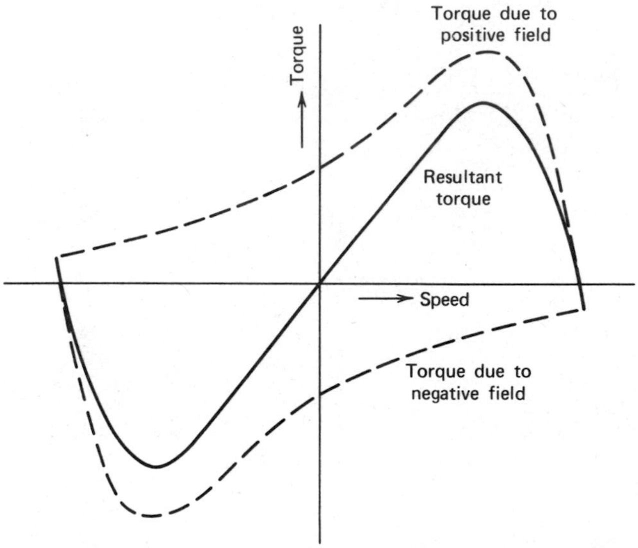

(b)

Figure 6.25. (*a*) Equivalent circuit of a single-phase induction motor, based on revolving-field theory. (*b*) Torque-speed characteristics of a single-phase induction motor based on revolving field theory.

1. Input current

$$I=\frac{V}{Z_e}=\frac{210}{124\angle 55^\circ}=1.7\angle -55^\circ A$$

2. Power factor $=\cos 55=0.573$ lagging
3. Developed power,

$$P_d=\left(\frac{0.5r_2}{s}I_f^2-\frac{0.5r_2}{2-s}I_b^2\right)(1-s)$$

$$\cong\left[\frac{V_f^2}{0.5r_2/s}-\frac{V_b^2}{0.5r_2/(2-s)}\right](1-s)$$

since $s=0.03$ (small). But $V_f=IZ_f=1.7(59.2+j89)=182$ V and $V_b=IZ_b=1.7\times 2.96=5.04$ V. Or

$$P_d=\left(\frac{182^2}{194}-\frac{5.04^2}{2.96}\right)(1-0.03)=156\ W$$

4. Shaft power, $P_s=P_d-P_{rot}=156-7=149$ W
5. Input power $=VI\cos\Theta=210\times 1.7\times 0.573=204$ W output power $=P_s-P_{iron}=149-35.5=113.5$ W efficiency $=\frac{113.5}{204}=55.6\%$.

6.11.2 STARTING OF SINGLE-PHASE MOTORS

We already know that because of the absence of a rotating magnetic field, when the rotor of a single-phase induction motor is at standstill, it is not self-starting. The two methods of starting a single-phase motor are either to introduce a commutator and brushes, such as in a repulsion motor, or to produce a rotating field by means of an auxiliary winding, such as by split-phasing. We shall consider this latter method in the following.

From the theory of the polyphase induction motor, we know that in order to have a rotating magnetic field we must have at least two mmf's which are displaced from each other in space and carry currents having different time phases. Thus, in a single-phase motor, a starting winding on the stator is provided as a source of the second mmf. The first mmf arises from the main stator winding. The various methods to achieve the time and space phase-shifts between the main winding and starting winding mmf's are summarized below.

1. **Split-phase motors.** This type of a motor is schematically represented in

Fig. 6.26*a*, where the main winding has a relatively low resistance and a high reactance. The starting winding, however, has a high resistance and a low reactance and has a centrifugal switch as shown. The phase angle α between the two currents I_m and I_s is about 30 to 45°, and the starting torque T_s is given by

$$T_s = KI_mI_s \sin\alpha \tag{6.55}$$

where K is a constant. When the rotor reaches a certain speed (about 75% of its final speed), the centrifugal switch comes into action and disconnects the starting winding from the circuit. The torque-speed characteristic of the split-phase motor is of the form shown in Fig. 6.26*b*. Such motors find applications in fans, blowers, and so forth, and are rated up to 1/2 hp.

A higher starting torque can be developed by a split-phase motor by inserting a series-resistance in the starting winding. A somewhat similar effect may be obtained by inserting a series inductive-reactance in the main winding. This reactance is short-circuited when the motor builds up speed.

2. **Capacitor-start motors.** By connecting a capacitance in series with the starting winding, as shown in Fig. 6.27, the angle α in (6.55) can be increased. The motor will develop a higher starting torque by doing this. Such motors are not restricted merely to fractional horsepower ratings, and may be rated up to 10 hp. At 110 V, a 1-hp motor requires a capacitance of about 400 μF whereas a 70 μF is sufficient for a 1/8-hp motor. The capacitors generally used are inexpensive electrolytic types and can provide a starting torque that is almost four times that of the rated torque.

As shown in Figure 6.27, the capacitor is merely an aid to starting and is disconnected by the centrifugal switch when the motor reaches a predetermined speed. However, some motors do not have the centrifugal switch. In such a motor, the starting winding and the capacitor are meant for permanent opera-

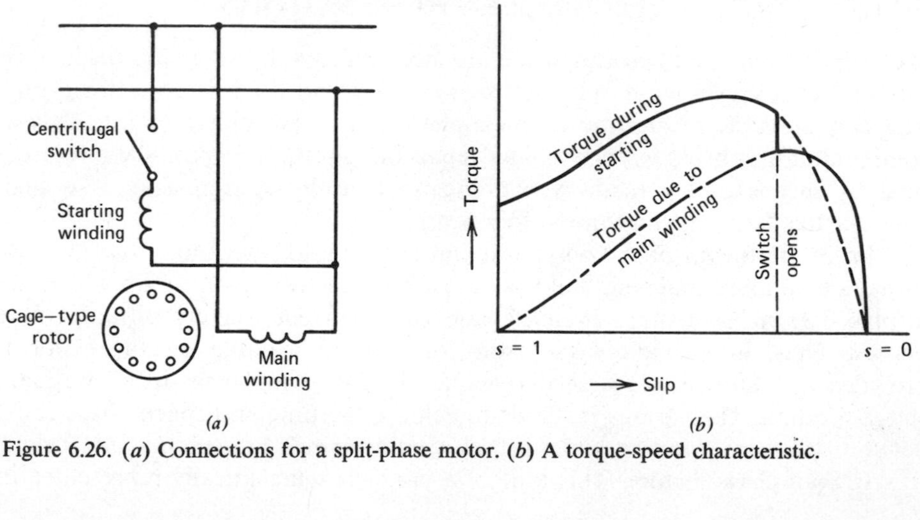

Figure 6.26. (*a*) Connections for a split-phase motor. (*b*) A torque-speed characteristic.

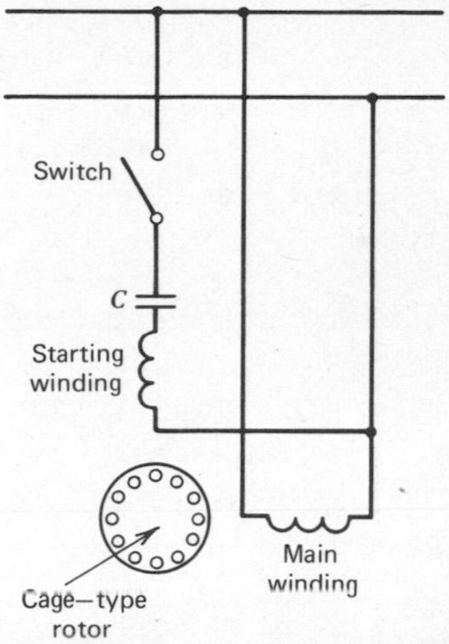

Figure 6.27. A capacitor-start motor.

tion and the capacitors are much smaller. For example, a 110 V 1/2-hp motor requires a 15 μF capacitance.

A third kind of the capacitor motor uses two capacitors—one which is left permanently in the circuit along with the starting winding and the other which gets disconnected by a centrifugal switch. Such motors are, in effect, unbalanced two-phase induction motors.

3. **Shaded-pole motors.** Another method of starting very small single-phase induction motors is to use a shading band on the poles, as shown in Figure 6.28, where the main single-phase winding is also wound on the salient poles. The shading band is simply a short-circuited copper strap wound on a portion of the pole. Such a motor is known as the shaded-pole motor. The purpose of the shading band is to retard (in time) the portion of flux passing through it in relation to the flux coming out of the rest of the pole face. Thus, the flux in the unshaded portion reaches its maximum before that located in the shaded portion. Thus, we have a progressive shift of flux from the direction of the unshaded portion to shaded portion of the pole, as shown in Figure 6.28. The effect of the progressive shift of flux is similar to that of a rotating flux, and because of it, the shading band provides a starting torque. Shaded-pole motors are the least expensive of the fractional horsepower motors and are generally rated up to 1/20 hp.

In summary, in this chapter we have studied the steady-state characteristics of the polyphase and single-phase induction motors operating under balanced conditions. Further topics relating to the induction motor, such as its dynamics and unbalanced operations, are given in Chapter 9. Solid-state control of induction motors is introduced in Chapter 8.

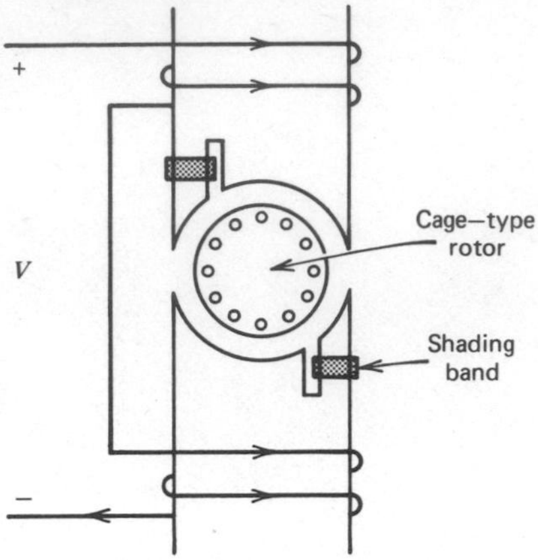

Figure 6.28. A shaded-pole motor.

REFERENCES

1. A. S. Langsdorf, *Theory of Alternating Current Machinery*, McGraw-Hill Book Company, New York, 1955.
2. D. C. White, and H. H. Woodson, *Electromechanical Energy Conversion*, John Wiley & Sons, Inc., New York, 1959.
3. S. Seely, *Electromechanical Energy Conversion*, McGraw-Hill Book Company, New York, 1962.
4. A. F. Puchstein, T. C. Lloyd, and A. G. Conrad, *Alternating-current Machines*, Third Edition, John Wiley & Sons, Inc., New York, 1954.

PROBLEMS

6.1. A four-pole 60-Hz induction motor runs at 1710 rpm. Calculate: (a) the slip in percent (b) the frequency of rotor currents and (c) the speed of the rotating magnetic field produced by (i) the stator and (ii) the rotor, with respect to the stator, in rpm and in rad/s.

6.2. A two-pole 60-Hz wound rotor induction motor has 127 V-per-phase across its stator. The voltage induced in the rotor is 3.81 V-per-phase. Assuming that the stator and the rotor have equal effective numbers of turns-per-phase, calculate (a) the motor speed and (b) the slip.

6.3. A cage-type induction motor consists of 42 bars, each having a resistance of 4.12×10^{-5} ohm (including the resistances of the two end rings). The stator winding has the following data: six-poles; three-phase; 36 slots; 144 turns-per-phase; 0.836 winding factor. Calculate the equivalent rotor resistance-per-phase that is referred to the stator.

6.4. The power crossing the air gap of an induction motor is 24.3 kW. If the developed electromagnetic power is 21.9 kW, what is the slip? The rotational loss at this slip is 350 W. Calculate the output torque, if the synchronous speed is 3600 rpm.

6.5. Using the circuit shown in Figure 6.17*a*, obtain an expression of the slip at which the motor develops the maximum torque. Derive an expression for the maximum torque.

6.6. A three-phase 230 V 60-Hz Y-connected two-pole induction motor operates at 3% slip while taking a line current of 22 A. The stator resistance and leakage reactance-per-phase are 0.1 and 0.2 ohm, respectively. The rotor leakage reactance is 0.15 ohm-per-phase, respectively. Calculate (a) the rotor resistance, (b) power crossing the air gap, and (c) developed power. Neglect X_m.

6.7. The per-phase constants of a three phase 600-V 60-Hz four-pole Y-connected wound rotor induction motor are:

$$r_1 = 0.75 \text{ ohm} \qquad X_1 = X_2 = 2.0 \text{ ohms}$$

$$r_2' = 0.80 \text{ ohm} \qquad X_m = 50.0 \text{ ohms}$$

Neglect the core losses. (a) Calculate the slip at which the maximum developed torque occurs; (b) find the value of the maximum torque; (c) specify the range of speed for a stable operation of the motor; and (d) compute the starting torque and compare it with the maximum torque.

6.8. Repeat Problem 6.7, Parts (a), and (b), by considering the rotor circuit only. Assume $r_1 = X_1 = 0$, in which case 600 V may be taken as the applied line voltage to the rotor.

6.9. A 440 V 25-Hz Two-pole Y-connected motor has a magnetizing reactance of 10.5 ohms-per-phase and rotor leakage reactance of 0.12 ohm-per-phase. Using the rotor circuit only, determine the slip for maximum electromagnetic torque at a per-phase rotor resistance of (a) 0.03 ohm, (b) 0.06 ohm, and (c) 0.1 ohm. Hence show the effect of rotor resistance on the torque-speed characteristics of the motor. Neglect r_1 and x_1.

6.10. For the motor of Problem 6.7, determine the per-phase value of the resistance that must be inserted in the rotor circuit to obtain the maximum torque from the motor at starting.

6.11. Sketch qualitatively the torque-speed characteristics of an induction motor, comparing it with normal characteristics and showing the effects of the following:

(a) Varying the frequency, while keeping the applied voltage constant.

(b) Varying the applied voltage, while keeping the frequency constant.

6.12. A 220 V, three-phase, 60-Hz, four-pole, Y-connected induction motor has a per-phase stator resistance of 0.25 Ω. The no-load and blocked-rotor test data on this motor are No-load test-stator voltage = 220 V

input current = 3.0A

input power = 600 W

friction and

windage loss = 300 W

Blocked-rotor test: stator voltage = 34.6 V

input current = 15.0 A

input power = 720 W

(a) Obtain the approximate equivalent circuit for the machine.

(b) If the machine runs as a motor with 5% slip, calculate the dfeveloped power, developed torque, and efficiency.

(c) Determine the slip at which maximum torque occurs and calculate the maximum torque.

Chapter 7
Synchronous Machines

Synchronous machines are among three of the most common types of electric machines. The other two, the dc commutator machine and the polyphase induction machine, have been considered in the preceding chapters. The bulk of electric power for everyday use is produced by polyphase synchronous generators, which are the largest single-unit electric machines in production. For instance, synchronous generators with power ratings of several hundred megavolt-amperes (MVA) are fairly common, and it is expected that machines of several thousand MVA will be in use in the 1980s[1]. These are called synchronous machines because they operate at constant speeds and constant frequencies under steady-state. Like most rotating machines, a synchronous machine is capable of operating both as a motor and as a generator. However, because very few industrial drives run at fixed speeds, synchronous motors are not as commonly used as induction or dc motors. Rather, they are widely used as generators, several of them operating in parallel in a power station. While operating in parallel, the generators share the load with each other and at a given time one of the generators, it is allowed to "float" on the line as a synchronous motor on no-load. We shall discuss such a no-load operation of a synchronous motor in a later section.

The operation of a synchronous generator is based on Faraday's law of electromagnetic induction, and an ac synchronous generator works very much like a dc generator, in which the generation of emf's is by the relative motion of conductors and magnetic flux. Clearly, however, a synchronous generator does not have a commutator as does a dc generator. The two basic parts of a synchronous machine are the magnetic field structure, carrying a dc-excited winding, and the armature. The armature often has a three-phase winding in which the ac emf is generated. Almost all modern synchronous machines have stationary armatures and rotating field structures. The dc winding on the rotating field structure is connected to an external source through slip-rings and brushes. (Recall the construction of the slip-ring type induction motor, Chapter 6.) Some field structures do not have brushes, but rather have brushless excitation by rotating diodes. In some respects the stator carrying the armature windings is similar to the stator of a polyphase induction motor, which was studied in the last chapter. In the next section we discuss some of the constructional features of synchronous machines.

7.1 CONSTRUCTIONAL FEATURES OF SYNCHRONOUS MACHINES

Some of the factors that dictate the form of construction of a synchronous machine are the following.

1. Form of excitation. We recall from the preceding remarks that the field structure is usually the rotating member of a synchronous machine and is supplied with a dc-excited winding to produce the magnetic flux. This dc excitation may be provided by a self-excited dc generator mounted on the same shaft as the rotor of synchronous machine. Such a generator is known as the *exciter*. The dc thus generated is fed to the synchronous machine field winding shown in Figure 7.1. In slow-speed machines with large ratings, such as hydroelectric generators, the exciter may not be self-excited. Rather a *pilot exciter*, which may be self-excited or may have a permanent magnet, activates the exciter (Figure 7.7). The maintenance problems of direct-coupled dc generators impose a limit on this form of excitation at about 100 MW rating.

An alternative form of excitation is provided by silicon diodes and thyristors, which do not present excitation problems for large synchronous machines. The two types of solid-state excitation systems are:

(a) Static systems that have stationary diodes or thyristors, in which the current is fed to the rotor through slip-rings.

(b) Brushless systems that have shaft-mounted rectifiers which rotate with the rotor, thus avoiding the need for brushes and slip-rings. Figure 7.2 shows a brushless excitation system.

2. Field structure and speed of machine. We have already mentioned that the synchronous machine is a constant-speed machine. This speed, known as

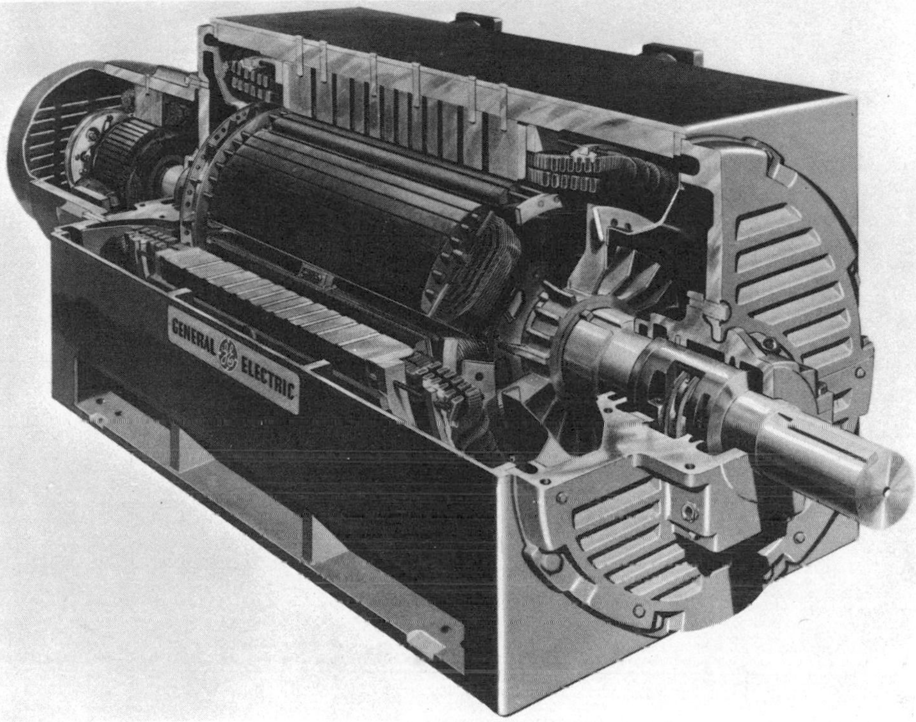

Figure 7.1. Cutaway of a salient pole synchronous machine. (Courtesy General Electric Company.)

synchronous speed, n_s, is given by (6.9). The equation is repeated below for convenience.

$$n_s = \frac{120f}{p} \tag{7.1}$$

Thus, a 60-Hz two-pole synchronous machine must run at 3600 rpm, whereas the synchronous speed of a 12-pole 60-Hz machine is only 600 rpm. The rotor field structure consequently depends upon the speed rating of the machine. Thus, turbogenerators, which are high-speed machines, have *round* or cylindrical rotors. (See Figures 7.3 and 7.4.) Hydroelectric and diesel-electric generators are low-speed machines and have *salient pole* rotors, as depicted in Figures 7.5 and 7.6. Such rotors are less expensive to fabricate than round rotors. They are not suitable for large high-speed machines, however, because of the excessive centrifugal forces and mechanical stresses that develop at speeds around 3600 rpm.

Another feature in the construction of a synchronous machine stems from the mounting of the rotor. For example, a round-rotor turbine-driven machine

Figure 7.2. Rotor of a 3360-kVA 6-kV brushless synchronous generator, with rotating diodes. (Courtesy Brown Boveri Company.)

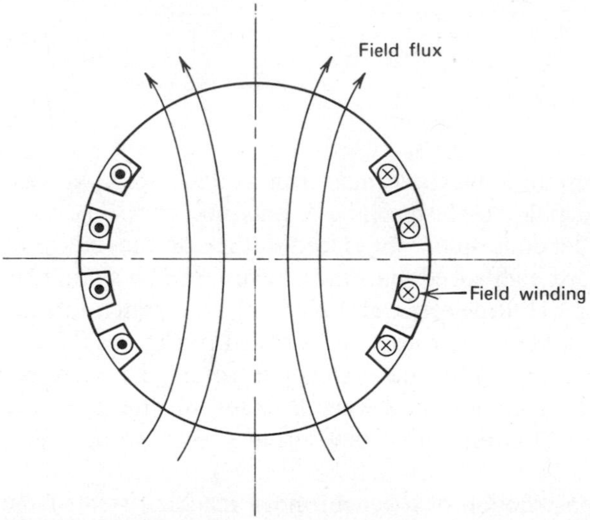

Figure 7.3. Field winding on a wound rotor.

Figure 7.4. Turbine rotor with direct water cooling during the mounting of damper hollow conductors. (Courtesy Brown Boveri Company.)

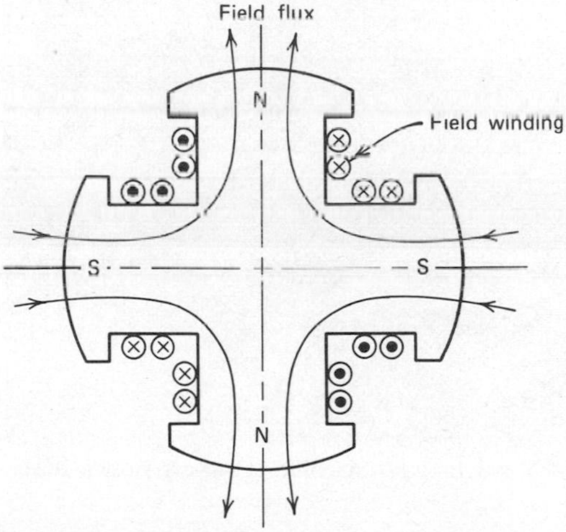

Figure 7.5. Field winding on a salient rotor.

Figure 7.6. Salient rotor of a 152.5-MVA 13.8-kV synchronous machine. (Courtesy Brown Boveri Company.)

(Figure 7.4) or a salient rotor diesel engine-driven machine (Figure 7.1) has a horizontally mounted rotor. A water wheel-driven machine (Figure 7.7) invariably has a vertically mounted salient pole rotor.

3. Stator. The stator of a synchronous machine is similar to that of a polyphase induction motor. (See Figure 7.8.) There is essentially no difference between the stator of a round-rotor machine and that of a salient rotor machine. The stators of waterwheel generators, however, usually have a large diameter armature compared to other types of generators (Figure 7.9). The stator core consists of punchings of high-quality laminations having slot-embedded lap windings.

4. Cooling. Because synchronous machines are often built in extremely large sizes, they are designed to carry very large currents. A typical armature current-density may be of the order of 10 A/mm^2 in a well-designed machine. Also, the magnetic loading of the core is such that it reaches saturation in many regions. The severe electric and magnetic loadings in a synchronous machine produce heat which must be appropriately dissipated. Thus, the manner in which the active parts of a machine are cooled determines its overall physical structures. In addition to air, some of the coolants used in synchronous machines include water, hydrogen, and helium. Figure 7.4 shows a turbine rotor with direct water cooling during the mounting of damper hollow conductors. The cooling arrangement of the stator of a turbine generator is shown in Figure 7.9.

5. Damper bars. So far we have mentioned only two electrical windings of a synchronous machine: the three-phase armature winding and the field winding. We also pointed out that, under steady-state, the machine runs at a constant speed, namely, the synchronous speed. However, like other electric machines, a synchronous machine undergoes transients during starting and abnormal conditions. During transients the rotor may undergo mechanical oscillations and its speed deviates from the synchronous speed, which is an undesirable phenomenon. To overcome this, an additional set of windings, resembling the cage of an induction motor, is mounted on the rotor. This winding is called damper winding and is shown in Figures 7.1, 7.4, and 7.6. When the rotor speed is different from the synchronous speed, currents are induced in the damper winding. The damper winding acts like the cage-rotor of an induction motor, producing a torque to restore the synchronous speed. Also, the damper bars provide a means of starting the machine as a synchronous motor, which is otherwise not self-starting.

In summary, Figures 7.1 to 7.9 show the various structural features of different types of synchronous machines. In principle, the machine has three electrical windings—the armature, the field, and the damper—located on the stator and the rotor as shown in Figure 7.1. In the following section we begin the study of a synchronous machine in its simplest form.

Figure 7.7. Mounting the rotor of a hydro-electric generator. (Courtesy Brown Boveri Company.)

Figure 7.8. Mounting stator conductors in slots of one stator half of a synchronous machine. (Courtesy Brown Boveri Company.)

Figure 7.9. End-winding region of a 722-MVA 22-kV turbine generator. (Courtesy Brown Boveri Company)

7.2 PRINCIPLE OF OPERATION

We have mentioned earlier that a synchronous machine is capable of operating either as a motor or as generator. In this section we shall consider the qualitative aspects of the action of a synchronous machine first as a motor and then as a generator.

7.2.1 OPERATION OF A SYNCHRONOUS MOTOR

From the discussions of Section 7.1 we notice a resemblance between the salient pole rotor of a synchronous machine and the rotor of a reluctance motor discussed in Chapter 4. Thus, from the energy-storage and energy-conversion principles developed in Chapter 4, we can show how a polyphase synchronous machine operates as a motor. We know that the stator of a three-phase synchronous machine is similar to that of a three-phase induction motor. We have demonstrated in the last chapter that a three-phase excitation, such as that found in the stator of an induction motor, produces a rotating magnetic field in the air gap of the machine. Referring to Figure 7.10*a*, we shall have a rotating magnetic field in the air gap of the salient pole machine when its stator (or armature) windings are fed from a three-phase source. The rotor will then have a tendency to align with the field at all times, in order to present the path of least reluctance. Thus, if the field is rotating, the rotor will tend to rotate with the field. From Figure 7.10*b*, we see that a round rotor will not tend to follow the rotating magnetic field, because the uniform air gap presents the same reluctance all around the air gap and the rotor does not have any preferred direction of alignment with the magnetic field. This torque, which we have in Figure 7.10*a* but not in Figure 7.10*b*, is called the *reluctance torque*. It is present by virtue of the variation of the reluctance around the periphery of the machine.

Next, let the field winding, Figure 7.10*a* or *b*, be fed by a dc source that produces the rotor magnetic field of definite polarities. By the principle of alignment of fields (Chapter 4), we conclude that when the rotor is excited, it will tend to align with the stator field, and will tend to rotate with the rotating magnetic field. We observe that for an excited rotor, a round rotor and a salient rotor both will tend to rotate with the rotating magnetic field, although the salient rotor will have an additional reluctance torque because of the saliency. In a later section we shall derive expressions for the electromagnetic torque in a synchronous machine attributable to field excitation and to saliency.

So far we have indicated the mechanism of torque production in a round-rotor and in a salient rotor machine. To recapitulate, we might say that the stator rotating magnetic field has a tendency to "drag" the rotor along, as if a north pole on the stator "locks in" with a south pole of the rotor. However, if the rotor is at a standstill, the stator poles will tend to make the rotor rotate in one direction and then in the other, as they rotate and sweep across the rotor

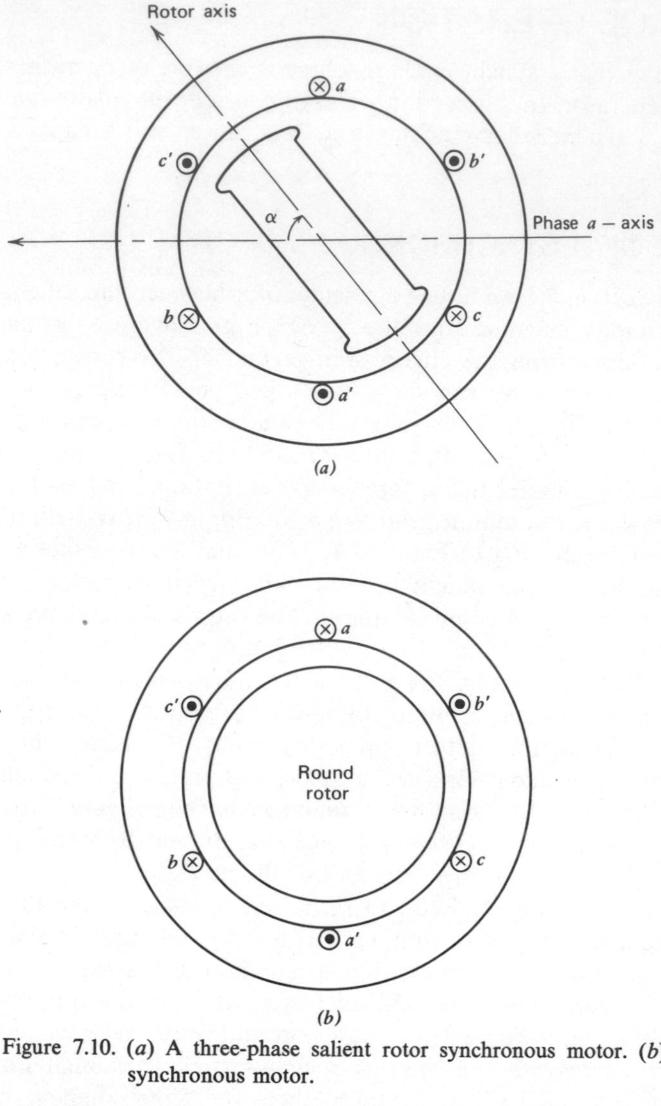

Figure 7.10. (*a*) A three-phase salient rotor synchronous motor. (*b*) A three-phase round rotor synchronous motor.

poles. Therefore, a synchronous motor is not self-starting. In practice, as we mentioned earlier, the rotor carries damper bars which act like the cage of an induction motor and thereby provide a starting torque. Once the rotor starts running and almost reaches the synchronous speed, it locks into position with the stator poles. The rotor pulls into step with the rotating magnetic field and runs at the synchronous speed; the damper bars go out of action. Any departure from the synchronous speed results in induced currents in the damper bars, which tend to restore the synchronous speed. Machines without damper bars, or very large machines with damper bars, may be started by an auxiliary motor. We shall discuss the operating characteristics of synchronous motors later, but first we shall consider the operation of a synchronous machine as a generator.

7.2.2 OPERATION OF A SYNCHRONOUS GENERATOR

Like the dc generator, a synchronous generator functions on the basis of Faraday's law. If the flux linking the coil changes in time, a voltage is induced in a coil. Stated in alternate form, a voltage is induced in a conductor if it cuts magnetic flux lines. Considering the machine shown in Figure 7.10*a* and neglecting the effects of winding distributions (i.e., we assume concentrated coils for each phase), we find the flux density distribution in the air gap produced by the field winding to be

$$B(\theta) = B_m \cos\theta \tag{7.2}$$

The flux linking an N-turn coil, of radius r and axial length l, is found as follows:

$$\text{flux linkage per pole} = \lambda = N\int_{-\pi/2}^{\pi/2} B(\theta)\, lr\, d\theta \tag{7.3}$$

And (7.2) and (7.3) yield

$$\lambda = 2NB_m lr \tag{7.4}$$

From Figure 7.10*a* it is clear that when $\alpha = 0$, λ is maximum and it is zero at $\alpha - 0$. The value of λ for any rotor position α is, thus, from (7.4)

$$\lambda(\alpha) = 2NB_m lr \cos\alpha \tag{7.5}$$

The voltage induced in the N-turn coil corresponding to phase a in our case is obtained from Faraday's law and from (7.5) is given by

$$v_a = -\frac{d\lambda}{dt} = -\frac{d\lambda}{d\alpha}\frac{d\alpha}{dt} = 2NB_m lr\omega \sin\alpha \tag{7.6}$$

where $d\alpha/dt = \omega$, the speed of the rotor. Also, we may put $\alpha = \omega t$ and $2B_m lr = \phi$ in (7.6) to obtain the final form for the voltage expression as

$$v_a = \omega N\phi \sin\omega t = V_m \sin\omega t \tag{7.7}$$

where $V_m = \omega N\phi$. The obvious conclusion is that for a sinusoidal flux density distribution produced by the field rotor winding, the voltage induced in the phase 'a' is sinusoidal and its frequency $f=2\pi/\omega$ is dependent on the rotor speed.

Next we consider the round-rotor machine of Figure 7.10*b*, where the flux density distribution produced by the rotor mmf in the air gap is uniform. According to the "flux-cutting" rule, the voltage induced in the N-turn coil at a given instant is

$$v_a = BlU_\perp \tag{7.8}$$

where $U_\perp$, B, and l are mutually perpendicular. The linear velocity U is related to the angular velocity by $U=r\omega$ and, from Figure 7.10*b*, its vertical component $U_\perp = U \sin\alpha$. Substituting these in (7.7), with $\alpha=\omega t$, yields

$$v_a = Blr\omega \sin\omega t = V_m \sin\omega t \tag{7.9}$$

which also implies a sinusoidal voltage generation in phase a. Because phases b and c are displaced from phase a by 120°, the corresponding voltages may be written as

$$v_b = V_m \sin(\omega t - 120°) \tag{7.10}$$

$$v_c = V_m \sin(\omega t + 120°) \tag{7.11}$$

The above voltages are expressed graphically in Figure 7.11.

7.3 CERTAIN REALISTIC CONSIDERATIONS

Before considering the methods of analysis used with synchronous machines that lead to the determination of their operational characteristics, we shall review some of the assumptions made or implied in the preceding discussions. First of all, while deriving the voltage equation in Section 7.2.2, we represented a phase winding by an N-turn coil. In reality, this is not so, as discussed in Chapter 6. We recall from Section 6.1 that a phase winding is distributed around the periphery of the stator. The coil-pitch of the winding is a fraction of the pole-pitch, which means that the winding may have a fractional pitch. To account for the winding distribution and fractional pitch, we use the winding factor, $k_w < 1$, as given by

$$k_w = k_d k_p \tag{7.12}$$

by which the voltage in a full-pitch, concentrated N-turn coil will be reduced. In (7.12) k_d = distribution factor and k_p = pitch factor, which is defined in Section 6.1. When distributing a phase winding in the slots around the stator, the resulting mmf distribution is ideally designed to be as close to sinusoidal as

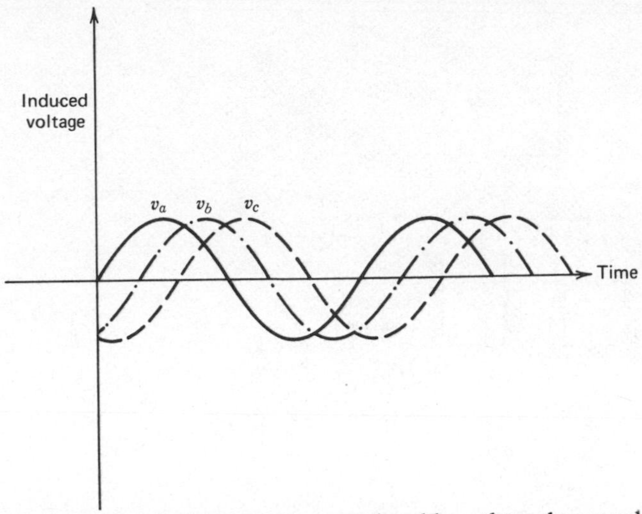

Figure 7.11. A three-phase voltage produced by a three-phase synchronous generator.

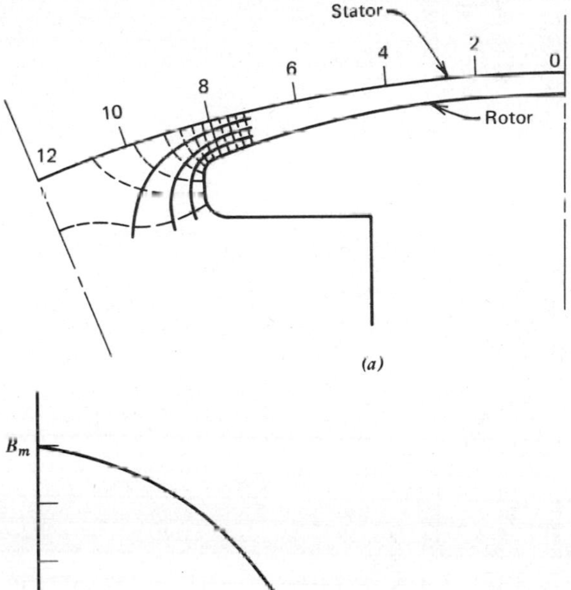

(a)

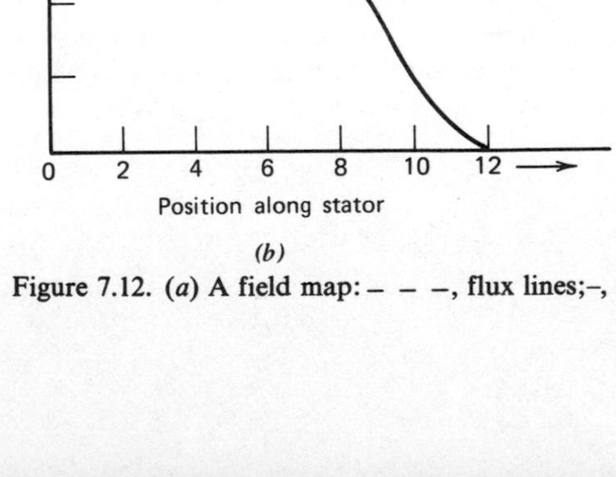

(b)

Figure 7.12. (*a*) A field map: – – –, flux lines;–, equipotentials. (*b*) Flux density distribution.

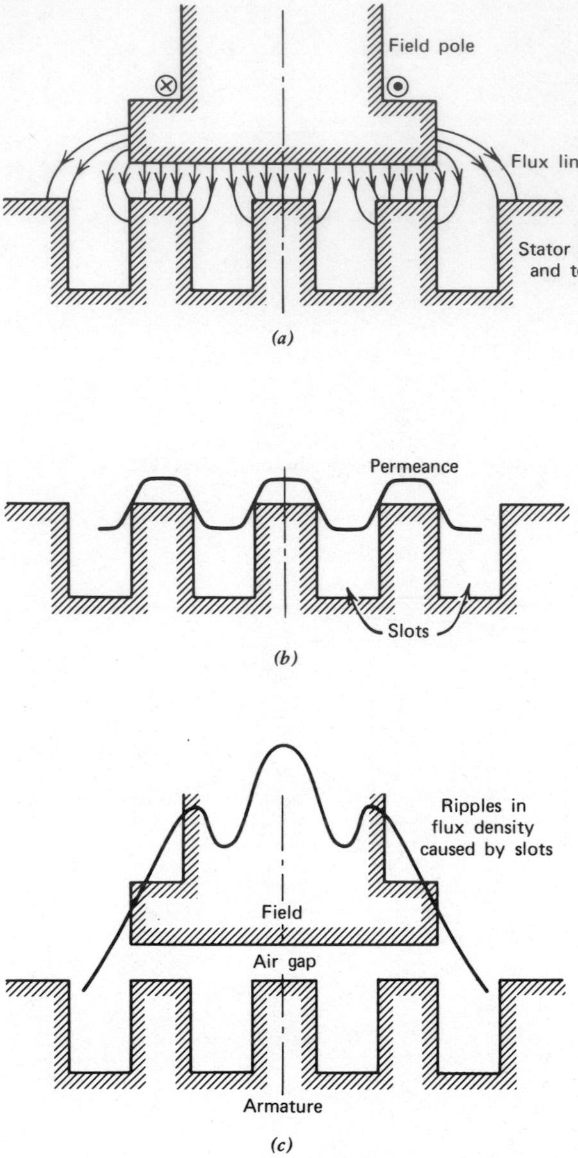

Figure 7.13. (*a*) Flux density distribution near tooth-tips. (*b*) Permeance variation along the armature surface. (*c*) Distortion in flux density distribution caused by the armature slots.

possible. However, in reality an armature winding does contain harmonies, as depicted by the Fourier series representation of the mmf distribution. For further details see References 2 and 3.

Second, for a salient pole machine, we assumed a sinusoidal flux density distribution in the air gap. Again, this is an idealistic assumption. In practice, the field distribution is nonsinusoidal and approaches a form shown in Figure 7.12*b*. The field distribution is obtained by the graphical field-mapping technique illustrated in Figure 7.12*a*. Neglecting the reluctance of the iron portion of the magnetic circuit, we find that the permeance of each flux tube is proportional to the ratio of its mean cross section and to its mean length. Thus, the permeance distribution around the stator periphery is known. Knowing the mmf distribution, the flux density distribution can be obtained.

Third, in our discussions so far, we have not given any consideration to the presence of slots and teeth. From Figures 7.13*a* and *b* it is clear that there is a permeance variation for the air gap flux because of the slots and teeth. This will cause the field form to contain ripples, as shown in Figure 7.13*c*, which are called tooth ripples. Such ripples create harmonics in the induced emf's which are of higher frequencies than 60 Hz (for instance, 875 Hz). The frequencies, in turn, give rise to magnetic noise. Proper design, such as having a large number of slots with small openings, compared to the air gap, or having skewed slots, or having the number of slots a nonintegral multiple of the number of poles, reduces the tooth ripples in a synchronous machine. Finally, the effect of saturation on the open-circuit voltage is shown in Fig. 7.14.

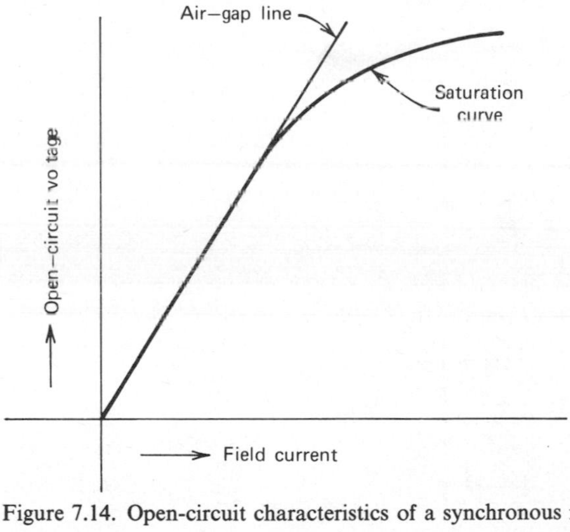

Figure 7.14. Open-circuit characteristics of a synchronous machine.

7.4 OPERATING CHARACTERISTICS OF SYNCHRONOUS MACHINES

We shall now look into some of the steady-state operating characteristics of synchronous machines from a quantitative viewpoint. For convenience, we shall consider generator operation separately from motor operation. The details of the method of analysis for the round-rotor machine will be somewhat different from the procedure for a salient pole machine. In all cases, the analysis makes use of the machine parameters, which must be identified before proceeding with the analytical details.

Since for the present we are considering only the steady-state behavior of the machine, circuit constants of the field and damper windings need not be considered. The presence of the field winding will be denoted by the flux produced by the field excitation. Turning to the armature winding, we shall represent it on a per-phase basis (as we did with the induction motor in Chapter 6). Obviously, the armature winding has a resistance. But the ohmic value of this resistance must include the effects of the operating temperature and the alternating currents flowing in the armature conductors (causing skin-effect, for ins-

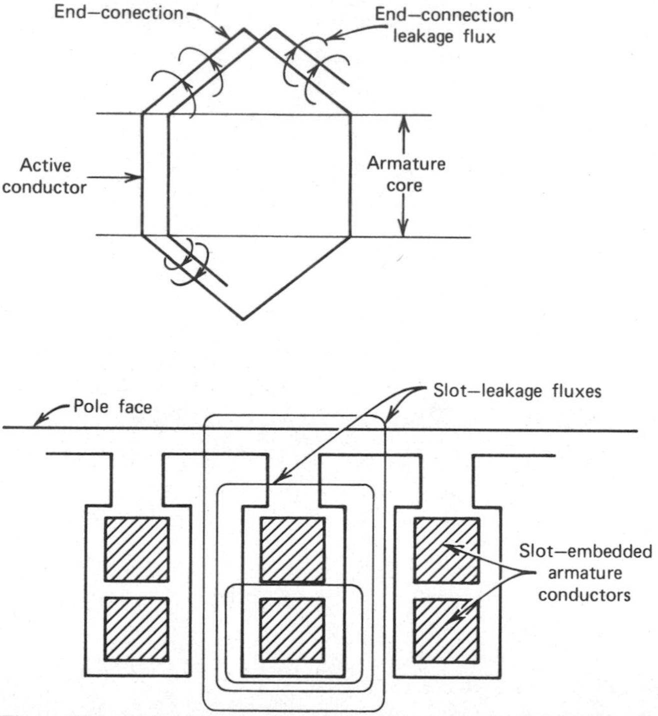

Figure 7.15. (*a*) End-connection leakage flux path. (*b*) Slot-leakage flux paths.

tance). As a consequence, the value of the armature resistance becomes larger, as compared to its dc resistance. The larger value of the resistance is known as the *effective resistance* of the armature resistance and is denoted by r_a. An approximate value of r_a is 1.6 times the dc resistance.

Next, we consider the reactances pertaining to the armature winding. First, the leakage reactance is caused by the leakage fluxes linking the armature conductors only, because of the currents in the conductors. These fluxes do not link with the field winding. As in an induction motor, for convenience in calculation, the leakage reactance is divided into: (a) end-connection leakage reactance, (b) slot-leakage reactance, (c) tooth-top and zig-zag leakage reactance, and (d) belt-leakage reactance. All of these components are not significant in every synchronous machine. In most large machines, the last two reactances are a small portion of the total leakage reactance.

Flux paths contributing to the end-connection leakage and slot-leakage reactances are shown in Figures 7.15*a* and *b*, respectively. We denote the total leakage reactance of the armature winding per-phase by X_a. Now, to proceed with the analysis, let us first consider a round-rotor synchronous generator. We shall also introduce the concept of *synchronous reactance*, the most important parameter in determining the steady-state characteristics of a synchronous machine.

7.4.1 PERFORMANCE OF ROUND-ROTOR SYNCHRONOUS GENERATOR

At the outset we wish to point out that we shall study the machine on a per-phase basis, implying a balanced operation. Thus, let us consider a round-rotor machine operating as a generator on no-load. Let the open-circuit phase voltage be V_0 for a certain field current I_f. Here, V_0 is the internal voltage of the generator. We assume that I_f is such that the machine is operating under unsaturated condition. Next, we short-circuit the armature at the terminals, keeping the field current unchanged (at I_f), and measure the armature phase current I_a. In this case, the entire internal voltage V_0 is dropped across the internal impedance of the machine. In mathematical terms,

$$V_0 = I_a Z_s \tag{7.13}$$

and Z_s is known as the *synchronous impedance*. One portion of Z_s is r_a and the other a reactance, X_s known as *synchronous reactance*; that is,

$$Z_s = r_a + jX_s \tag{7.14}$$

In (7.14) X_s is greater than the armature leakage reactance X_a, discussed earlier.

Where does the additional reactance come from? We shall attempt to answer this question in the following discussion.

Let the generator supply a phase current I_a to a load at unity power factor and at a terminal voltage, V_t, volt/phase. This is shown in the phasor diagram (Figure 7.16), where V_a is the phasor sum of V_t and the drop due to the armature resistance and armature leakage reactance. Notice that we now have two mmf's —F_a attributable to the armature current and F_f attributable to the field current present in the machine. To find the mmf F_r that produces the voltage V_a, refer to the open-circuit characteristic of the generator, in Figure 7.17, which shows F_r corresponding to V_a. The flux produced by an mmf is in phase with the mmf. As dictated by $e = -N\, d\phi/dt$, however, the voltage induced by a certain flux is behind the mmf by 90°. Therefore, we lay F_r ahead of V_a by 90° and F_a in phase with I_a, as depicted in Figure 7.16. The mmf F_a is known as the *armature reaction* mmf. Sufficient mmf must be supplied by the field to overcome F_a such that we have a net F_r to produce V_a. Because of F_a, an equal and opposite component, F_a of the field mmf is laid off from F_r, as shown in Figure 7.16. The phasor F_f is then the total field mmf in the machine. Corresponding to this mmf, the open-circuit voltage of the generator is V_0, as found from Figure 7.17. This open-circuit voltage is known as the *nominal induced emf*, and is also shown in Figure 7.16. From the geometry of the phasor diagram, it is clear that the triangles OST and OQR are similar. We also see that QR is perpendicular to OP and must pass through P because $OP = V_t + I_a r_a$. Thus, QRP is a continuous straight line and we have

$$V_0 = V_t + I_a(r_a + jX_s) \tag{7.15}$$

where X_s, the synchronous reactance, agrees with the definitions of (7.12) and (7.13). The "extra" reactance, in addition to X_a (Figure 7.16), is introduced by the armature reaction. Therefore, the synchronous reactance is the sum of the armature leakage reactance and armature reaction reactance.

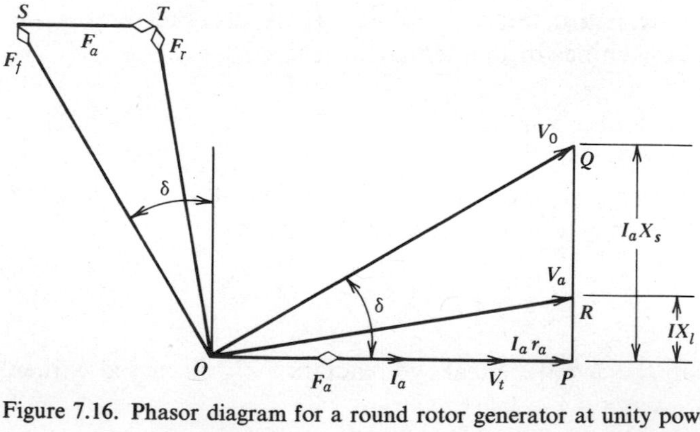

Figure 7.16. Phasor diagram for a round rotor generator at unity power factor (to define X_s).

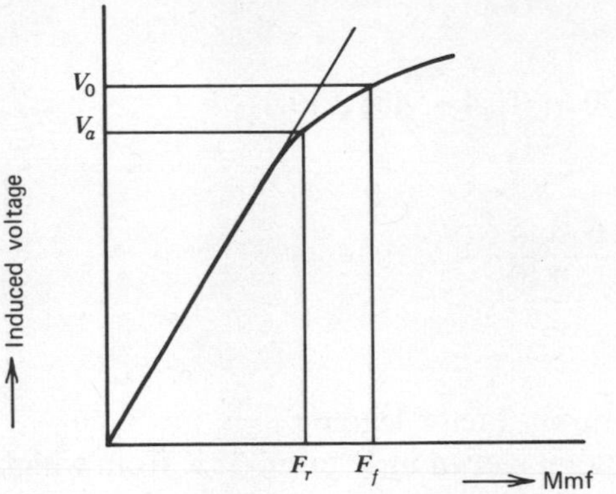

Figure 7.17. Open-circuit characteristics of a synchronous generator.

In an actual synchronous machine, except in very small ones, we almost always have $X_s \ll r_a$ in which case and $Z_s \cong jX_s$. We shall use this restriction in most of the analysis. Among the steady-state characteristics of a synchronous generator, its voltage regulation and power-angle characteristics are the most important ones. As for a transformer and a dc generator, we define the voltage regulation of a synchronous generator at a given load as

$$\text{percent voltage regulation} = \frac{V_0 - V_t}{V_t} \times 100 \tag{7.16}$$

where V_t is the terminal voltage on load and V_0 is the no-load terminal voltage. Clearly, for a given V_t, we can find V_0 from (7.14) and hence the voltage regulation as illustrated by the following example.

Example 7.1

Calculate the percent voltage regulation for a three-phase Y-connected 2500-kVA 6600-V turboalternator operating at full load and 0.8 power factor lagging. The per-phase synchronous reactance and the armature resistance are 10.4 and 0.071 ohms, respectively.

Clearly, we have $X_s \ll r_a$. The phasor diagram for the lagging power factor, neglecting the effect of r_a, is shown in Figure 7.18a. The numerical values are as follows:

$$V_t = \frac{6600}{\sqrt{3}} = 3810 \text{ volt}$$

$$I_a = \frac{2500 \times 1000}{\sqrt{3} \times 6600} = 218.7 \text{ A}$$

From (7.14) we have

$$V_0 = 3810 + 218.7(0.8 - j0.6)j10.4 = 5485\angle 19.3°$$

And

$$\text{percent regulation} = \frac{5485 - 3810}{3810} \times 100 = 44\%$$

Example 7.2
Repeat the above calculations with 0.8 power factor leading.

In this case we have the phasor diagram shown in Figure 7.18*b*, from which we get

$$V_0 = 3810 + 218.7(0.8 + j0.6)j10.4 = 3048\angle 36.6°$$

and

$$\text{percent voltage regulation} = \frac{3048 - 3810}{3810} \times 100 = -20\%$$

We observe from the last two examples that the voltage regulation is dependent on the power factor of the load. Unlike what happens in a dc generator, the voltage regulation for a synchronous generator may even become negative. The angle between V_0 and V_t is defined as the *power angle*, δ. To justify this definition, we reconsider Figure 7.18*a*, from which we obtain

$$I_a X_s \cos\phi = V_0 \sin\delta \tag{7.17}$$

From (7.16) and (7.17) we get

$$P_d = \frac{V_0 V_t}{X_s} \sin\delta \tag{7.18}$$

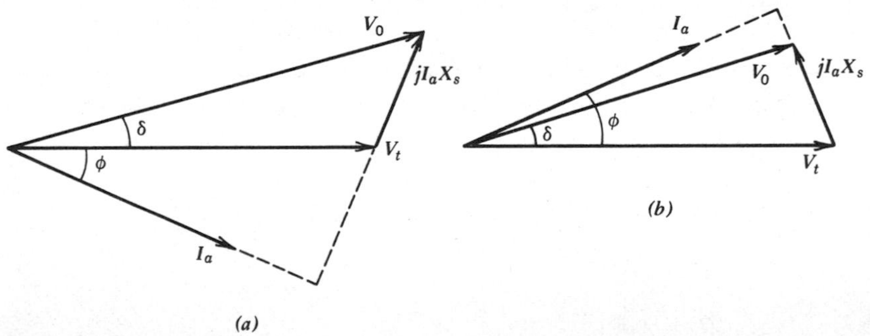

Figure 7.18. Phasor diagrams. (*a*) Lagging power factor. (*b*) Leading power factor.

which shows that the internal power of the machine is proportional to $\sin\delta$. Equation (7.18) is often said to represent the power angle characteristic of a synchronous machine.

7.4.2 PERFORMANCE OF ROUND-ROTOR SYNCHRONOUS MOTOR

Except for some precise calculations, we may neglect the armature resistance as compared to the synchronous reactance. Therefore, the steady-state per phase equivalent circuit of a synchronous machine simplifies to the one shown in Figure 7.19*a*. Notice that this circuit is similar to that of a dc machine, where the dc armature resistance has been replaced by the synchronous reactance. In Figure 7.19*a* we have shown the terminal voltage V_t, the internal excitation voltage V_0 and the armature current I_a going "into" the machine or "out of" it, depending upon the mode of operation—"into" for motor and "out of" for generator. With the help of this circuit and (7.18) we shall study some of the steady-state operating characteristics of a synchronous motor. In Figure 7.19*b* we show the power-angle characteristics as given by (7.18). Here positive power and positive δ imply the generator operation while a negative δ corresponds to a motor operation. Because δ is the angle between V_0 and V_t, V_0 is ahead of V_t in a generator, whereas in a motor, V_t is ahead of V_0. The voltage-balance equation for a motor is, from Figure 7.19*a*

$$V_t = V_0 + jI_aX_s \tag{7.19}$$

If the motor operates at a constant power, then (7.17) and (7.18) require that

$$V_0 \sin\delta = I_aX_s\cos\phi = \text{constant} \tag{7.20}$$

We recall that V_0 depends upon the field current, I_f. Consider two cases: (1)

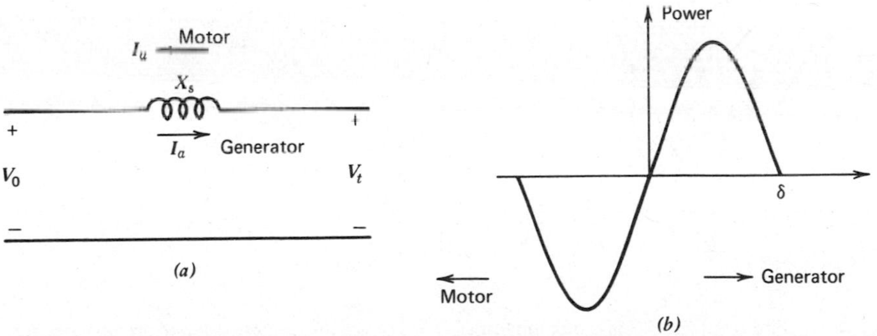

Figure 7.19. (*a*) An approximate equivalent circuit. (*b*) Power-angle characteristics of a synchronous machine.

when I_f is so adjusted that $V_0 < V_t$ and the machine is *underexcited*; and (2) when I_f is increased to a point that $V_0 > V_t$ and the machine becomes *overexcited*. The voltage-current relationships for the two cases are shown in Figure 7.20*a*. For $V_0 > V_t$ at constant power δ is greater than the δ for $V_0 < V_t$, as governed by (7.20). Notice that an underexcited motor operates at a lagging power factor (I_a lagging V_t), whereas an overexcited motor operates at a leading power factor. In both cases the terminal voltage and the load on the motor are the same. Thus, we observe that the operating power factor of the motor is controlled by varying the field excitation, hence altering V_0. This is a very

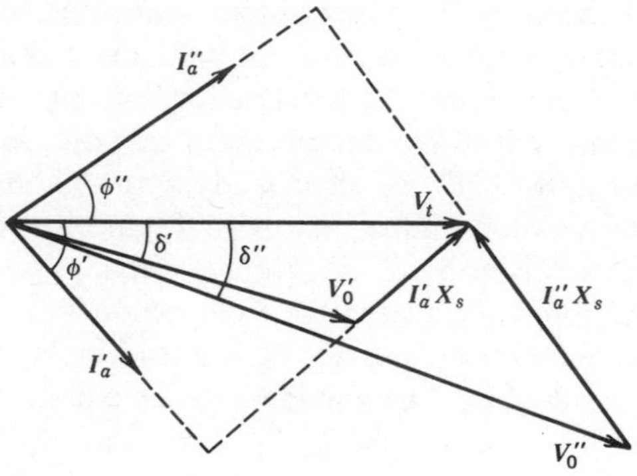

(*a*)

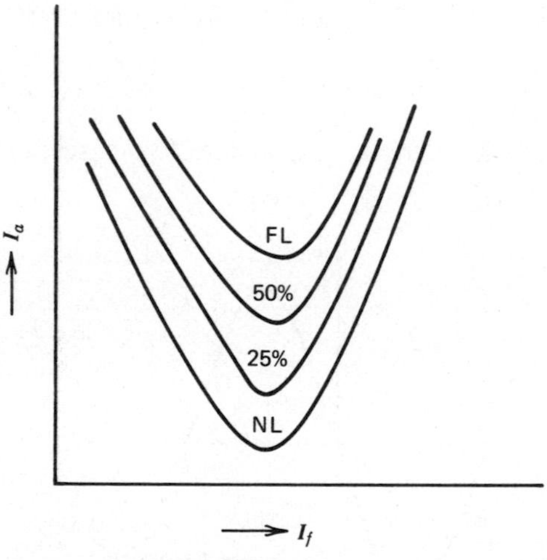

(*b*)

Figure 7.20. (*a*) Phasor diagram for motor operation (V_0', I_a', ϕ', and δ') correspond to underexcited operation. (V_0'', I_a'', ϕ'', and δ'') correspond to overexcited operation. (*b*) *V*-curves of a synchronous motor.

important property of synchronous motors. The locus of the armature current at a constant load, as given by (7.20), for varying field current is also shown in Figure 7.20*a*. From this we can obtain the variations of the armature current I_a with the field current, I_f (corresponding to V_0), and this can be done for different loads as shown in Figure 7.20*b*. These curves are known as the *V curves* of the synchronous motor. One of the applications of a synchronous motor is in power factor correction, as demonstrated by the following example.

Example 7.3
A three-phase, Y-connected load takes 50 A current at 0.707 lagging power factor at 220 V between the lines. A three-phase Y-connected round-rotor synchronous motor, having a synchronous reactance of 1.27 ohm per phase, is connected in parallel with the load. The power developed by the motor is 33 kW at a power angle of 30°. Neglecting the armature resistance, calculate (a) the reactive kVA of the motor, and (b) the overall power factor of the motor and the load.

The circuit and the phasor diagram, on a per-phase basis, are shown in Figure 7.21. From (7.18) we have

$$P_d = \frac{1}{3} \times 33{,}000 = \frac{220}{\sqrt{3}}\,\frac{V_0}{1.278}\sin 30°$$

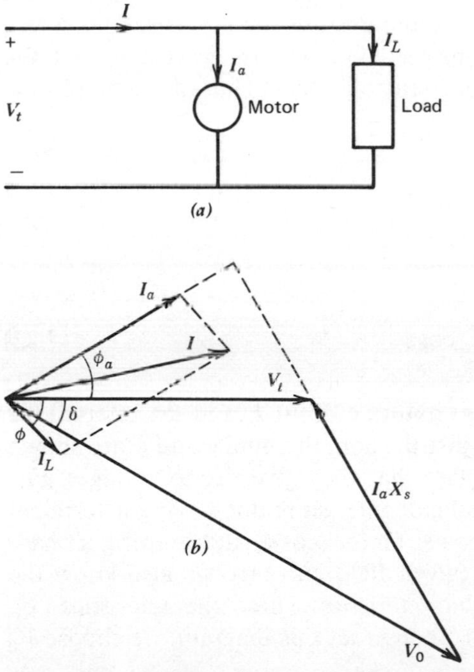

Figure 7.21. (*a*) Circuit diagram. (*b*) Phasor diagram.

which yields $V_0=220$ volt. From the phasor diagram, $I_aX_s=127$ or $I_a=127/1.27$ $=100$ A and $\phi_a=30°$. The reactive kVA of the motor $=\sqrt{3}\times V_tI_a\sin\phi_a=\sqrt{3}$ $\times\frac{220}{1000}\times100\times\sin 30=19$ kvar.

The overall power-factor angle ϕ is given by

$$\tan\phi=\frac{I_a\sin\phi_a-I_L\sin\phi_L}{I_a\cos\phi_a+I_L\cos\phi_L}=0.122$$

or $\phi=7°$ and $\cos\phi=0.992$ leading.

7.4.3 SALIENT POLE SYNCHRONOUS MACHINES

In the preceding, we have analyzed the round-rotor machine and made extensive use of the machine parameter, which we defined as synchronous reactance. Because of saliency, the reactance measured at the terminals of a salient rotor machine will vary as a function of the rotor position. This is not so in a round-rotor machine (see also Chapter 4, Reluctance Motor).

To overcome this difficulty, we use the *two-reaction theory* proposed by André Blondel (see also Section 9.5). The theory proposes to resolve the given armature mmf's into two mutually perpendicular components, with one located along the axis of the rotor salient pole, known as the direct (or d-) axis and with the other in quadrature and known as the quadrature (or q-) axis. The d-axis component of the mmf, F_d, is either magnetizing or demagnetizing and the q-axis component, F_q, results in a crossmagnetizing effect. Thus, if the amplitude of the armature mmf is F_a then

$$F_d=F_a\sin\psi \tag{7.21}$$

and

$$F_q=F_a\cos\psi \tag{7.22}$$

where ψ is the phase angle between the armature current I_a and the internal (or excitation) voltage V_0. In terms of space distribution, the mmf's and ψ are shown in Figure 7.22. The effects of F_d and F_q are that they give rise to voltages. For the sake of illustration, we consider a salient pole generator having a terminal voltage, V_t, supplying a load of lagging power factor ($\cos\phi$) and drawing a phase current I_a. For the operating condition (given field current), we also know the no-load voltage, V_0, from the no-load characteristics. These characteristics (V_t, V_0, I_a, and θ) are shown in Figure 7.23. To construct this diagram, we choose V_0

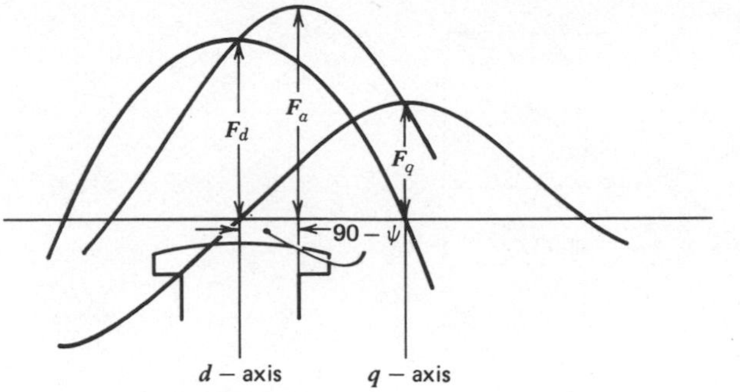

Figure 7.22. Armature mmf and its d- and q- components.

as the reference phasor and neglect the armature resistance. Now, on load because of armature reaction alone, V_0 will be reduced to V_a as determined by V_d and V_q, which are caused by F_d and F_q, respectively. From V_a we obtain V_t by subtracting the armature leakage reactance drop I_aX_a. Ignoring V_a, we might say that the difference between V_0 and V_t is attributable to the armature reaction reactance and the armature leakage reactance, that is, the synchronous reactance. We can resolve I_a into its d- and q-axis components, I_d and I_q respectively, so also the I_aX_s drop shown as I_dX_d and I_qX_q in Figure 7.23. Thus, the phasor diagram is complete. (See the next example.) We can associate physical meanings to the reactances X_d and X_q. These are, respectively, the direct-axis and quadrature-axis reactances and the maximum and minimum values of the synchronous reactance of a salient pole machine. These reactances can be measured experimentally, as we shall discuss in a later section. The preceding discussions are valid on a per-phase basis for a balanced machine. We now show the details of some of the calculations by the following example.

Example 7.4

A 20-kVA, 220-V, 60-Hz, Y-connected three-phase salient pole synchronous generator supplies rated load at 0.707 lagging power factor. The phase constants of the machine are $r_a = 0.05$ ohm and $X_d = 2X_q = 4.0$ ohms. Calculate the voltage regulation at the specified load.

$$V_t = \frac{220}{\sqrt{3}} = 127 \text{ volt}$$

$$I_a = \frac{20{,}000}{\sqrt{3} \times 220} = 52.5 \text{ A.}$$

$$\phi = \cos^{-1} 0.707 = 45°$$

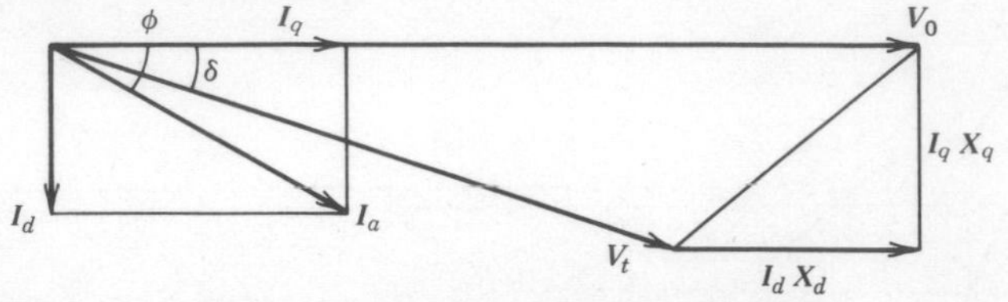

Figure 7.23. Phasor diagram of a salient pole machine.

From Figure 7.23 we have

$$I_d = I_a \sin(\delta + \phi)$$

$$I_q = I_a \cos(\delta + \phi)$$

$$V_t \sin \delta = I_q X_q = I_a X_q \cos(\delta + \phi)$$

or

$$\tan \delta = \frac{I_a X_q \cos \phi}{V_t + I_a X_q \sin \phi}$$

$$= \frac{52.5 \times 2 \times 0.707}{127 + 52.5 \times 2 \times 0.707} = 0.37$$

or

$$\delta = 20.6°$$

$$I_d = 52.5 \sin(20.6 + 45) = 47.5 \text{ A.}$$

$$I_d X_q = 47.5 \times 4 = 190.0 \text{ V}$$

$$V_0 = V_t \cos \delta + I_d X_d$$

$$= 127 \cos 20.6 + 190 = 308$$

$$\text{And percent regulation} = \frac{V_0 - V_t}{V_t} \times 100\%$$

$$= \frac{308 - 127}{127} \times 100 = 142\%$$

The previous example shows how the phasor diagram Figure 7.23 can be used to determine the voltage regulation of a salient pole synchronous generator. In fact, the phasor diagram depicts the complete performance characteristics of the machine. For instance, to obtain the power-angle characteristics of a salient pole machine, operating either as a generator or as a motor, we refer to Figure 7.23. From this Figure we have, neglecting r_a and the internal losses,

$$\text{power output} = V_t I_a \cos\phi = \text{developed power} = P_d \tag{7.23}$$

and

$$I_q X_q = V_t \sin\delta$$
$$I_d X_d = V_0 - V_t \cos\delta \tag{7.24}$$

Also

$$I_d = I_a \sin(\delta + \phi)$$
$$I_q = I_a \cos(\delta - \phi) \tag{7.25}$$

Substituting (7.25) in (7.24) and solving for $I_a \cos\phi$ yields

$$I_a \cos\phi = \frac{V_0}{X_d}\sin\delta + \frac{V_t}{2X_q}\sin 2\delta - \frac{V_t}{2X_d}\sin 2\delta \tag{7.26}$$

Finally, substituting (7.26) and (7.23) gives

$$P_d = \frac{V_t V_0}{X_d}\sin\delta + \frac{1}{2}V_t^2\left(\frac{1}{X_q} - \frac{1}{X_d}\right)\sin 2\delta \tag{7.27}$$

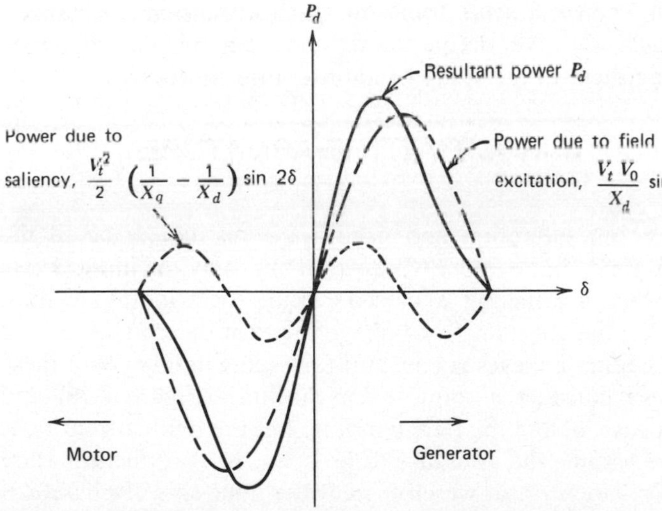

Figure 7.24. Power-angle characteristics of salient pole machine.

This variation of the developed power, P_d, as a function of the power angle δ, is shown in Figure 7.24. Notice that the resulting power is composed of power due to saliency—the second term in (7.27)—and of power due to field excitation—the first term in (7.27). Clearly, when $X_d = X_q$ the machine has no saliency and only the first term in (7.27) is nonzero, which represents the power-angle characteristics of a round-rotor machine. On the other hand, if there is no field excitation, implying $V_0 = 0$, the first term in (7.27) reduces to zero. We then have the power-angle characteristics of a reluctance machine as given by the second term, also discussed in Chapter 4. As in a round-rotor machine discussed earlier, the power-angle characteristics given by (7.27) reflect the generator as well as motor operation. The term, δ, is positive for the former and negative for the latter.

7.5 TRANSIENTS IN SYNCHRONOUS MACHINES

In the preceding sections we focused our attention on the steady-state behavior of synchronous machines. In this section we shall briefly review some cases involving transients in synchronous machines. Of particular interest are (a) the sudden short circuit at the armature terminals of a synchronous generator, and (b) mechanical transients caused by a sudden load change on the machine. There are numerous other cases involving transients in synchronous machines, but these will not be considered at this point.

We know from earlier considerations that the performance of a machine, for a given operating condition, can be determined if the machine parameters are known. For instance, we have already expressed the steady-state power-angle characteristics of a salient pole synchronous machine in terms of the *d*-axis and *q*-axis reactances. Similarly, the constants, by which transient behavior of a synchronous machine is known, are the transient and subtransient reactances, and pertinent time constants. We define these quantities in the following subsection, while relating them to the study of an armature short-circuit.

7.5.1 SUDDEN SHORT-CIRCUIT AT THE ARMATURE TERMINALS

At the outset we assume no saturation and neglect the resistances of all the windings—the armature, field, and damper windings. Thus, only the inductances remain, implying that the flux linking a closed circuit (or winding) cannot change instantaneously, as the constant flux linkage theorem determines. Stated differently, the sum of the flux linkages is constant for each winding. With these assumptions in mind, we consider a round-rotor machine (Figure 7.25) and focus our attention on phase 'a' and the field winding. Let the field current be I_f at $t = 0$. Prior to $t = 0$ we assume the armature to be open. At $t = 0$ the armature winding is suddenly short-circuited, at which instant the mmf axis of phase *a* is

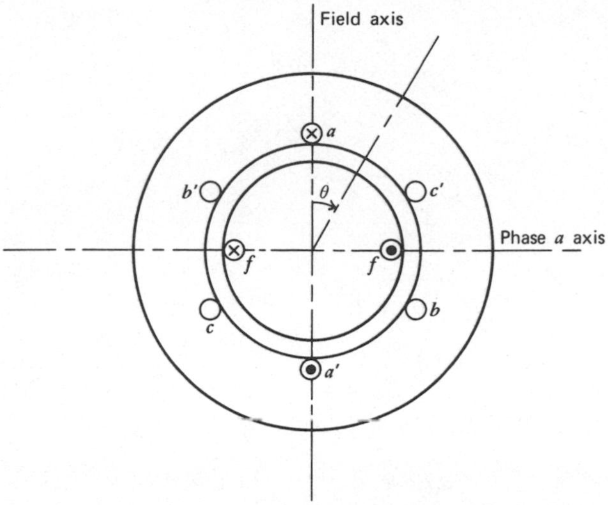

Figure 7.25. Three-phase round-rotor machine (only phase a and field winding carry currents).

at right angles to the mmf axis of the field winding. This occurs so that there is no mutual coupling between the two windings. Clearly, at $t=0$, the flux that links the armature is $\lambda_a=0$. But, the flux that links the field winding is $\lambda_f=L_f I_f$, where $L_f=$ field-winding inductance. We may divide L_f such that

$$L_f = L_l + L_{ad} \tag{7.28}$$

where $L_l=$ field-leakage inductance and $L_{ad}=$ mutual inductance between the field and armature windings. Thus, L_{ad} also corresponds to the armature reaction reactance. We may now rewrite λ_f, using (7.28) as

$$\lambda_f = (L_l + L_{ad})I_f = L_{ad}(1+\tau_f)I_f \tag{7.29}$$

where $\tau_f = L_l / L_{ad}$ = field-leakage coefficient.

After a time, t, let the rotor rotate through angle θ (see Figure 7.25), in which case i_a and $(i_f + I_f)$ will flow through the armature and field windings to maintain the flux linkages with these windings. Therefore, for the armature winding we have

$$\lambda_a = i_a L_{ad}(1+\tau_a) + (i_f + I_f)L_{ad}\sin\theta = 0 \tag{7.30}$$

where $\tau_a = X_l / \omega L_{ad}$ = armature leakage coefficient. Likewise, for the field winding we have

$$\lambda_f = (i_f + I_f)L_{ad}(1+\tau_f) + i_a L_{ad}\sin\theta$$
$$= I_f L_{ad}(1+\tau_f) \tag{7.31}$$

Solving (7.30) and (7.31) for i_a and i_f yields

$$i_a = \frac{\left[(1+\tau_f)\sin\theta\right]I_f}{\sin^2\theta - (1+\tau_a)(1+\tau_f)} \tag{7.32}$$

$$i_f = -\frac{(\sin^2\theta)I_f}{\sin^2\theta - (1+\tau_a)(1+\tau_f)} \tag{7.33}$$

The maximum values of these currents occur at $\theta = \pi/2$. In this case we have

$$(i_a)_{\max} = -\frac{(1+\tau_f)I_f}{\tau_a + (1+\tau_a)\tau_f} \tag{7.34}$$

$$(i_f)_{\max} = \frac{I_f}{\tau_a + (1-\tau_a)\tau_f} \tag{7.35}$$

Multiplying the numerator and denominator of (7.34) by ωL_{ad} yields

$$(i_a)_{\max} = -\frac{(1+\tau_f)V_0}{X_l + (1+\tau_a)X_f} \tag{7.36}$$

where $V_0 = \omega L_{ad} I_f$ = (internal) induced voltage, $X_f = \omega L_{ad}\tau_f$ = field leakage reactance and $X_l = \omega L_{ad}\tau_a$ = armature leakage reactance. The circuit corresponding to (7.36) is shown in Figure 7.26 and the input reactance of this circuit is the direct-axis transient reactance, x_d'.

Next, to include the effect of damper windings in the direct axis, we observe that the effect of the winding is indistinguishable in impact from that of the field winding, except for the current carried. Thus in the d-axis, we now have the armature, field, and damper windings all in parallel, as shown in Figure 7.27, where x_{Dd} = damper winding leakage reactance. From Figure 7.27 we have the subtransient reactance, x_d'', given by

$$x_d'' = x_l + x_{Dd}\frac{\tau_f}{\tau_{Dd} + \tau_f(1+\tau_{Dd})} \tag{7.37}$$

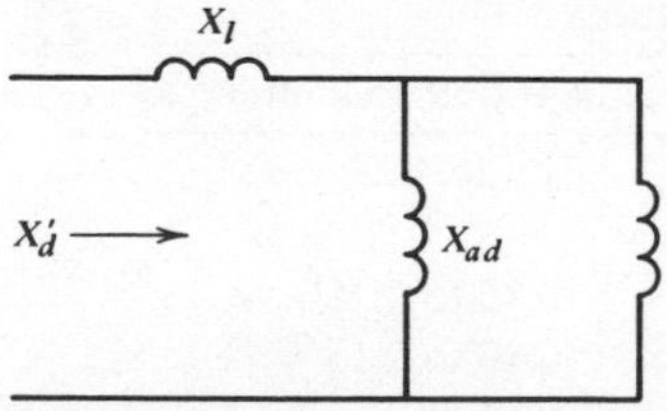

Figure 7.26. Equivalent circuit for the transient reactance.

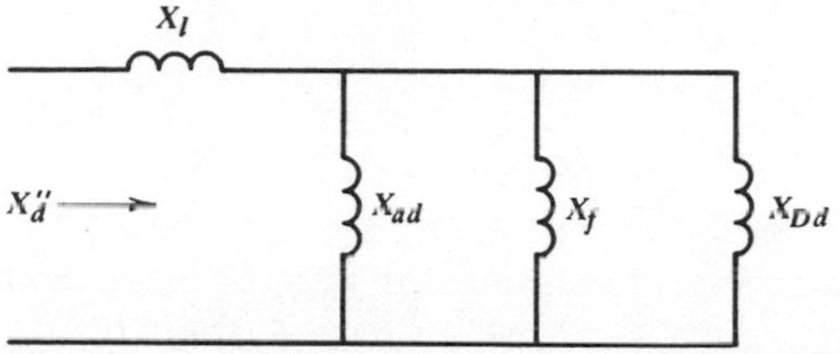

Figure 7.27. Equivalent circuit for the subtransient reactance.

where

$$\tau_{Dd} = \frac{x_{Dd}}{x_{ad}}.$$

Having defined x_d' to account for the presence of the field winding and x_d'' to denote the effect of the damper winding, we can now consider the presence of the various resistances in determining the short-circuit current in the armature. To obtain an explicit solution for the short-circuit current, including the effects of x_d' and x_d'', is a cumbersome exercise and beyond our present scope. However, we can arrive at some useful and important conclusions about the transient-current waveform by the following reasoning. We assume that a sudden short-circuit at the armature terminals occurs when the steady-state current passes through zero. In such a case the current waveform would be like that shown in Figure 7.28. Here, the rate of decrease of the consecutive peaks is determined by the time constants of the windings. The current decays over time because of the presence of resistances. First of all, the damper winding has the smallest time constant. Thus, the maximum current is V_0/x_d''. The effect of the damper lasts for only the first few cycles. Next, the maximum current is determined by x_d' and is V_0/x_d'. Finally, the steady-state current is limited by the synchronous reactance X_d.

Typical per unit values of various reactances of synchronous machines are given in Table 7.1.

Table 7.1 **Per Unit Synchronous Machine Reactances**

Reactance	Round-rotor machine	Salient pole machine
X_d	1.0 to 1.25	1.0 to 1.2
X_q	0.65 to 0.80	
x'_d	0.35 to 0.40	0.15 to 0.25
x''_d	0.20 to 0.30	0.10 to 0.15
x''_q	0.20 to 0.3	0.10 to 0.15

In Table 7.1, the per unit values are based on the machine rating.

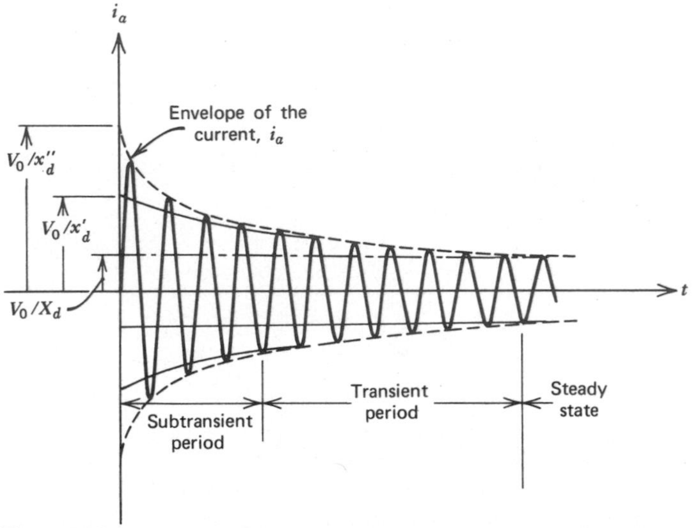

Figure 7.28. Armature current versus time, in a short-circuited generator.

7.5.2 MECHANICAL TRANSIENTS

The mechanical equation of motion of the synchronous machine is

$$J\ddot{\Theta}_m + b\Theta_m = T_e + T_m \tag{7.38}$$

where T_e = torque developed by the machine, T_m = externally applied torque, J = moment of inertia of the rotating system (including the load or prime mover), and b = friction coefficient, including electrical damping.

For the sake of illustration, let us consider a two-pole cylindrical-rotor machine, and assume that the frequency of mechanical oscillations is small, so that the steady-state power-angle characteristics can be used. Note that this analysis is only approximate. The per-phase power developed by the machine is given by

$$P_d = \frac{V_0 V_t}{X_s} \sin\delta \tag{7.39}$$

which also represents the electrical torque on a different scale. Let the changes in Θ_m, T_e, and T_m caused by a sudden load change be represented by $\Delta\Theta_m$, ΔT_e, and ΔT_m, respectively, so that (7.39) modifies to the following:

$$(Jp^2 + bp)\Delta\Theta_m = \Delta(T_e + T_m) \tag{7.40}$$

The change in the electrical torque is, from (7.39)

$$\Delta T_e = \Delta\left(\frac{V_0 V_t \sin\delta}{\omega_m X_s}\right) \tag{7.41}$$

where ω_m = mechanical velocity of the rotor and is the same as synchronous speed under steady-state conditions. In (7.41), for constant voltages, only sin δ changes for load changes. For small variations $\Delta(\sin\delta) \cong \Delta\delta$. Also $\Delta\Theta_m = \Delta\delta$; the number of poles are two. Therefore, (7.41) becomes

$$\Delta T_e = -K_e \Delta\delta \tag{7.42}$$

We see from Figure 7.24 that ΔT_e is negative for generator operation, so that in (7.42), $K_e = -(V_0 V_t/\omega_m X_s)$. From (7.40) and (7.42) we have

$$(Jp^2 + bp + K_e)\Delta\delta = \Delta T_m \tag{7.43}$$

which is a linear second-order differential equation in terms of the power angle δ. If we compare this with the second-order differential equation of a mechanical system, the natural frequency of oscillation, and the damping ratio are, respectively (see also Chapter 4),

$$\omega_n = \sqrt{\frac{K_c}{J}} \tag{7.44}$$

$$\zeta = \frac{b}{2\sqrt{K_e J}} \tag{7.45}$$

Example 7.5

A 30 hp, 220 V, three-phase, Y-connected, 60 Hz, 3600 r/min, cylindrical rotor machine, on no-load is brought up to the rated speed by an auxiliary motor and is then suddenly connected to a 220 V, three-phase source with the proper phase sequence. Study the mechanical transient from the following data:

synchronous reactance/phase = 2.0 Ω

excitation voltage V_0 of (7.39) = 150 V/phase

moment of inertia of rotating parts = 1.5 MKS units

damping torque b of (7.40) = 12 N-m/rad/s

Denoting $\Delta\delta$ by δ', we find that the equation of motion is, from (7.43)

$$J\frac{d^2\delta'}{dt^2} + b\frac{d\delta'}{dt} + K_e\delta' = 0 \tag{7.46}$$

In (7.46) K_e is known as the *synchronizing torque*. For motor operation, K_e (for the three-phase machine) is given by

$$K_e = \frac{V_0 V_t}{\omega_m X_s} \times 3 \tag{7.47}$$

For the given machine, $V_0 = 150$ V, $V_t = 220/\sqrt{3} = 127$ V, $X_s = 2.0$ Ω, and $\omega_m = 120\pi$ rad/s. Substituting these in (7.47), we obtain

$$K_e = \frac{150 \times 127 \times 3}{120\pi \times 2} = 756 \text{ N-m/rad}$$

Equation (7.46) therefore becomes

$$(1.5p^2 + 12p + 756)\delta' = 0 \tag{7.48}$$

From (7.44), (7.45) and (7.48) we have

$$\omega_n = \sqrt{\frac{756}{1.5}} = 22.5 \text{ rad/s}$$

$$\zeta = \frac{12}{2\sqrt{756 \times 1.5}} = 0.178$$

In terms of cycles per second, the natural frequency of oscillation is obtained from

$$2\pi f_n = \omega_n$$

or

$$f_n = \frac{\omega_n}{2\pi} = \frac{22.5}{2\pi} = 3.6 \text{ c/s}$$

In most machines,

$$0.2 < f_n < 2$$

$$\zeta \cong 0.2$$

Knowing ζ and ω_n, we can obtain the mechanical behavior from the equation

$$\frac{\delta'}{\delta'_{ss}} = 1 - \frac{1}{\sqrt{1-\zeta^2}} e^{-\zeta\omega_n t} \sin\left(\sqrt{1-\zeta^2}\, \omega_n t + \phi\right) \tag{7.49}$$

where

$$\phi = \tan^{-1} \frac{\sqrt{1-\zeta^2}}{\zeta} \tag{7.50}$$

and δ'_{ss} = steady-state power-angle. Note that (7.49) is the solution to (7.46) for $\zeta < 1$.

7.6 DETERMINATION OF MACHINE REACTANCES

The synchronous reactance X_s of a cylindrical-rotor machine can be obtained from the open-circuit and short-circuit tests on the machine. The open-circuit saturation curve and the steady-state armature current are shown in Figure 7.29 on a per-phase basis. For a 2-A field current, the short-circuit current is 25 A, whereas the open-circuit voltage is 57 V. Consequently, the synchronous impedance is $57/25 = 2.48\ \Omega$. Neglecting armature resistance, $Z_s \cong X_s = AC/BC = 2.48$. As shown in Figure 7.29 X_s varies with saturation.

For the salient pole machine, it is necessary that we know the values of both X_d and X_q. The physical significance of these reactances was discussed earlier, and these are the maximum and minimum values of the armature reactance, respectively, for different rotor positions. These reactances are determined by the *slip test*. In this test, the machine is excited by a three-phase source (for a three-phase machine) and driven mechanically at a speed slightly different from the synchronous speed. The field winding is unexcited and open-circuited. Oscillograms are taken of the armature current, armature voltage, and the induced field voltage. These are shown in Figure 7.30. The ratio of maximum to minimum armature current yields the ratio X_d/X_q. For instance, from the

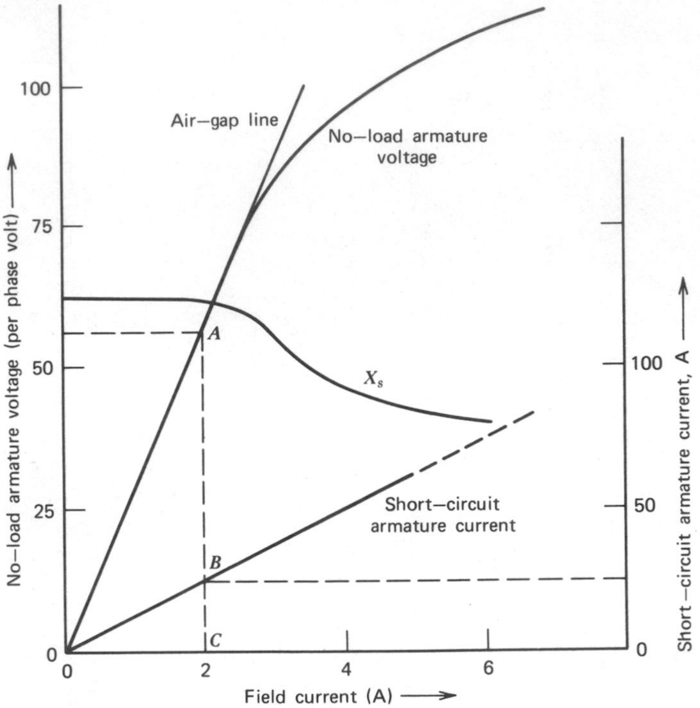

Figure 7.29. Test data for determining X_s.

diagram we find that $X_d/X_q = 1.6$. Knowing X_d from the open-circuit and short-circuit tests described above for the cylindrical-rotor machine, we can calculate X_q. There are other methods available for the determination of these reactances. The interested reader should consult References 2 and 3.

The transient and subtransient reactances, x_d' and x_d'', are determined by recording the three-phase currents when a sudden short circuit is applied to the machine running on no-load and at rated speed. Generally, the currents of the various phases are not symmetrical about the time axis. However, x_d' and x_d'' can be determined either (1) by eliminating the dc component, or (2) from an oscillogram such as the one shown in Figure 7.28.

7.7 SMALL SYNCHRONOUS MOTORS

The three-phase synchronous machines discussed so far are assumed to be large (of the order of several hundred kW, or even larger), because they have been considered in terms of possible applications in electric power systems. However, there are numerous applications which require synchronous motors of small (i.e., fractional horsepower) ratings. Most often, such motors are designed to operate

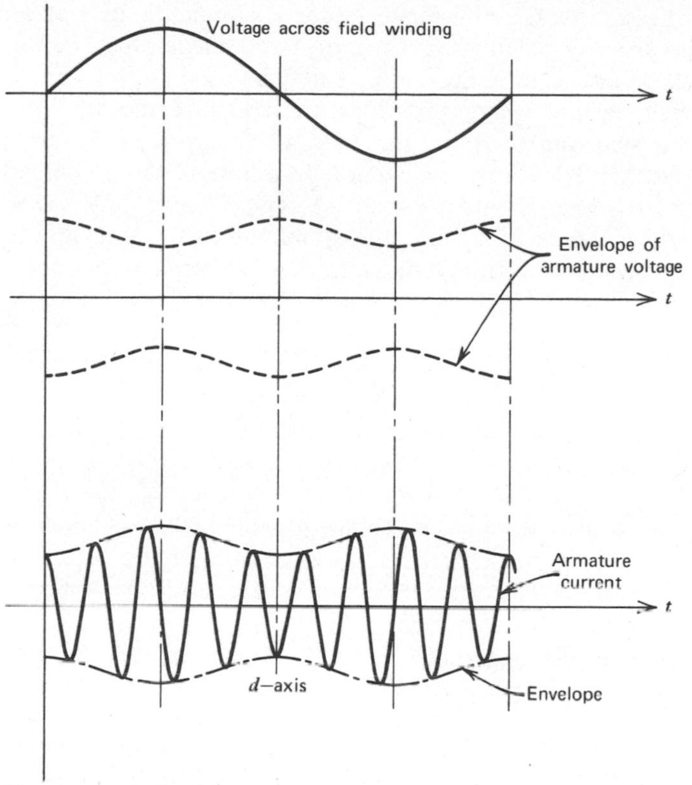

Figure 7.30. Oscillograms from slip test.

on a single-phase supply and do not require a dc excitation or the use of a permanent magnet. In this regard the fractional-horsepower synchronous motor is considerably different from its three-phase counterpart which has a relatively large rating. The two types of small synchronous motors are the *reluctance motor* and the *hysteresis motor*. These motors are used in clocks, timers, turntables, and so forth.

7.7.1 THE RELUCTANCE MOTOR

We are somewhat familiar with the reluctance motor, as may be recalled from Example 4.2. We know that the torque in a reluctance motor is similar to the torque arising from saliency in a salient pole synchronous motor. We may also recall that the time-average torque of a reluctance motor is nonzero only at one speed for a given frequency, and the power-angle characteristics of the motor are as discussed in Example 4.2.

A reluctance motor starts as an induction motor, but normally operates as a synchronous motor. The stator of a reluctance motor is similar to that of an induction motor (single-phase or polyphase). Thus, to start a single-phase motor, almost any of the methods discussed in Section 6.11 may be used. A three-phase reluctance motor, is self-starting when started as an induction motor. After starting, to pull it into step and then to run it as a synchronous motor, a three-phase motor should have low rotor resistance. In addition, the combined inertia of the rotor and the load should be small. A typical construction of a four-pole rotor is shown in Figure 7.31. Here, the aluminum in the slots and in spaces where teeth have been removed serves as the rotor of an induction motor for starting.

7.7.2 THE HYSTERESIS MOTOR

Like the reluctance motor, a hysteresis motor does not have a dc excitation. Unlike the reluctance motor, however, the hysteresis motor does not have a salient rotor. Instead the rotor of a hysteresis motor has a ring of special magnetic material, such as chrome, steel, or cobalt mounted on a cylinder of aluminum or some other nonmagnetic material, as shown in Figure 7.32. The stator of the motor is similar to that of an induction motor, and the hysteresis motor is started as an induction motor.

In order to understand the operation of the hysteresis motor, we may consider the hysteresis and eddy-current losses in the rotor. We observe that, as in an induction motor, the rotor has a certain equivalent resistance. The power dissipated in this resistance determines the electromagnetic torque developed by the motor, as discussed in Chapter 6. We may conclude that the electromagnetic torque developed by a hysteresis motor has two components—one by virtue of

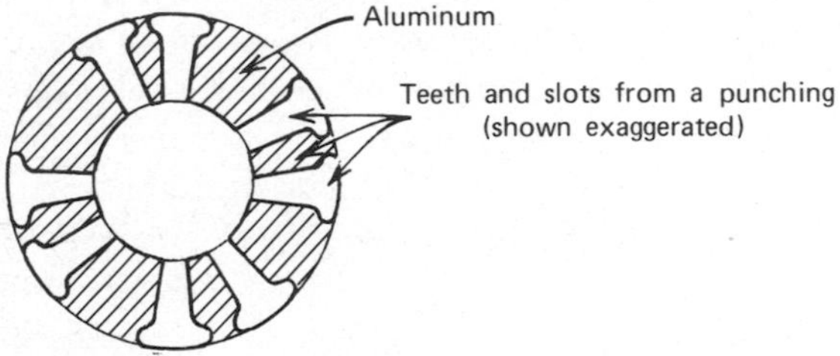

Figure 7.31. Rotor of a reluctance motor.

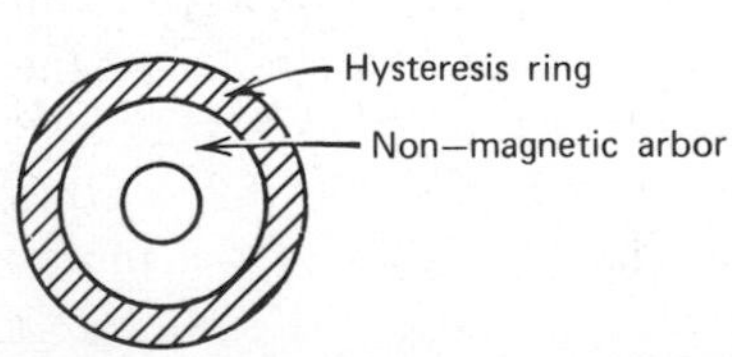

Figure 7.32. Rotor of a hysteresis motor.

the eddy-current loss and the other because of the hysteresis loss. We know that the eddy-current loss can be expressed as

$$p_e = K_e f_2^2 B^2 \tag{7.51}$$

where K_e = a constant; f_2 = frequency of the eddy currents; and B = flux density. In terms of the slip s, the rotor frequency f_2 is related to the stator frequency f_1 by

$$f_2 = s f_1 \tag{7.52}$$

Thus (7.51) and (7.52) yield

$$p_e = K_e s^2 f_1^2 B^2 \tag{7.53}$$

And the torque T_e is related to p_e by (see Chapter 6)

$$T_e = \frac{p_e}{s\omega_s} \tag{7.54}$$

so that (7.53) and (7.54) give

$$T_e = K' s \tag{7.55}$$

where $K' = K_e f_1^2 B^2 / \omega_s$ = a constant.

Next, for the hysteresis loss, p_h, we have

$$p_h = K_h f_2 B^{1.6} = K_h s f_1 B^{1.6} \tag{7.56}$$

and for the corresponding torque, T_h, we obtain

$$T_h = K'' \tag{7.57}$$

where $K'' = K_h f_1 B^{1.6} / \omega_s$ = a constant.

Notice that the component, T_e, as given by (7.55), is proportional to the slip and decreases as the rotor picks up speed. It is eventually zero at synchronous speed. The component of the torque aids in the starting of the motor. The second component, T_h, as given by (7.57) remains constant at all rotor speeds and is the only torque when the rotor achieves the synchronous speed. The physical basis of this torque is the hysteresis phenomenon, which causes a lag of the magnetic axis of the rotor behind that of the stator. In Figures 7.33*a* and *b*, respectively, the absence and the presence of hysteresis are shown measured by the shift of the rotor magnetic axis. The angle of lag δ, shown in Figure 7.33*b*, causes the torque arising from hysteresis. As mentioned above, this torque is independent of the rotor speed (shown in Figure 7.33*c*) until the breakdown torque.

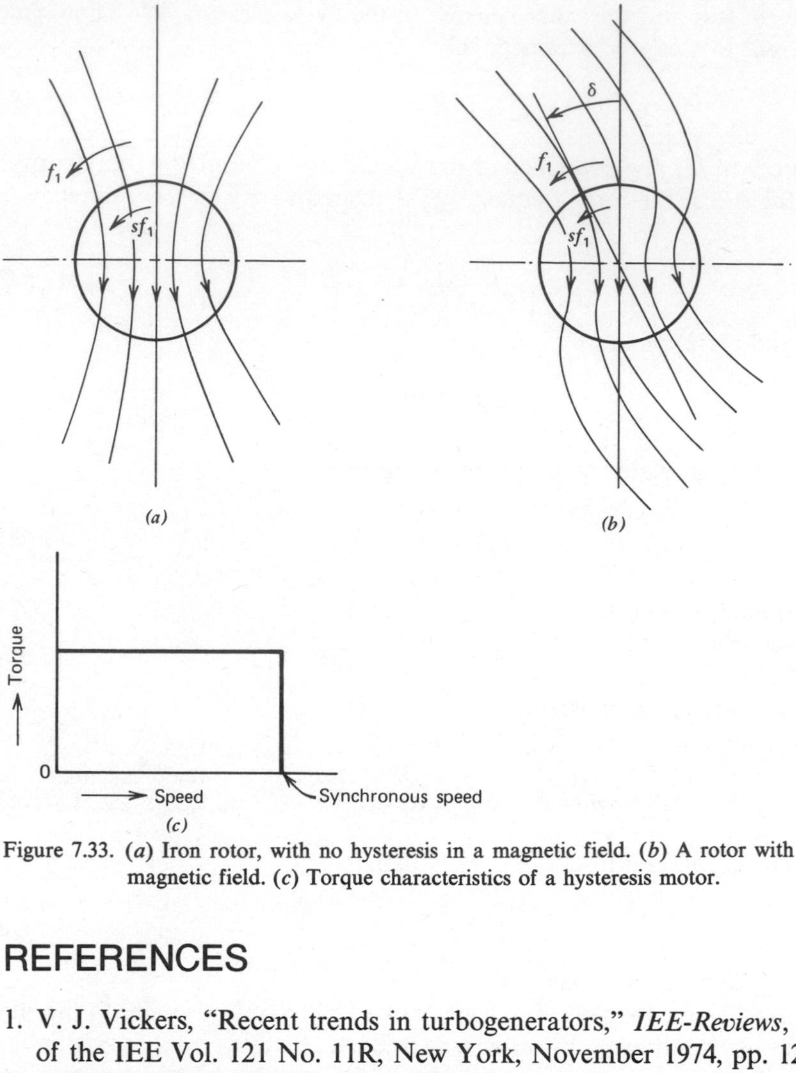

Figure 7.33. (*a*) Iron rotor, with no hysteresis in a magnetic field. (*b*) A rotor with hysteresis in a magnetic field. (*c*) Torque characteristics of a hysteresis motor.

REFERENCES

1. V. J. Vickers, "Recent trends in turbogenerators," *IEE-Reviews*, Proceedings of the IEE Vol. 121 No. 11R, New York, November 1974, pp. 1273–1306.
2. A. S. Langsdorf, *Theory of Alternating Current Machinery*, McGraw-Hill Book Company, New York, 1955.
3. M. Liwschitz-Garik and C. C. Whipple, *Alternating Current Machines*, Reinhold Company, New York, Van Nostrand, 1961.

PROBLEMS

7.1. A 60-Hz synchronous generator feeds an eight-pole induction motor running with a slip of 2%. What is the speed of the motor? At what speed must the generator run if it has (a) two poles and (b) six poles?

7.2. The flux density distribution produced by the field mmf of a two pole salient rotor synchronous machine is sinusoidal, having an amplitude of 0.75 tesla. If the rotor runs at 3600 rpm, calculate the frequency and amplitude of the voltage induced in a 150-turn coil on the armature. The axial length of the armature is 12 cm and its inner diameter is 10 cm.

7.3. Let the field winding of a two-pole synchronous machine be excited by an ac source such that the air-gap flux density distribution is $B(\theta,t)=B_m \cos\omega_1 t \cos\theta$. The armature has a bore $2r$ and length l. Obtain an expression for the voltage induced in an N-turn coil on the armature if the rotor (or field) rotates at ω_2 rad/s. Study the special case when $\omega_1=\omega_2=\omega$.

7.4. A 30 kVA 230 V Y-connected round-rotor synchronous motor operates at full-load at a leading power factor of 0.707. A three-phase Y-connected inductive load having an impedance of $(4+j3)$ ohm per phase is connected in parallel with the motor. Calculate (a) the overall power factor of the motor and the inductive load, (b) active and reactive power for (i) the motor and (ii) for the load, and (c) line current for the motor and inductive load combination.

7.5. In the text we have obtained an expression for the power-angle characteristics of round-rotor synchronous machine, neglecting the armature resistance. If this resistance has a value r_a, derive the modified expression for the power-angle characteristics of the machine.

7.6. Draw the phasor diagram of a salient pole synchronous generator supplying a load having a leading power factor. From this diagram obtain an expression for the power angle δ in terms of the machine constants, X_d and X_q, the armature current I_a and the load power-factor angle ϕ. Neglect armature resistance.

7.7. A salient pole synchronous generator is Y-connected and operates at 220 V (line-to-line), and at a power angle of 30°. The machine constants per phase are $X_d=5$ ohm, $X_q=3$ ohm, and $r_a=$negligible. If the generator develops a total of 16 kW power, calculate the voltage regulation for the given operating conditions.

7.8. A 20 kVA, 220 V, 60 cycle, Y-connected, three-phase salient pole synchronous generator supplies rated load at 0.707 leading power factor. The per-phase constants of the machine are armature resistance $R^s=0.05\ \Omega$; direct-axis reactance $X_d^s=4.0\ \Omega$; quadrature-axis reactance $X_q^s=2.0\ \Omega$. Calculate the developed power and percent voltage regulation at the specified load.

7.9 A synchronous motor is delivering 0.5 N-m torque at 1800 rpm. The load torque is suddenly reduced to zero and the torque angle is observed to oscillate initially over a 12° range with a period of 0.3 s. After 7.5 s, the oscillations have decreased to 4.42°. Calculate the synchronizing torque, J, and b for the zero-load condition.

Chapter 8
Electronic Control Of Electric Motors

This chapter is primarily concerned with the theory and applications of electronic devices and circuits in the control of electric motors.

The characteristics of a number of types of electromechanical devices have been described in the preceding chapters but none of these characteristics is realizable without the use of additional components and circuitry that can be lumped under the heading, *machine controls*. Such equipment may be as simple as an on/off switch or as complex as the assembly of semiconductors, circuit components, and mechanical devices that control the one-second reversal from full speed of a 10,000 horsepower rolling mill motor used in the fabrication of steel. Controls are applied to both ports of rotating machines, that is, to the electrical terminals and to the mechanical shaft, often with a connecting link —such as a feedback circuit—between the two ports.

There is a great diversity and variety of components and systems used to control rotating machines. In terms of practical applications and theoretical analysis, no other aspect of rotating machines has undergone such dramatic changes in recent years or holds more potential for improving machine characteristics in the future than does the electronic control of electric machines.

Many mechanical and fluid devices are used in motor control, such as mechanical clutches and brakes, hydraulic couplings, fluid control systems, and so forth. There are also many electromechanical devices used in motor control—relays, contactors, starters, solenoids, magnetic amplifiers, amplidynes, and so forth. These devices are still the most common type of control found in many household and automotive motor applications. Since the mechanical, hydraulic, and electromechanical control systems are well-described in the literature and have been in common usage from close to the beginning of motor applications, these devices will be described or discussed only as needed when they play an integral role in an otherwise electronic control system.

8.1 SOME GENERAL CONSIDERATIONS OF MOTOR CONTROL

The purpose of a motor control system is to control one or more of the motor output parameters, that is, shaft speed, angular position, acceleration, shaft torque, and mechanical output power. The control of temperatures at various points in the motor is also a frequent objective of motor control systems. Since it is the mechanical parameters of the motor that are being controlled by the input electrical parameters, the peculiar characteristic of the individual machine—that is, the characteristics that relate input electrical quantities to output mechanical quantities—are of vital importance in the design and analysis of electronic control. It is customary to treat the control of dc commutator motors separately from that of synchronous and induction motors for this reason. The nature of the load and power supply will also influence the nature of the control. Figure 8.1 illustrates the basic arrangement of electronic motor control. This figure illustrates a total motor system including load and power source. The feedback loops are shown by dotted lines, since many motor control schemes are "open-loop" schemes. The load box represents a very general concept of load and may be a pure inertia load. The power source box is also generalized and is meant to cover *all* power or excitation sources required by the motor, such as the field and armature power supplies in a dc commutator motor. Figure 8.1 has no analytical value and should not be confused with the block diagram describing control signal flow such as those in control theory. Here we simply illustrate the

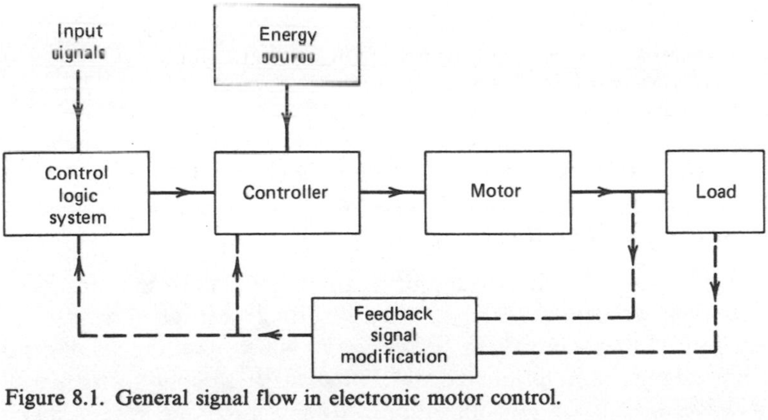

Figure 8.1. General signal flow in electronic motor control.

general topological layout of the principal elements of a general motor control system. The primary concern of this chapter will be with the box labeled "controller."

The materials and structural features of power semiconductors used in motor control are much different from those of electromechanical devices. As would be expected, the operating and environmental characteristics are also much different. It is essential to recognize these differences in the design and use of electronic motor control in which the two different types of components are used in a common environment and are subjected to the same voltage and current values. The principal differences may be summarized as follows.

1. The thermal characteristics of semiconductor devices are much different from those of electromechanical devices such as motors. Motors have large thermal capacities and can sustain thermal overloads for time periods measured in minutes. Semiconductors have very short thermal time constants, lasting often less than a second. They also have poor natural thermal conduction paths and require heat sinks in most applications.
2. The current overload characteristics of the two types of devices are quite different, partly as a result of the thermal differences noted above. Semiconductor devices have relatively little overcurrent capability. Thyristors have an I^2t or surge current rating, which is a nonrecurrent rating. Device characteristics are usually altered for short time periods following such a surge. Too frequent surges, even within the surge rating, will result in a degradation of the device characteristics. Motors can generally sustain overcurrents as long as thermal ratings are not exceeded and can be designed for periodic high surge current operations, such as during "plugging"—a means of rapidly braking a machine.
3. Many semiconductors are limited by the rate of change of current, as will be explained for thyristors below. There is no equivalent limitation in a motor.
4. Thyristors are also limited by a rate-of-change of voltage characteristic, which is not a factor in motor operation.
5. Semiconductor circuits are very susceptible to electromagnetic interference, both conductive and inductive, whereas motors are in no way similarly affected.
6. Semiconductors are much more sensitive to shock and vibration than motors. However, with proper mounting and packaging, semiconductors can be applied in most levels of vibration.
7. Motors have traditionally been operated in manners that result in continuous current and voltage waveforms, such as steady direct current or sinusoidal alternating current. When controlled by semiconductor systems, waveforms are usually less regular and often discontinuous with steeply rising wavefronts. Such waveforms have various implications concerning motor losses

and excitation characteristics and, in some cases, suggest modification of the motor design.

8. Electromagnetic motors are inherently inductive devices and large magnitudes of inductive energy storage are associated with their normal operation. The management of inductive energy during the rapid switching action of semiconductor devices, to prevent voltage spikes and excessive dv/dt that may damage the devices, is a principal design problem in electronic motor control.

These inherent differences between semiconductor control devices and electric motors are significant in almost every aspect of the design and operation of motor electronic control.

8.2 VOLTAGE AND CURRENT WAVEFORMS

One of the chief characteristics of motor control systems that use semiconductor devices is the nature of voltage and current waveforms. These are generally irregular, and the standard formulas based on continuous sinusoidal or steady dc waveforms are inadequate. The Fourier series method is often used in the analysis of electromagnetic devices having nonlinearities caused by the characteristics of the magnetic material. The nonlinear exciting currents, voltages, or magnetic flux of transformers and induction machines, which often require Fourier series representation, have been discussed in previous chapters. However, when a motor or transformer is excited through an electronic control system, the *applied* waveforms are nonsinusoidal and often discontinuous. Further distortion or modification of the waveform will result from the nonlinear characteristics of various elements of the motor.

Another interesting characteristic of electronic control is that the waveforms may change appreciably as a function of the level of motor torque or motor speed. As a result, both load and exciting components of voltage and current in motor electronic control systems frequently have nonsinusoidal waveforms. This characteristic results in many implications for those involved with the design, analysis, or operation of motor electronic control systems as follows.

1. The measurement of voltages and currents must be performed with instruments capable of accurately indicating the type of waveforms being measured. Thermocouple instruments are adequate for measuring power components in most electronic motor control systems. Oscilloscopes are usually essential for waveform analysis of both the power and control signal parameters.

2. Loss measurements should be performed, if possible, with the motor excited as it is to be used, rather than with standard sinusoidal or dc excitation. Magnetic material loss data, such as that discussed in Chapter 2, is obtained with sinusoidal excitation and is often incorrect for other types of excitation.

The measurement of core losses is difficult when the waveforms are like those described above. In that case, special wattmeters, such as electronic-multiplier, Hall-effect, or thermal-type instruments, should be used.

3. Standard circuit theory based upon single frequency sine wave parameters is inadequate in the analysis of electronic motor control circuits. It is frequently necessary to evaluate the instantaneous time variation of both power and control signal currents and voltages. Fourier methods are also useful, as noted above.
4. The standard numerical values for the relationships between average, rms, and maximum values of current and voltages are seldom applicable.
5. The range of frequencies of the voltage and current components in an electronic motor control system is always much greater than the fundamental frequency applied to the motor. This is readily apparent if one considers the Fourier components of a nonsinusoidal periodic function. The fundamental frequency results from the switching action of power semiconductors in the control system and is usually in the power or low audio range of frequencies, seldom more than 3000 Hz. The range of frequencies in various currents and voltages may easily be 100,000 Hz or higher, however. This fact must be recognized when the choice of instrumentation used in the laboratory, is made, when considering audible and electromagnetic noise interference that results from the control system, when designing filters, and when protecting the control logic circuitry used to switch the power devices.

The calculation of average and rms values of voltage and current is quite important in electronic control systems for calculating motor power and torque, for heating of wires and other components, and for sizing components and instrumentation. To make these calculations, it is often necessary to return to the definitions of average and rms values that are given in Section 1.3.

Table 8.1 catalogues several of the common waveforms characteristic of electronic control systems and shows their average and rms values. In all of the waveforms shown in Table 8.1, T_0 is the time duration or "on-time" of the pulse, and T_p is the length of a full period of the signal. The average rms formulas are also applicable if $T_0 = T_p$, that is, if there is no "off-time" in the signal. The fundamental frequency of the signal referred to above is

$$f_p = \frac{1}{T_p} \tag{8.1}$$

Example 8.1
Calculate the rms and time-average values of the waveform shown in Table 8.1, item 4.

It is first necessary to derive an equation for the instantaneous variation of the parameter during one time period. For waveform item 4, this is a simple

TABLE 8.1 **Motor Controller Waveforms**

		RMS	Average
1.	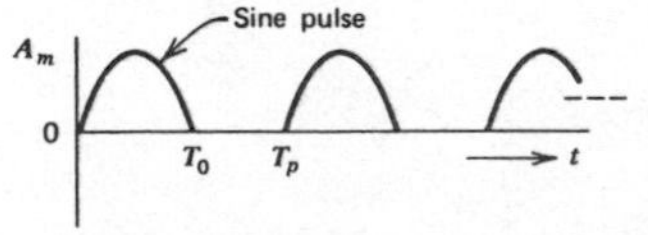	$A_m\sqrt{\frac{T_0}{2T_p}}$	$\frac{2}{\pi}\cdot\frac{T_0}{T_p}\cdot A_m$
2.	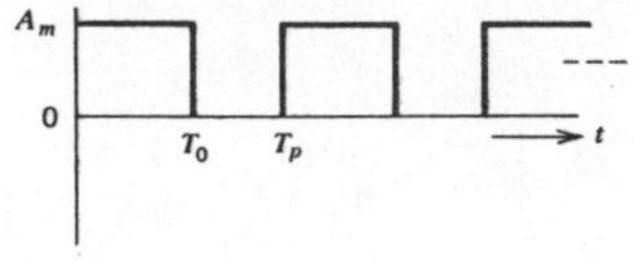	$A_m\sqrt{\frac{T_0}{T_p}}$	$\frac{T_0}{T_p}\cdot A_m$
3.	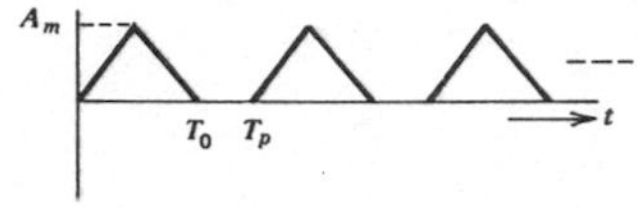	$A_m\sqrt{\frac{T_0}{3T_p}}$	$\frac{1}{2}\cdot\frac{T_0}{T_p}\cdot A_m$
4.	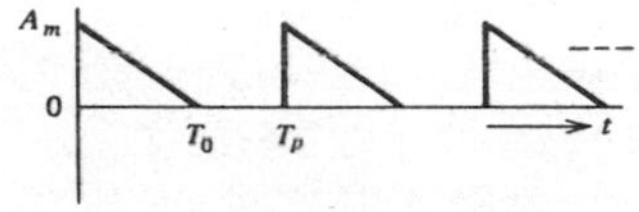	Same as No. 3	Same as No. 3
5.	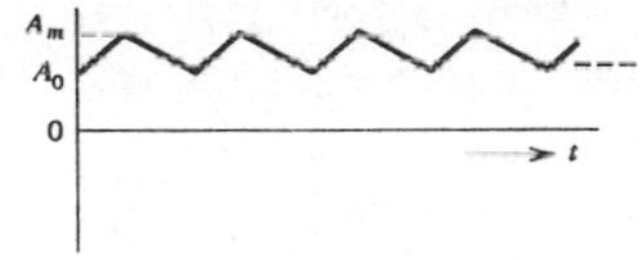	$A_m\sqrt{\frac{1}{3}(1+\zeta+\zeta^2)}$ $(\zeta=\frac{A_0}{A_m})$	$\frac{1}{2}A_m(1+\zeta)$ $(\zeta=\frac{A_0}{A_m})$
6.	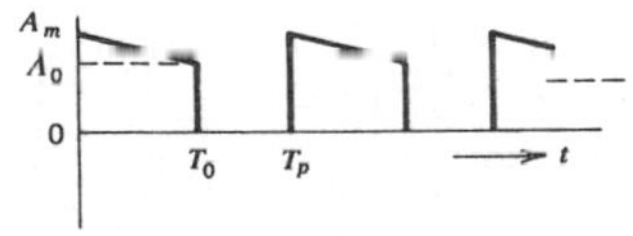	$A_m\sqrt{\frac{T_0}{T_p}\cdot\frac{1}{3}(1+\zeta+\zeta^2)}$ $(\zeta=\frac{A_0}{A_m})$	$\frac{1}{2}\cdot\frac{T_0}{T_p}A_m(1+\zeta)$ $(\zeta=\frac{A_0}{A_m})$
7.	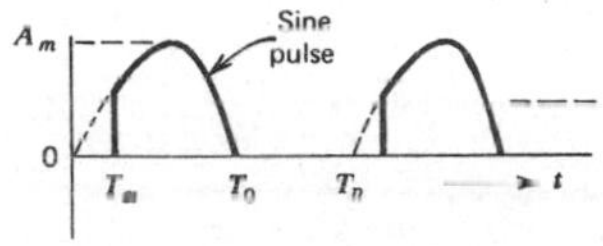	$A_m\sqrt{\frac{1}{T_p}[\frac{T_0-T_\sigma}{2}+\frac{T_0\sin\left(2\pi\frac{T_\sigma}{T_0}\right)}{4\pi}]}$	$\frac{T_0}{\pi T_p}\cdot A_m\cdot(1+\cos\frac{\pi T_\sigma}{T_0})$

Note: The symbol, A_m, represents a maximum value of a current or voltage.

straight line equation,

$$a=A_m-\frac{A_m}{T_0}t$$

The rms value is found by taking the square root of the mean (average) of the square of the function. The mean of the square over one full period is

$$\frac{1}{T_p}\int_0^{T_p}a^2dt=\frac{1}{T_p}\int_0^{T_0}\left[A_m^2-2\frac{A_m^2}{T_0}t+\left(\frac{A_m}{T_0}\right)^2t^2\right]dt=\frac{T_0}{T_p}\frac{A_m^2}{3}$$

The square root of this term is the rms value and equals

$$A = A_m \sqrt{\frac{1}{3}\frac{T_0}{T_p}}$$

The time-average (not the half-wave average) is found from (1.9):

$$\frac{1}{T_p}\int_0^{T_p} a\,dt = \frac{1}{T_p}\int_0^{T_0} a\,dt = \frac{1}{T_p}\int_0^{T_0}\left(A_m - \frac{A_m t}{T_0}\right)dt$$

$$A_{\text{av}} = A_m\left(\frac{1}{2}\frac{T_0}{T_p}\right)$$

Note that the same expression for both rms and time-average values would result for the saw-tooth pulse with positive slope.

The waveforms of the type shown in Table 8-1 will be referred to as "chopped" waves. In electronic motor control circuits, portions of the circuit are energized for a short time interval and then disconnected by a solid-state switching device for a short interval. This process is then repeated. In many circuits, the on-time pulse is of a regular shape or can be closely approximated by a regular shape. This is true of the first four examples of Table 8.1. In such cases, a generalized expression for rms and average values is available:

$$A = A_m \sqrt{K_f \frac{T_0}{T_p}} \tag{8.2}$$

$$A_{\text{av}} = A_m K_A \frac{T_0}{T_p} \tag{8.3}$$

In these expressions, K_f and K_A are factors that depend only upon pulse shape. For example, in a sine pulse, $K_f = 1/2$, $K_A = 2/\pi$; in a triangular pulse, $K_f = 1/3$, $K_A = 1/2$; in a square pulse, $K_f = K_A = 1$.

Example 8.2

Find the time-average, half-wave average, and rms values of the waveform of item 1 in Table 8.1 if each alternate sine pulse is negative.

From (8.2), the rms value is

$$A = A_m \sqrt{\frac{1}{2}\cdot\frac{T_0}{T_p}}$$

The half-wave average, from (1.8), is

$$A_{HWA} = \frac{1}{T_p}\int_0^{T_p} \sin\omega t\, dt = \frac{1}{T_p}\int_0^{T_0} \sin 2\pi\left(\frac{1}{2T_0}\right) t\, dt$$

$$= \frac{1}{T_p}\cdot\frac{T_0}{\pi}\left[-\cos\frac{\pi}{T_0}t\right]_0^{T_0} = \frac{ct_0}{T_p}\cdot\frac{2}{\pi}$$

The time-average *over a full period* is zero,

$$A_{\text{av}} = 0$$

The differences between the average and half-wave average are brought out in these examples, since both are of value in the analysis of chopped and non-sinusoidal waveforms. In the waves with pulses of only one polarity, as shown in Table 8.1, the half-wave average has no real meaning. The factor, K_A, in (8.3) is actually the half-wave average of the pulse itself.

8.2.1 RESET OF INDUCTIVE COMPONENTS AND CAPACITORS

The unidirectional waveforms of the type shown in Table 8.1 are found in many portions of motor control circuits as well as in switching regulators and similar systems. Such waveforms can often lead to unwanted operation of inductors, transformers, and capacitors, since the polarity of these devices is not reversed naturally by an alternating waveform. Inductive devices with ferromagnetic cores will be operating on a minor hysteresis loop. For square-loop cores, which are commonly used in switching circuits, this means that the signal causes the core to be traversed across the flat portion of its hysteresis loop and the device is essentially "noninductive" Likewise, a capacitor would be fully charged after a few pulses, which would greatly distort the circuit voltages for normal operation.

Normal performance of inductive components under unidirectional waveforms often requires an auxiliary winding coupled to the core. The function of this winding is to "reset" the core to its opposite magnetic state during the off-time (T_0 to T_p in Table 8.1) of the main signal. Capacitor voltages are often required to be reversed during this off-time by auxiliary ringing circuits. Several problems at the end of this chapter explore the application of these useful circuits.

8.3 POWER SEMICONDUCTOR DEVICES

Many types of semiconductors are now available for electronic motor-control applications. The type to be used in a specific application will depend upon a number of factors, primarily the power, voltage, and current requirements of the motor to be controlled. Other factors include environmental considerations, particularly the ambient temperature of this semiconductor environment; the control modes or control philosophy to be used; and overall system cost considerations. In this section, we shall summarize the principal characteristics of the more important *power* semiconductors used in motor control.

The power semiconductors and their associated circuitry are often referred to as the power circuit of a motor control system. The box labeled "controller" in Figure 8.1 contains the power circuit of the motor control system. There is another grouping of circuits and components required in a complete motor control system that operates at very low levels of voltage and current compared to the power circuit. This group is often referred to as the "control logic," "gate-control circuitry," or some similar designation. Its function is to implement the control philosophy in order to cause the motor to operate in a prescribed manner, which includes circuitry for turning the power semiconductors on and off.

Table 8.2 lists the principal devices used in motor control systems. Also shown in Table 8.2 are standard symbols for these devices in circuit diagrams and the "state-of-the-art" maximum voltage, current, and time-response capabilities of each class of device. These capabilities are seldom achievable concurrently in a single device. Usually only one of these maximum capabilities can be realized in one device. This fact and the various trade-offs available in the design of semiconductor devices for specific applications will be discussed below. The time or speed parameter in Table 8.2 generally refers to the minimum turnoff time achievable in each device class. In the column showing device symbols in Table 8.2, "A" refers to anode; "*K*", cathode; "G", gate; "E", emitter; "C", collector; and "B", base. The Zener diode is not a true power-controlling device in the same sense as the other devices listed, but it is included because it is widely used as a voltage control and sensing device in many motor controllers.

The devices listed in Table 8.2 may generally be categorized according to structure: two-layer devices, of which the silicon rectifier is the best known; three-layer devices, typified by the power transistor; and four-layer devices, "whose bistable action depends upon *p-n-p-n* regenerative feedback,"[1] or thyristor. The term *thyristor* is used to describe all four-layer devices, including the silicon controlled switch (SCS), the Triac, the Schockley diode, gate turnoff devices (GTO), and the silicon controlled rectifier (SCR). The latter device has found by far the greatest application in electronic motor control, mainly because until recently it has been the only controlled semiconductor device that could be built to handle the large voltage and current levels required in motor control. However, power transistors and Darlington devices presently can be built in voltage and current ratings capable of handling many motor control applications, as can be seen from Table 8.2. It is expected that the capabilities of these devices will be increased further in the future.

8.3.1 SILICON RECTIFIER

Silicon rectifiers were among the first power semiconductor devices to be developed and have, along with germanium rectifiers, almost totally supplanted

Table 8.2 **Power Semiconductors**

	Device	Symbol	Maximum Capabilities		
			Volts	Rms amp.	Speed (μs)
	Silicon rectifier		5000	7500	
SCR	Silicon-controlled rectifier	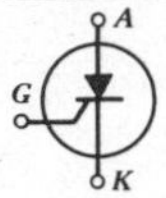	5000	3000	1
TRIAC	Bidirectional switch	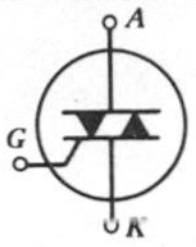	1000	2000	1
GTO GCS	Gate Turn-off SCR Gate controlled switch	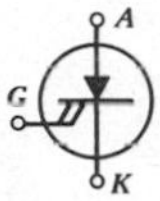	400 1000	 200	0.2 2
	Power transistor	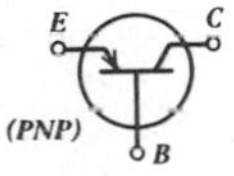	3000	500	0.2
	Darlington	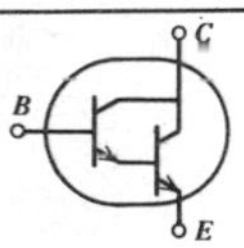	1000	200	1
	Zener diode		500		100

vacuum and gas tube power rectifiers. Figure 8.2 shows two silicon rectifier configurations in common service today. One is a stud mounted rectifier, so named because of the threaded stud on the bottom of the device, which, electrically, is the cathode and is mounted on a plate or heat sink. The other device in Figure 8.2 is known variously as a "press-pak," "hockey-puck," or "disk" configuration. The two electrical contacts are on the top and bottom of the disk, and it must be pressure-mounted between two flat conducting surfaces to achieve good electrical contact. The principal advantage of the disk configuration is an improved heat transfer path, with low thermal impedance running through both electrical terminals. In the stud-mount design low thermal impedance runs through only one terminal.

The principal parameters of a silicon rectifier are the repetitive peak reverse voltage (PRV) or blocking voltage, average forward current, and maximum

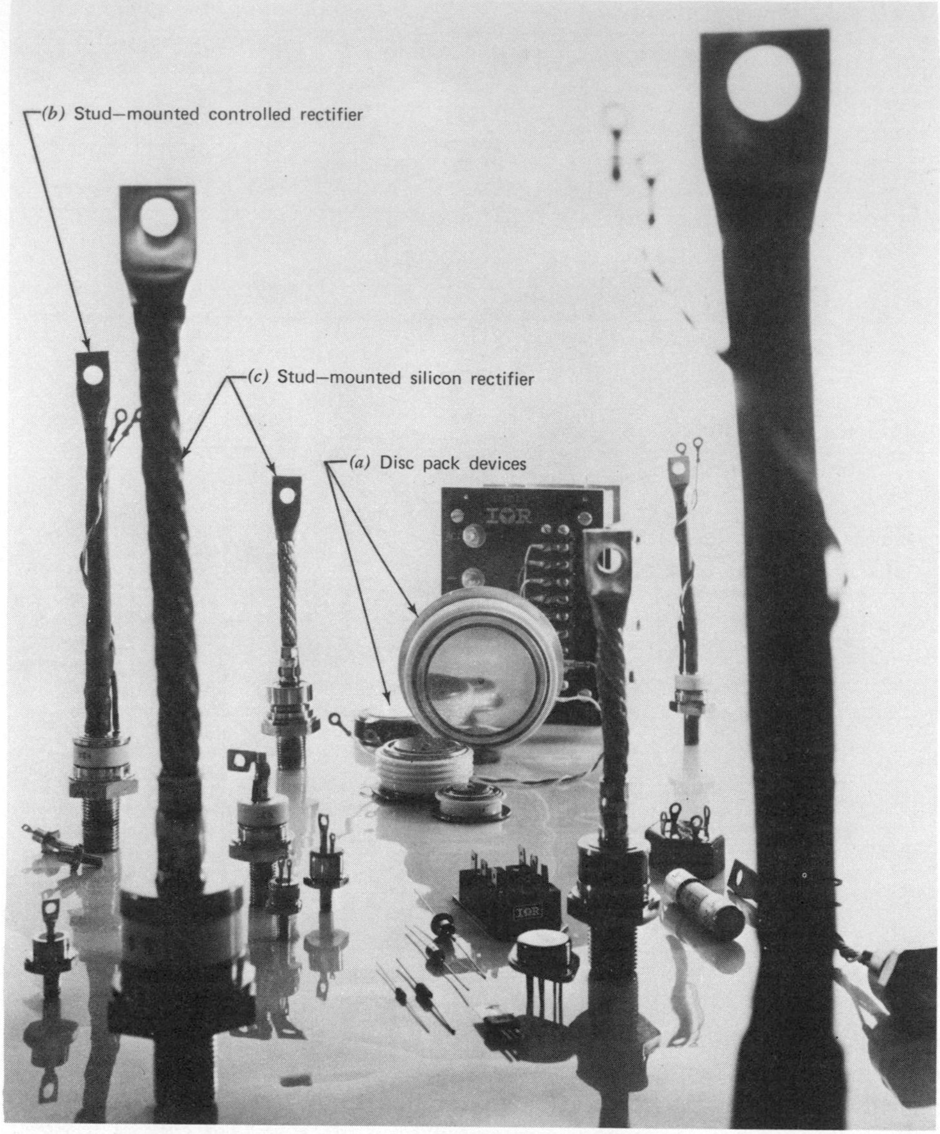

Figure 8.2. Silicon rectifiers and silicon controlled rectifiers (SCR's) illustrating (*a*) disc (hockey puck) devices; (*b*) stud-mounted controlled rectifier; (*c*) stud-mounted (silicon rectifier). (Courtesy International Rectifier Corporation).

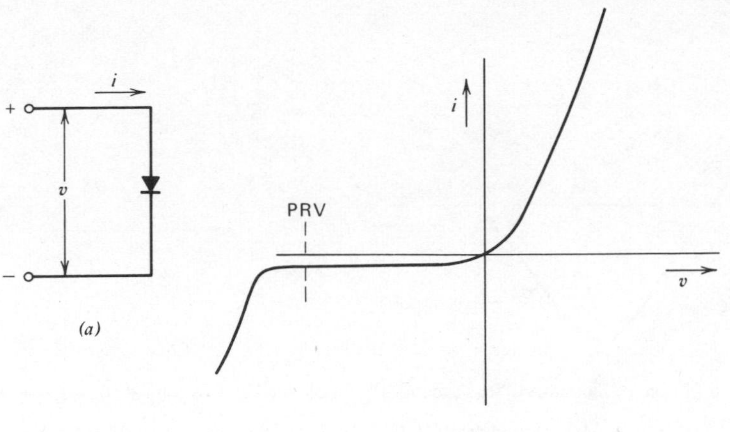

Figure 8.3. Silicon rectifier characteristics. (*a*) Symbols. (*b*) vi-characteristics.

operating junction temperature. The latter is 125°C for most silicon devices. The v-i characteristic is shown in Figure 8.3. In a forward conducting mode, the silicon rectifier is not a perfect conductor but has a forward voltage drop of approximately one volt at all current levels within its rating. There are also surge current and overvoltage ratings associated with a rectifier that may be significant in a given application.

Another important characteristic in certain applications is the rectifier recovery time, that is, the time following the end of forward conduction before the rectifier assumes its full reverse blocking capability. This characteristic is known as the "reverse recovery" performance of a diode and is related to the time required to "sweep out" minority charge carriers stored in the diode following the end of conduction. This time period is in the range of a few microseconds and requires current flow in the blocking direction, that is, opposite to normal diode current flow. This phenomenon can be the source of electromagnetic noise or ringing in a motor control system and must be considered in the design of these systems. The recovery characteristics determine the rate at which blocking voltage can be reapplied to the rectifier and, hence, the frequency of operation of the rectifier. Many motor control applications of rectifiers require very short recovery times and for this purpose fast-recovery devices have been developed in which the recovery time can be only several hundred nanoseconds. A comparison of the recovery characteristics of conventional and fast-recovery silicon rectifiers is shown in Figure 8.4.

One common function supplied by silicon rectifiers in motor control systems is that of the "free wheeling" diode. This function provides a path for continuation of motor current following the switching-off of another power device between the motor terminals and the energy source. The free wheeling diode will appear in many of the control systems discussed later in this chapter.

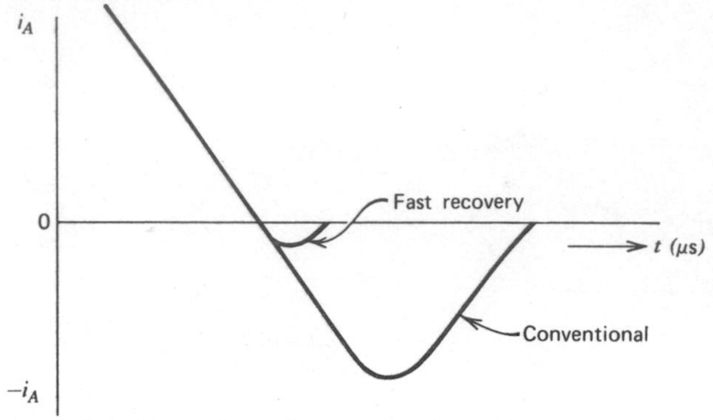

Figure 8.4. Comparison of conventional and "fast recovery" silicon rectifiers in the reverse-recovery region.

8.3.2 SILICON CONTROLLED RECTIFIER

This device is also frequently termed a *thyristor*, although we prefer to use thyristor as the name of the general family of four-layer devices. The silicon controlled rectifier, or SCR, has three terminals: anode, cathode, and gate. In the reverse or blocking direction, the SCR functions very much like the silicon rectifier described in the previous section. In the forward direction, conduction can be controlled to a limited extent by the action of the gate, which is connected to low-signal circuitry that can be electrically isolated from the power circuitry connected to the anode and cathode. Control is limited, since the gate can turn on the device, that is, initiate the conditions for forward current flow but cannot stop current flow. Control by the gate is only momentary, during which carriers are injected into the lower *p*-base material to cause breakdown of the *p-n* junction voltages. This process is called avalanche breakdown. When current flow begins through the anode, the SCR latches on and remains in the conductive state until "turned off" by external means. The turn-off process is known as commutation of the SCR and will be discussed in detail in Section 8.4. Figure 8.2 shows the two principal physical configurations of SCRs, the stud-mount and disk configurations. Except for the gate lead, both configurations are similar in appearance to silicon rectifier configurations. Figure 8.5 illustrates a typical disc-type SCR "package," including heat sink and fusing.

Figure 8.6 is a copy of a commercial power SCR data sheet. The principal parameters (and magnitudes for this particular device, No. C419PN) for applying SCRs to motor control systems are

1. Repetitive peak reverse voltage (PRV); (1800 V).
2. Maximum value of average on-state current; this parameter is related to the heating within the semiconductor; ($2/\pi \times 850 = 540$ A).

3. Maximum value of rms on-state current; this is the current rating of metal conductor portions of the device, such as the anode pigtail in stud devices; (850 A).
4. Peak one-cycle on-state current; this is the surge current limit (6500 A).
5. Critical rate-of-rise of forward blocking voltage; there are usually two ratings: initial (when the device is first turned on) and reapplied (following commutation); (200 V/μs).
6. Turnoff time; the off-time required following commutation before forward voltage can be reapplied (see Section 8.4); (40 μs).
7. Maximum rate of rise of anode current during turnon; too high di/dt may result in local hot-spot heating, a main cause of device failure; (300 A/μs).
8. Maximum operating junction temperature; (125°C).

There are additional parameters of importance in applying SCRs as witnessed by the data sheet of Figure 8.6. Thermal management of the SCR is extremely critical in all applications. Most of the parameters just mentioned vary considerably as a function of device temperature, which also characterises other types of semiconductors. Therefore, much of the engineering required in the application of an SCR is in the design of its heat sink, mounting method, and auxiliary cooling (if required). The use of SCRs in series or parallel electrical connections generally assists in meeting the thermal requirements.

There are many classes of SCRs. For motor controller use, there are two broad classifications, sometimes termed *inverter type* and *chopper type*. The former are applicable to inverters, cycloconverters, and brushless dc motor systems while the latter apply to choppers, phase controlled rectifiers, regulators, and so forth. The main difference between the two types is primarily that of time response, related to the fifth, sixth, and seventh parameters listed earlier. Inverter types are generally more costly than chopper type SCRs. There are further ways to classify SCRs, especially in the smaller ratings used in control and communication applications. Many SCRs in the 35-A (rms) rating and below can be packaged in plastic cases, which results in a lower device cost than metallic SCRs of similar ratings. Another classification relates to the gate signal. A very useful class of SCRs for position sensing in motor control systems is the light-activated class, or LASCR. In place of energy injection by electrical current, the LASCR is triggered on by photon energy. Light activation is available for all of the other three and four terminal devices listed in Table 8.2, as well as for the SCR.

The forward voltage drop of an SCR varies considerably during anode current conduction and may be very large during the first few microseconds

Figure 8.5. Disc-type thyristor showing heat sink and fuses. (Courtesy Westinghouse Electric Corporation).

Maximum Allowable Ratings

Types	Repetitive peak off-state voltage, V_{DRM^a} $T_j = 40$ to $+125°C$	Repetitive peak reverse voltage, V_{RRM^a} $T_j = 0.40$ to $+125°C$	Nonrepetitive peak reverse voltage, V_{RSM^a} $T_j = +125°C$
C449PN	1800 volts	1800 volts	2040 volts
C449PS	1700	1700	1920
C449PM	1600	1600	1790
C449PE	1500	1500	1700

[a]Half sinewave waveform, 10 ms max pulse width. Consult factory for lower-rated voltage devices.

Peak one cycle surge (nonrepetitive) on-state current, I_{TSM}	6500 amperes
Critical rate-of-rise of on-state current, Nonrepetitive	500 A/μs
Critical rate-of-rise of on-state current, repetitive	300 A/μs
Average gate power dissipation, $P_{G(av)}$	5 watts
Storage temperature, T_{stg}	-40 to $+150°C$
Operating temperature, T_j	-40 to $+125°C$
Mounting force required	3000 lb + 500 lb − 0 lb 13.3 KN + 2.2 KN − 0 KN

Figure 8.6. SCR data sheet. (Courtesy General Electric Company.)

after turn-on. This high voltage drop results in a power loss known as the switching loss of an SCR. The *average* forward voltage drop of an SCR is 1.5 to 2.0 V.

8.3.3 TRIAC

The TRIAC, often called a bidirectional switch, is approximately equivalent to a pair of back-to-back or antiparallel SCRs fabricated on a single chip of semiconductor material. Triggered conduction may occur in both directions, that is, the TRIAC is a quasi-bilateral device. TRIAC applications include light-dimming, heater control, and ac motor speed control. The same parameters listed as important in the application of SCRs in the previous section generally apply to the TRIAC. However, it should be noted that the TRIAC is a three-terminal device with only one gate, which has an effect on its time-response as compared with that of two distinct SCRs connected in antiparallel position. The turn-off time of a TRIAC is in the same order of magnitude as that of an SCR. This implies that a time period approximately equal to the turn-off time must be observed before applying *reverse* voltage to a TRIAC. In an antiparallel pair, however, reverse voltage can be immediately applied after cessation of forward current in one SCR. TRIACs are not available in as high voltage and current ratings as SCRs at the present time and, therefore, are used in control of motors of relatively low power ratings.

Characteristics

Test	Symbol	Min.	Types	Max.	Units	Test conditions
Repetitive peak reverse and on-state current	I_{RRM} and I_{DRM}	—	10	25	mA	$T_j = +25°C$, $V = V_{DRM} = V_{RRM}$
Repetitive peak reverse and off-state blocking Current	I_{RRM} and I_{DRM}	—	45	60	mA	$T_j = +125°C$, $V = V_{DRM} = V_{RRM}$
Thermal resistance	$R_{\theta JC}$	—	—	.04	°C/watt	Junction-to-case—Double-side cooled
Critical linear rate-of-Rise of off-state voltage (Higher values may cause	dv/dt	200	—	—	V/μsec	$T_J = +125°C$, $V_{DRM} = 0.80$ rated, gate -open. Exponential or linear rising wave form. Exponential $di/dt = 0.8\ V_{DRM}\ (0.632)/\tau$
device switching)	Higher minimum dv/dt selections available—consult factory					
Gate trigger current	I_{GT}	—	—	200	mA dc	$T_C = +25°C$, $V_D = 6$ V dc, $R_L = 3$ ohms
		—	-	150		$T_C = +125°C$, $V_D = 6$ V dc, $R_L = 3$ ohms
Gate trigger voltage	V_{GT}	—	—	3	V dc	$T_C = 25$ to $+125°C$, $V_D = 6$ V dc $R_L = 3$ ohms
		—	—	5,...,3		$T_C = -40$ to $25°C$, $V_D = 6$ V dc, $R_L = 3$ ohms

Figure 8.6 (Cont'd)

Peak on-state voltage	V_{TM}	—	—	2.8	volts	$T_C = 25°C$, $I_T = 2000$ A peak Duty cycle $\leqslant 0.01\%$
Conventional circuit Commutated turnoff Time (with reverse voltage)	t_q				μsec	(1) $T_C = +125°C$ (2) $I_{TM} = 500$ A (3) $V_R \geqslant 50$ volts (4) 80% of V_{DRM} reapplied
C449–60		—	—	60		(5) Rate-of-rise of off-state voltage =
C449–60		—	—	40		200 V/μs (6) Gate bias = open during turnoff Interval, 0 Volts, 100 Ohms (7) Duty cycle $\leqslant .01\%$
Conventional circuit Commutated turnoff Time (with feedback diode)	t_q				μsec	(1) $T_C = +125°C$ (2) $I_{TM} = 500$ A (3) $V_R = 2$ volts, minimum (4) 80% of V_{DRM} reapplied
C449–60		—	60	*b*		(5) Rate-of-rise of off-state voltage =
C449–40		—	40	*b*		200 V/μs (6) Gate bias = open during turnoff interval (7) Duty cycle $\leqslant 0.01\%$

[b]Consult factory for maximum turn-off time.

Figure 8.6 (Cont'd.) (Courtesy General Electric Company)

8.3.4 GATE TURN-OFF SCR OR GATE-CONTROLLED SWITCH

Both of these names apply to a three-terminal, four-layer semiconductor device in the thyristor family with characteristics very similar to those of the SCR. In addition, the GTO or GCS may be turned off by an appropriate gate signal. The merits of this feature will be further appreciated after reading Section 8.4 of this chapter on SCR commutation. At the present time this class of thyristor is available in voltage and current ratings much lower than those of SCRs.

In early versions of this device, the gate current required to turn off the device was almost equal to the anode current being turned off. This equality meant that the gate circuitry would have to have been of equivalent current-carrying ability to that of the anode circuitry. Recently, however, the required gate current for device turn-off has been reduced considerably. The ratio of gate current for anode turn-off to anode current being turned off is termed turn-off gain, A_{ITO}, and is similar to equivalent turn-off gains used in power transistor terminology. Present GTO and GCS device turn-off gains vary from 2 to 1000, the latter figure being available in devices of low anode current ratings (5 A or less). The turn-off gate pulse is of the opposite polarity to that of the turn-on gate pulse. In the range of currents required in many motor control applications, 50 to 200 A, the GCS turn-off time is considerably lower than that of an equivalent SCR. Therefore, these devices are used in inverters, choppers, and cycloconverters requiring high device-switching frequencies. The use of GCSs in place of SCRs will frequently result in a lower weight controller, because of the elimination of SCR commutation circuitry.

8.3.5 POWER TRANSISTOR

When used in motor control circuits, power transistors are almost always operated in a switching mode. The transistor is driven into saturation and the linear gain characteristics are not used. The common-emitter configuration is the most common, because of the high power gain in this connection. The collector-emitter saturation voltage, $V_{CE(SAT)}$ for typical power transistors is from 0.2 to 0.8 V. This range is considerably lower than the on-state anode-to-cathode voltage drop of an SCR. Therefore, the average power loss in a power transistor is lower than that in an SCR of equivalent power rating. The switching times of power transistors are also generally faster than those of SCRs, and the problems associated with turning off or commutating an SCR are almost nonexistent in transistors. However, a power transistor is more expensive than an SCR of equivalent power capability. In addition, the voltage and current ratings of power transistors available are much lower than those of existing SCRs. It has already been stated that the maximum ratings listed in Table 8.2 are generally unobtainable concurrently in a single device. This is particularly true of power transistors. Devices with voltage ratings of 1000 V or above have limited current ratings of 10 A or less. Similarly, the devices with higher current ratings, 50 A and above, have voltage ratings of 200 V or less. There has been relatively little operating experience in practical motor control circuits with transistors whose voltage or current capabilities approach the upper limits listed in Table 8.2. For handling motor control requiring large current ratings at 200 V or below, it has been common to parallel transistors of lower current rating. This requires great care to assure equal sharing of collector currents and proper synchronization of base currents among the paralleled devices.

Typical specifications for a group of relatively low power switching transistors is shown in Figure 8.7. Ratings of significance for motor control application (some of which are not given in Figure 8.7) include:

1. Breakdown voltage, specified by the symbols BV_{CEO}—collector to emitter breakdown voltage with the base open, and BV_{CBO}—collector to base breakdown voltage with emitter open.
2. Collector saturation voltage, $V_{CE(SAT)}$.
3. Emitter-base voltage rating, V_{EBO}.
4. Maximum collector current, I_c, average and peak.
5. Forward current transfer ratio, H_{FE}, the ratio of collector to base current in the linear region.
6. Power dissipation.
7. Maximum junction temperature, typically 150 to 180°C.
8. Switching times: rise time, t_r; storage time, t_s; and fall time, t_f. Sometimes these switching times are related to a maximum frequency of switching, as in Figure 8.7.

The thermal impedances and temperature coefficients are also important parameters. In paralleling power transistors, the variation of device characteristics with temperature becomes especially significant. The I_c-V_{BE} characteristic is extremely temperature-sensitive.

The switching of a power transistor is controlled through the base circuit. A continuous base current is required to keep the transistor in a saturated or "on" condition and the power dissipation associated with the base circuit may be a significant portion of the total device loss. This factor should be included in determining power efficiencies of motor controllers. The rise time in turning on a transistor may be reduced by overdriving the base current. As the collector current reaches its maximum value and the device is in saturation, the base current is reduced to the minimum value required to hold the device in saturation. To initiate turn-off, the excess saturation charge—which is proportional to the excess base current—must be removed from the base region of the

6 TO 30 AMPERE ALLOY TRANSISTORS—HIGH SAFE OPERATING AREA

Style	A500		A510		A670	
Ic Rating	6 Amperes	6 Amperes	7.5 Amperes	7.5 Amperes	20 Amperes	20 Amperes
Type No.	151	152	153	154	163	164
V_{ceo}(sus) 40	151-04	152-04	153-04	154-04	163-04	164-04
50	151-05	152-05	153-05	154-05	163-05	164-05
60	151-06	152-06	153-06	154-06	163-06	164-06
70	151-07	152-07	153-07	154-07	163-07	164-07
80	151-08	152-08	153-08	153-08	163-08	164-08
90	151-09	152-09	153-09	154-09	163-09	164-09
100	151-10	152-10	153-10	154-10	163-10	164-10
120	151-12	152-12	153-12	154-12	163-12	164-12
140	151-14	152-14	153-14	154-14	163-14	164-14
160	151-16	152-16	153-16	154-16	163-16	164-16
180	151-18	152-18	153-18	154-18	163-18	164-18
200	151-20	152-20	153-20	154-20	163-20	164-20
220	151-22	152-22	153-22	154-22	163-22	164-22
h_{FE} min	11 + 1.5 A	18-+-1.5 A	15 + 1.5 A	25 + 1.5 A	15 + 5 A	25 + 5 A
V_{ce}(sat) max	1.3 V	1.25 V	1.3 V	1.25 V	1.1 V	1.0 V
f_T typical	0.5 MHz	0.5 MHz	0.5 MHz	0.5 MHz	0.5 MHz	0.5 MHz
Maximum ratings	V_{EB}=25 V, I_B=3 A $T_{J\&TSG}$= −65 to +150°C P_T= 100 W + 80°C $R\theta_{JC}$=0.7°C/W		V_{EB}=25 V, I_B=3 A $T_{J\&TSG}$= −65 to +175°C P_T=200 W + 25°C $R\theta_{JC}$=0.75°C/W		V_{EB}=25 V, I_B=7.5 A $T_{J\&TSG}$= −65to +175°C P_T=200 W + 75°C $R\theta_{JC}$=0.5°C/W	

Style	A67H			
Ic Rating	30 Amperes			
Type No.	110	110	110	110
V_{ceo}(sus) 50	2N2757	2N2763	2N2769	2N2775
100	2N2758	2N2764	2N2770	2N2776
150	2N2759	2N2765	2N2771	2N2777
200	2N2760	2N2766	2N2772	2N2778
250	2N2761	...	...	...
h_{FE} min	10 + 10 A	10 + 15 A	10 + 20 A	10 + 25 A
V_{ce}(sat) max	1.5 V	1.5 V	1.5 V	1.5 V
f_T typical	0.5 MHz	0.5 MHz	0.5 MHz	0.5 MHz
Maximum ratings	V_{FB}=25 V, I_B=7.5 A $T_{J\&TSG}$= −65 to +175°C P_T=200 W + 75°C $R\theta_{JC}$=0.5°C/W			

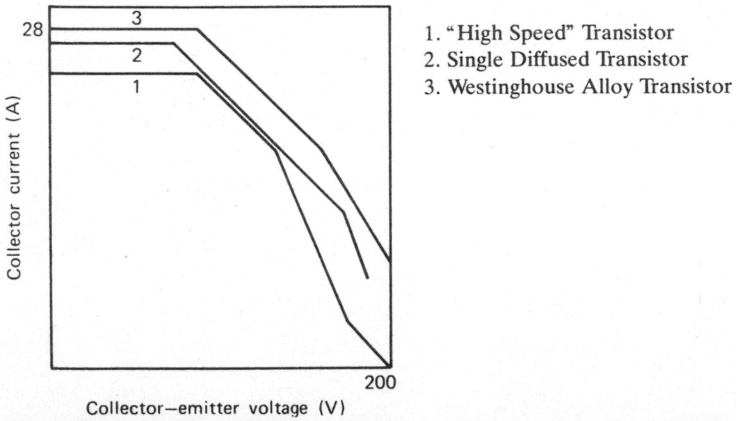

Figure 8.7. Low-power switching transistor applications. (Courtesy Westinghouse Electric Corp.)

transistor before the transistor can begin to switch back into the off state. This results in the storage delay time, t_s. This time can be minimized by applying a reverse current to the base during turn-off. The fall time, t_f, is also reduced by overdriving the base with reverse current during turn-off. During reverse bias of the base circuit which occurs at turn-off, the base voltage rating, V_{EBO}, must not be exceeded or device failure may result. "Slow" turn-offs are obtained by reducing the base current to zero with no reverse base current.

Overdriving the base to reduce switching times can result in transistor failure by a mechanism known as "secondary breakdown." Secondary breakdown is a thermal runaway process which is induced by current concentration at the emitter junction. This concentration produces the possibility of "hot spot" formation, which often causes a continuous thermal buildup that results in a catastrophic failure of the device. The safe operating limits of collector current and collector-emitter voltage are often defined by marking off a region on a graph of I_c versus V_{CE}, known as the "safe operating area" or *SOA*. Operation outside of this area may initiate the secondary breakdown situation. Motor circuits contain many inductances and during transistor turn-off, the energy stored in the inductive components in series with the collector will increase the collector-emitter voltage voltage as a function of the rate of decrease of collector current. Therefore, rapid turn-off attributable to large reverse base current will result in high values of V_{CE}. Such high values both increase the loss known as the switching loss and add to heating in the transistor, which may result in secondary breakdown. An example of an *SOA* region is shown in Figure 8.7.

8.3.6 POWER DARLINGTON

This designation generally refers to the well-known Darlington-connected transistor pair fabricated on a single chip. The same characteristics are of course achievable through the use to two discrete transistors, albeit usually in a larger, more complex, and more costly package. The principal merit of the Darlington device is its high current gain. The operating parameters and failure modes discussed for transistors in the previous section are also applicable to the Darlington.

Darlington amplifiers are a recent entry into the area of motor controls but have met with considerable acceptance due to their potential for reducing the size, cost, and weight of motor controllers. These devices are used both in choppers for dc commutator motor control and in inverters for ac motor control, generally for lower power applications. Recently, larger devices have been developed with ratings as high as 200 A and 100 V or 100 A and 450 V and have been applied to the control of traction motors used in lift trucks and industrial electric vehicles. Current gains as high as 1600 have been achieved at these high current levels. However, with high current gain there is a higher on-state or saturation voltage, $V_{CD(SAT)}$. An example of analysis of a Darlington inverter used in motor control is given in Reference 9.

8.3.7 ZENER AND AVALANCHE DIODES

There is a group of two-terminal devices which have very useful switching properties for motor control and many other electronic circuit applications. These properties are not generally used for current or power control, but rather, for protecting other semiconductor devices from overvoltage conditions, for supplying a reference voltage, and for voltage regulation. The devices consist of two-or four-layer silicon chips and are similar in appearance to other silicon diodes and, theoretically, can be used in rectifying service. The best-known member of this family is the Zener diode.[10]

The voltage reference and voltage protection capabilities of the Zener diode are based upon the breakdown characteristics of a *pn*-junction. When the reversed-bias voltage applied to the diode exceeds a certain value, known as the Zener voltage, V_Z, breakdown occurs. The breakdown condition is characterized by a drastic decrease in device resistance and a sudden inrush of current. This breakdown mechanism is a combination of two phenomena, termed *Zener breakdown* and *avalanche breakdown*. The breakdown is nondestructive—as long as the power rating of the device is not exceeded—and the device quickly recovers its ability to block reverse voltage. Zener diodes are available in ratings of from 2.4 to 200 V. In the forward-biased direction, the Zener diode looks very much like a conventional silicon rectifier. Like other semiconductors, Zener diodes are temperature-sensitive and have a temperature coefficient defining the variation of the Zener voltage as a function of temperature. This must be considered in the application of Zener diodes in electronic circuits. Temperature-compensated diodes are available with temperature coefficients as low as 0.0002% per °C.

There are many devices having properties similar to those of the Zener diode. The Zener diode is often termed a *unidirectional breakdown diode. Bidirectional breakdown diodes* are available exhibiting the avalanche breakdown characteristic in both reverse and forward directions. The reverse switching rectifier (RSR) is a two-terminal, four-layer thyristor with breakdown characteristics similar to those of the Zener diode but with much higher voltage rating and very high inrush *di/dt*-capabilities. It is frequently used in continuous pulsed operation much like an SCR where high *di/dt* is required, but with the on-state initiated by circuit voltage rather than by the action of a gate. Turn-on by reversed-bias anode voltage is, of course, a property of all thyristors, including SCRs and TRIACs. However, in most SCR and TRIAC applications where the desired mode of operation is through a controlled gate signal, reversed-bias voltage turn-on is a detrimental operating mode and often leads to device failure.

8.4 SCR COMMUTATION TECHNIQUES

A major consideration in the use of SCRs and TRAICs in motor control is the type of circuitry required to commutate or turn-off the device and the associated cost, weight, and complexity of commutation circuitry. In this section, we shall examine the basic requirements of commutation and some of the principal methods used to achieve good commutation in practical motor control systems. More detailed development of commutation theory and practice may be found in References 1, 2, 4, 11, and 12.

The basic operation of SCRs has been presented briefly in Section 8.3.2. It was noted that the SCR is turned on by injection of energy into the device by means of the gate connection which, causes a rapid breakdown of junction voltages and buildup of anode current. Beyond a minimum value of anode current, the device "latches on" and the device loses control of anode current. From that time on, anode current is determined solely by conditions in the external circuit between anode and cathode until the device is once more brought into the blocking state. Commutation refers to the process of bringing the device from a forward conducting state to a forward blocking state, or "turning off" the SCR. Turn-off or commutation results from two possible conditions in the overall anode-cathode circuit:

1. Zero or very low anode current caused by opening a switch or some other similar mechanism in the external circuit or by a large increase in external circuit resistance. This condition is termed *natural or starvation commutation* and requires a time period much longer than the minimum turn-off times given in the device specification.
2. Reverse voltage bias across the anode-cathode, that is, the anode potential is below that of the cathode. This condition is termed *forced commutation* and is required in almost every practical application of SCRs.

Forced commutation is achieved in systems energized from dc sources by an arrangement of energy storage elements (capacitors and inductors) and by additional switching devices (usually SCRs). In systems energized from ac sources, forced commutation is obtained by means of the cyclic potential reversal of the energy source. The mechanism of forced commutation will be explained with the aid of Figures 8.8 to 8.10. Figure 8.8[1] shows the voltage and current relationships that must exist in an SCR during commutation for commutation to be successful, that is, for the SCR to turn off. This characteristic could represent the current and voltage in the SCR, S_1, of either Figure 8.9 or 8.10. The current, i_1, is shown in these diagrams. The voltage, v_1, is anode-cathode voltage across S_1. The means of achieving these voltage-current relationships in the circuits of Figures 8.9 and 8.10 will be explained in the next sections.

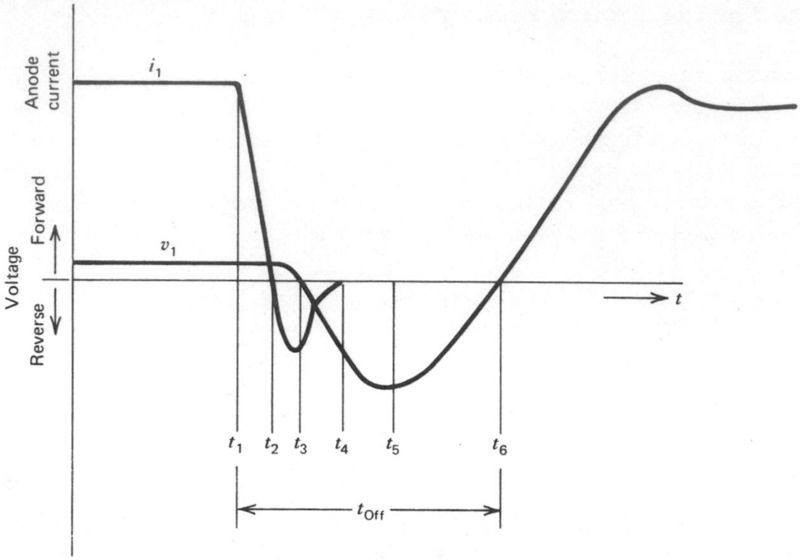

Figure 8.8. SCR voltage-current during commutation.

The SCR is in a certain conducting or on-state condition at the leftmost side of this Figure 8.8 and commutation is initiated at time, t_1, when a negative voltage is introduced into the external anode-cathode circuit. The anode-cathode voltage drop remains at the low on-state level (1.5 to 2.0 V) until anode current decreases to zero at time, t_2, at which time the voltage begins to go negative. The anode must be maintained at a negative (reversed-biased) potential for a prescribed time period, until time, t_6. The period, t_1 to t_6, is known as the circuit time-off period, T_a, and this time interval must be somewhat larger than the

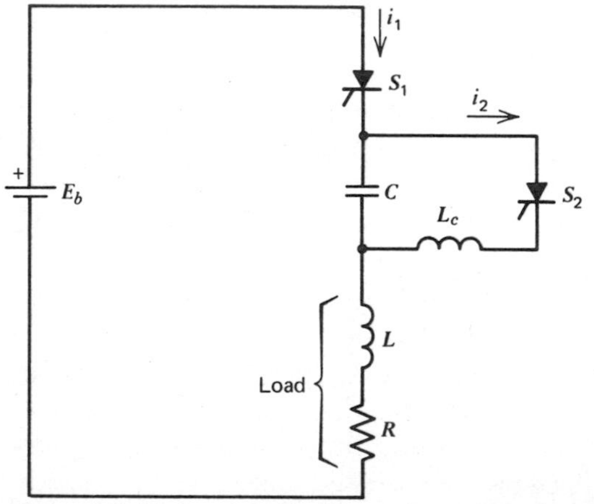

Figure 8.9. Motor controller illustrating series commutation.

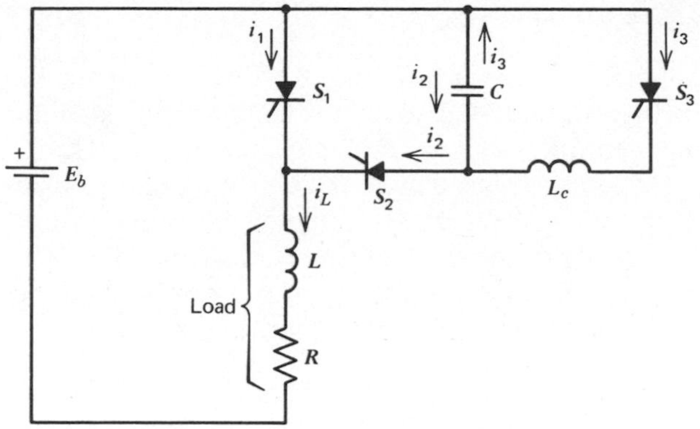

Figure 8.10. DC chopper showing parallel commutation.

device turn-off time, t_{OFF} (Figure 8.6), to assure safe device turn-off. Following zero anode current, there is a reverse recovery interval, t_2 to t_4 (usually 3 μs or less), during which reverse conduction occurs as explained for the silicon rectifier in Section 8.3.1. During the commutation process the following two characteristics of the device must be carefully controlled in order to achieve good commutation and to prevent the device failure.

1. From t_1 to t_4, the rate of change of anode current, $-di/dt$, may be very high. (There is no internal limit to the di/dt in this interval.) Even the slightest amount of inductance within the SCR or in series with the anode can result in unexpectedly high induced voltages between the anode and cathode terminals. To protect the device during this interval, the induced voltage can be limited by placing a capacitor-diode series circuit across the anode-cathode (called a "snubber network"), or by shunting the anode to ground through a "surge suppressor". (These protective schemes will be further discussed in Section 8.6.)
2. Following the forced turn-off time (after t_6 in Figure 8.8), there is a further limit on the rate at which forward voltage can be reapplied to the SCR. This is known as the "reapplied dv/dt" rating of an SCR. It is in the order of 20 to 50 V/μs for most SCRs.

Forced commutation schemes for motor controllers energized from dc sources are divided into two general classifications: series and parallel. This classification is based upon the circuit arrangement of the commutating capacitor, that is, whether it is electrically in series or parallel with the SCR which is being commutated. There are many circuit configurations in both classifications. The two basic circuit concepts will be presented here and further variations will appear in some of the control schemes discussed in later sections.

8.4.1 SERIES COMMUTATION

A simple series commutation circuit is shown in Figure 8.9.[13] In this circuit S_1 is the main SCR, which is being commutated and is supplying the load that consists of a resistive component, R, and an inductive component, L. The commutating capacitor is C; S_2 is the reversing SCR; and L_c is the reversing inductor. Operation is as follows: Load current is initiated by turning on the main SCR, S_1, by means of its gate circuit. Current flows from the source through the load and through the commutating capacitor and can be described, during this interval, by the differential equation of a series RLC-circuit

$$E_b = Ri_1 + L\frac{di_1}{dt} + \frac{1}{C}\int i_1 dt \tag{8.4}$$

The solution for the current, i_1, is, for the underdamped case $(L/C > R^2/4)$,

$$i_1 = \left(\frac{E}{Z_o} - \frac{\alpha}{\beta}I_{lo}\right)e^{-\alpha t}\sin\beta t + I_{lo}e^{-\alpha t}\cos\beta t \tag{8.5}$$

where

$E = E_b - V_{co}$

V_{co} = capacitor voltage at time, $t=0$; V_{co} is positive when the upper plate of C in Figure 8.9 is positive with respect to the lower plate

$Z_0 = \sqrt{L/C-(R/2)^2}$, the characteristic impedance in ohm

I_o = current at $t=0$

$\alpha = R/2L$, attenuation constant, S^{-1}

$\beta = \sqrt{1/LC-(R/2L)^2} = Z_o/L$, phase constant in rad/s

The solution for the voltage across the capacitor is

$$v_c = E_b - I_{10}R_o e^{-\alpha t}\left[\frac{\sin(\beta t + \epsilon + \Theta)}{\sin\Theta}\right] \tag{8.6}$$

where

$R_o = \sqrt{L/C}$

$Z_f = E/I_{10}$ = load impedance

$\sin\Theta = Z_o/\sqrt{Z_f^2 - RZ_f + (L/C)}$; $\Theta = \tan^{-1}Z_o/(Z_f - R/2)$

$\sin\epsilon = Z_o/\sqrt{L/C}$; $\epsilon = \tan^{-1}(\beta/\alpha) = \tan^{-1}(2Z_o/R)$

The solution for the load voltage (across L and R) is

$$v_{LR}=I_oR_oe^{-\alpha t}\left[\frac{\sin(\beta t+\epsilon+\Theta)}{\sin\Theta}\right] \tag{8.7}$$

In many cases of interest in series commutation, the current at the beginning of the SCR on-time is zero. Substituting $I_{10}=0$ in (8.5) to (8.7) gives

$$1_1=\frac{E}{Z_o}e^{-\alpha t}\sin\beta t \qquad (I_{10}=0) \tag{8.8}$$

$$v_c=E_b-\frac{E}{\sin\epsilon}e^{-\alpha t}\sin(\beta t+\epsilon), \qquad (I_{10}=0) \tag{8.9}$$

$$V_{LR}=\frac{E}{\sin\epsilon}e^{-\alpha t}\sin(\beta t+\epsilon) \qquad (I_{10}=0) \tag{8.10}$$

In the above equation, L should be interpreted as the *total* series circuit inductance, including the battery or source inductance and inductances associated with the SCR, anode and cathode leads, and the load inductance. Likewise, R should be interpreted as the *total* series circuit resistance.

If the SCR, S_1, were a bilaterial switch, the above equations would describe the typical transient response or "ringing" of an *RLC*-circuit energized from a dc source, which is sinusoidal in waveform. However, since reverse current cannot flow through the SCR, current flow ceases after the end of the first half-period of the sinusoidal response. This current will subsequently be referred to as a sine pulse. It is the characteristic response of an *RLC*-circuit excited from a dc source through an unilaterial device such as a diode, SCR, or transistor and occurs frequently in all types of electronic circuits. The pulse width of this sine pulse is given by

$$T_o-\frac{1}{\beta}(\pi-\Theta) \tag{8.11}$$

$$=\frac{\pi}{\beta} \qquad (I_{10}=0) \tag{8.12}$$

The time, T_m, after initial SCR turn-on (at $t=0$) when the sine pulse of current reaches its maximum value is

$$T_m=\frac{1}{\beta}(\epsilon-\Theta) \tag{8.13}$$

$$=\frac{\epsilon}{\beta} \qquad (I_{10}=0) \tag{8.14}$$

During this sine pulse, the inductance, L, is first charged to its maximum value of energy, which is completed at approximately $t = T_m$, or (if $I_o = 0$) at one-half of the sine-pulse period, T_o. This energy is subsequently returned to the capacitance, C. At the end of the sine pulse, when the current, $i = 0$, the inductance is fully discharged and the capacitance has been charged to a voltage given by

$$v_c(T_o) = E_b + e^{-\alpha T_o} I_{1o} R_o \left(\frac{\sin \epsilon}{\sin \Theta} \right) \tag{8.15}$$

$$= E_b (1 + e^{-\alpha T_o}) \pm V_{co} e^{-\alpha T_o} \qquad (I_{1o} = 0) \tag{8.16}$$

Let us assume that the capacitance is initially uncharged ($V_{co} = 0$) and the initial current is zero. For this case, from (8.16)

$$v_c(T_o) = E_b (1 + e^{-\alpha T_o}) = E_b (1 + e^{-R/2Z_o}) \tag{8.17}$$

For the underdamped circuit, the exponent, $R/(2Z_o)$, is much less than 1.0 and the capacitance voltage at the end of the current pulse is somewhat less than *twice the source voltage*, E_b. If the attenuation constant were zero, implying a lossless circuit ($R = 0$), the capacitor voltage at the end of the pulse would be exactly twice the source voltage. At the end of the sine pulse of current, neglecting the brief reverse recovery interval, the current and the voltage across the inductance are zero. The net voltage appearing between the anode-cathode terminals is the difference between the capacitor voltage and source voltage that reverse-biases the SCR. If this condition is maintained for a time interval larger than the SCR turn-off time, t_{off}, the SCR is commutated and reverts to a forward blocking state. This is the mechanism of series commutation. It is probably the simplest and safest means of commutating an SCR. However, this circuit, by itself, is of little value, since further operation of the circuit is not possible with the SCR in a reverse-biased situation.

The next step in the operation of such a circuit is to reverse the capacitance voltage through the circuit S_2 to L_c, which is initiated by turning on SCR, S_2. The resulting voltage and current relationships through this reversing circuit can be obtained using (8.5) to (8.17), with E_b and I_{1o} set equal to zero and the initial capacitance voltage, V_{co}, set equal to the capacitance voltage at the end of the current pulse through S_1.

From these equations and initial conditions, it is evident that another sine pulse of current flows through the circuit $C - S_2 - L_c$ with opposite direction in C as compared to the first pulse. At the end of this second pulse, the capacitance voltage is reversed in polarity with the upper plate in Figure 8.9, now negative with respect to the lower plate. The magnitude of the capacitance voltage depends upon the resistance of the inductor, L_c. If this resistance were zero, the voltage would be unchanged from the voltage at the end of the first pulse. The second pulse needed to reverse capacitance voltage can be initiated immediately following the termination of the first pulse. This is so provided that the second pulse period results in a reverse-biased condition that is maintained across the main SCR, S_1, for a time interval somewhat longer than the turn-off time of S_1. For the circuit as shown in Figure 8.9, this time interval, T_Q, can be obtained from the expression,

$$\sin(\xi - \beta T_Q) = \frac{E_b}{E} \sin\xi \tag{8.18}$$

where

$$\tan\xi = \frac{\sin\epsilon}{\dfrac{R_o}{Z_f} - \cos\epsilon}$$

$$= \tan\epsilon \qquad (I_{10} = 0)$$

Notice that T_Q is the interval following turn-on of S_2, during which negative voltage appears across the capacitance. If we assume that T_Q, which is the "circuit commutation time," is sufficiently large to turn off S_1, the circuit will then be ready for another operation following the reversal of capacitance voltage. Note that S_2 is commutated by a series commutation process, also, at the end of the reversing pulse. Also, at the end of the reversing pulse, the source and capacitance voltages are in series-aiding. If S_1 is now turned on, forward current flow is again possible through S_1. The maximum current of the pulse will be of larger magnitude in this third pulse due to the larger applied voltage across the series *RLC*-circuit, which is now the sum of source and capacitance voltage, $E_b + V_{co}$. Setting E equal to this value in (8.5) through (8.17) allows the circuit to be analyzed as during the previous pulses.

If the process of alternate operation of S_1 and S_2 is continued as described in the last paragraph, the capacitance voltage would build up to some value that could be much larger than the source voltage. If the circuit resistance were zero, the capacitance voltage would approach an infinite value by this technique. With finite circuit resistance, which always exists, the final or steady-state voltage across the capacitance at the instant S_1 is turned on is

$$V_{co} = E_b \coth \frac{\alpha T_0}{2} \tag{8.18a}$$

This process is useful in many types of voltage multplier circuits and in portions of series and parallel thyristor commutation circuits.

8.4.2 PARALLEL COMMUTATION

One of the earliest and most simple *parallel* commutation circuits is the Morgan circuit[1,14], in which turn-off is initiated by the switching action of a saturating transformer (see Chapter 3). As a result of declining costs of SCRs since the introduction of the Morgan circuit, the saturating transformer has been replaced by a commutating SCR and the reversing circuit shown in Figure 8.10. In Figure 8.10, S_1 represents the main SCR which is commutated and is supplying the load that consists of a resistance, R, and an inductance, L. S_2 is the commutating SCR; S_3, the reversing SCR; C, the commutating capacitor; and L_c, the reversing inductor.

In many parallel commutation circuits, S_2 can be replaced by a diode. If we assume that S_2 is an SCR, operation is as follows. Assume that the circuit of Figure 8.10 is initially in a discharged state. The SCRs S_1 and S_2 are turned on simultaneously. Currents i_1 and i_2 flow through both SCRs and the load with $i_L = i_1 + i_2$. Current i_2, which flows through the path $E_b - S_2 - C - R - L$, can be determined by the equations presented in the previous section. Current i_1 through the path $E_b - S_1 - R - L$ follows the typical current rise in an $R - L$ circuit that is energized from a dc source. The value of capacitance is chosen such that (a) the sine current pulse during this operation is relatively short compared to the desired current on-time pulse in the load, and (b) sufficient energy is stored in the capacitance to commutate S_2 at the end of the load on-time pulse. The SCR S_2 is commutated by means of the mechanism of series commutation, as described in the previous section, and $i_2 = 0$ at the completion of commutation.

While S_1 is still in an on-state condition, the voltage across the commutating capacitance is reversed through the reversing circuit, $S_3 - L_c$, with current i_3, as described in the previous section. At the end of this operation, the capacitance voltage is somewhat less than twice the source voltage, with the upper plate at a negative potential in relation to the lower plate. When it is desired to end the load current pulse through L and R, S_2 is again turned on. This causes the main SCR, S_1, to be reverse-biased by the capacitance voltage, and its current, i_1, is almost immediately transferred to the capacitance. If there is negligible inductance in S_1, C, and the interconnecting wiring between S_1 and C, this current transfer would occur instantaneously. If such a condition is maintained for a sufficient time interval, that is, an interval greater than the turn-off time of S_1, S_1 will be commutated. The voltage on C will again reverse, and S_2 eventually turns off by series commutation. The rapid transfer of anode current from the SCR undergoing commutation to a parallel capacitor is the situation illustrated in Figure 8.8.

The current through the capacitance and load during the commutation period can be calculated from the equations in the previous section by setting I_o equal to the current in S_1 at the initiation of commutation (when S_2 is turned on) and by setting V_{co} equal to the voltage on C (with a negative sign). This gives $E = E_b + V_{co}$. It is seen that this current during turn-off of S_1 is a sine pulse of period, T_o, determined by the values of C, L, R, and Θ. The value of Θ is determined by the values of I_o and V_{co}. These equations permit the design of a commutating circuit for commutating a given value of current, I_o, in an SCR of given specifications. The SCR specifications required for this design are the forward and reverse peak voltages (which set an upper limit on V_{co}), t_{off} (which determines the required circuit turn-off time, T_Q), and the reapplied term dv/dt. Thermal characteristics during commutation should also be examined.

In using the equations of Section 8.4.1 for designing SCR commutation circuitry, several aspects of these equations must be kept in mind:

1. These equations are based upon a linear inductance, L. Saturation of the inductance, if it has a ferromagnetic core, can easily occur during the current peak of the commutation pulse.
2. If the load is a motor, the equivalent load representation must also include an active voltage (such as the back emf in a dc commutator machine).
3. Many inductive loads, including many motors loads, are paralleled by a freewheeling diode which will alter the nature of the load impedance. This will be discussed with dc choppers later.
4. L and R must represent total circuit inductance and resistance. Battery and lead inductance may be significant in many designs.
5. Similarly, excessive circuit resistance can cause problems not only by increasing system losses but by reducing the charge on the commutating capacitor. As has been noted, a series commutation circuit must be an underdamped circuit to achieve the overcharging of the capacitor.

If the load R and L, the battery voltage, and the SCR specifications are known, the design of a commutation circuit consists of determining the value of the commutating capacitance, C, the value of the reversing inductance, L_c, and the ratings or specifications of the auxiliary SCRs required. This is done by assuming a value of the capacitance and V_{co} and determining the resulting circuit turnoff time, T_Q, given by (8.18). This process is best done by means of a computer simulation of (8.5) to (8.18), since a number of iterations are usually required and manual solutions of these equations are not simple. This is especially the case if it is necessary to use a nonzero value for I_o, which is required in the design of parallel commutation schemes. However, it is many times more desirable to estimate commutation characteristics without a complicated analysis, which can be done by making various simplifying assumptions about these equations. The most obvious simplifying assumption is to neglect circuit resistance, R, modifying equation parameters as follows:

$$\alpha=0; \qquad \beta=\frac{1}{\sqrt{LC}}; \qquad Z_o=\sqrt{\frac{L}{C}}; \qquad \epsilon=\frac{\pi}{2} \tag{8.19}$$

$$\Theta=\tan^{-1}\frac{Z_o}{Z_f}; \qquad \xi=\tan^{-1}\frac{Z_f}{Z_o}=\frac{\pi}{2}-\Theta \tag{8.20}$$

$$\cos(\Theta+\beta T_Q)=\frac{E_b}{E}\cos\Theta \tag{8.21}$$

$$\cos\beta T_Q=0 \qquad (I_o=0) \tag{8.22}$$

From (8.22) has come a "rule-of-thumb" for a first guess on the value of the commutating capacitor[1]: It is seen that $\beta T_Q=\pi/2$, or

$$T_Q=\sqrt{LC}\left(\frac{\pi}{2}\right)=\frac{\pi/2}{\beta}=T_m \tag{8.23}$$

which is often expressed approximately as

$$\sqrt{LC}\approx 0.6\, T_Q \tag{8.24}$$

The maximum value of current is, from (8.5) and (8.13)

$$I_{\max}=i(T_m)=I_o e^{-\alpha T_m}\left(\frac{\sin\epsilon}{\sin\Theta}\right) \tag{8.25}$$

$$=\frac{E}{Z_o}e^{-\alpha T_m}\sin\epsilon, \qquad (I_o=0) \tag{8.26}$$

By the use of the simplifying assumption that $R=0$, (8.25) becomes

$$I_{\max}=\frac{\sqrt{E^2+(I_oZ_o)^2}}{Z_o}=E_{PK}/Z_o \tag{8.27}$$

This is a very useful expression, since it gives the ratio of peak series-circuit voltage to maximum current including the effects of an initial current, I_o, and an initial capacitor voltage, V_{co}. Using the simplified expression for Z_o from (8.19) and substituting into (8.24) to eliminate L gives

$$C\approx\frac{0.6\, T_Q I_{\max}}{E_{PK}} \tag{8.28}$$

where

$$E_{PK}=\sqrt{(E_p \pm V_{co})^2+(I_o Z_o)^2}$$

I_{max} = peak capacitor current during commutation

Also, E_{pk} is generally taken to be the maximum voltage seen by the capacitor.
These relationships will now be illustrated by examples.

Example 8.3
A series circuit such as shown in Figure 8.9 consists of the following circuit elements: $R=0.4$, $L=125\ \mu H$, $C=100\ \mu F$, and $E_b=72$ V. Determine the maximum current, current pulse width, and time at which the current is maximum for the current pulse following S_1 being turned on.

Several of the parameters listed under (8.5) must first be calculated

$$Z_o=\sqrt{\frac{L}{C}-\left(\frac{R}{2}\right)^2}=1.1; \qquad \beta=\sqrt{\frac{1}{LC}-\left(\frac{R}{2L}\right)^2}=0.88\times 10^4$$

$$\epsilon=\tan^{-1}\frac{2Z_o}{R}=79.7°=1.39 \text{ rad}$$

Using these values, we find that the current pulse width, from (8.12), is

$$T_o=\frac{\pi}{.88\times 10^4}=358\ \mu s$$

From (8.14) we obtain

$$T_m=\frac{1.39}{(0.89\times 10^4)}=158\ \mu s$$

Substituting T_m into (8.26) and calculating α yields

$$\alpha=\frac{R}{2L}=\frac{0.4}{250\times 10^{-6}}=1.6\times 10^3$$

$$I_{max}=\frac{72}{1.1}(e^{-.252})(\sin 79.7°)=50.2 \text{ A}$$

Example 8.4
What is the value of voltage on the capacitance at the end of the current pulse of Example 8.3?

Using (8.16), we get

$$v_c(T_o)=72(1+e^{-.572})=72(1+0.565)=113\text{ V}$$

Example 8.5

In Figure 8.9, both S_1 and S_2 are now to be operated alternately. Use the same circuit constants as in Example 8.3; let $L_c = 20\ \mu h$; and assume that, in the steady state, the voltage on C is 180 V each time S_1 and S_2 are turned on. This implies zero resistance in the reversing circuit. Determine the maximum value of the new current pulse through S_1, the pulse width and maximum current of the pulse through S_2. Check to determine that the circuit commutation time, T_Q, is large enough to insure safe commutation of S_1.

The time parameters of the pulse through S_1 have not changed from Example 8.3, so the maximum current can be determined from

$$I_{\max}=\frac{(72+180)}{1.1}(e^{-0.252})(\sin 79.7°)=176\text{ A}$$

For the current pulse in the reversing circuit

$$Z_o=\sqrt{\frac{L}{C}}=\sqrt{\frac{20}{100}}=0.448;\qquad \beta=\sqrt{\frac{1}{LC}}=\sqrt{\frac{10^{12}}{2000}}=2.24\times10^4$$

$$\varepsilon=\frac{\pi}{2};\qquad \alpha=0$$

$$T_o=\frac{\pi}{\beta}=141\ \mu\text{s};\qquad T_m=\frac{\pi/2}{\beta}=70.5\ \mu\text{s}$$

$$I_{\max}=\frac{180}{0.448}=402\text{ A}$$

Note that this pulse has a narrow pulse width and high peak current. This is typical of the required current pulses in this type of circuit. The turn-off or circuit commutating time , [the time after S_2 is turned on] is, in this case, the time during which the capacitor voltage is still larger than the source voltage, in order to maintain a reverse bias on S_2. This can be found by setting v_c equal to 72 V in (8.9) and solving for time. Note that, in the reversing circuit calculation, E_b is zero.

$$72=0-\frac{180}{1.0}(1.0)\sin\left(\beta t_Q+\frac{\pi}{2}\right)$$

Solving for t_Q gives

$$\beta t_Q = 66.4° = 1.14 \text{ rad}; \qquad t_Q = \frac{1.14}{2.24 \times 10^4} = 51\ \mu\text{s}$$

This is adequate for many types of SCRs in this power range.

Example 8.6
Assuming no time delay in turning on the SCRs in Example 8.4, determine the rms current in the capacitor.

The current waveform in the capacitor has the characteristics shown in Figure 8.11. To determine rms current, Table 8.1 can be used

1. Due to S_1, $I_1 = 176\sqrt{\left[\dfrac{353}{2(353+141)}\right]} = 105$ A
2. Due to S_2, $I_2 = 402\sqrt{\left[\dfrac{141}{2(353+141)}\right]} = 152$ A
3. rms current in $C = \sqrt{105^2 + 152^2} = 185$ A

8.5 GATE AND BASE CIRCUIT TURN-ON TECHNIQUES

The control capabilities of thyristors and power transistors arise from their ability to switch from a high impedance state to a very low impedance state in a

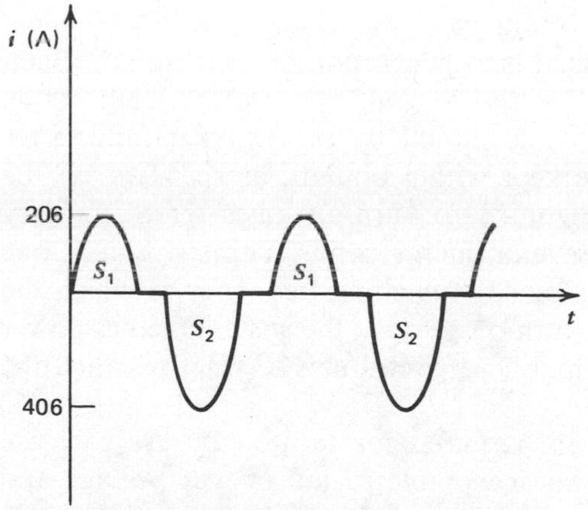

Figure 8.11. Current waveform in Example 8.6.

very short period of time. This action, which we have been calling device turn-on, is initiated by means of the gate terminal on thyristors and the base connection on the common-emitter transistor. The circuitry connected to the gate or base terminal is designed to translate the fundamental system control signals into correct turn-on action in the power device. The fundamental system control signals may come directly from some human operator or pretaped program, or they may come in part from feedback signals arising from motor operating conditions such as speed, torque, or temperature. Earlier in this chapter, we called this portion of the circuitry connected to the gate the logic control system. Our primary concern in this section is with the circuitry that is immediately adjacent electrically to the gate or base terminal, the function of which is to develop the proper electrical signal for device turn-on. It is often called the triggering circuit.

8.5.1 THYRISTOR GATE SIGNAL REQUIREMENTS

A thyristor gate signal for device turn-on is a pulse signal, generally lasting for only a few microseconds. Its function is to inject energy into the lower *p*-material layer of the device to start current flow between gate and cathode. This reduces the junction voltages in the device, causing it to lose its forward blocking capability and allowing anode current to begin to flow. The initial breakdown begins with a small spot around the gate load and then begins spreading radially outward from the gate connection. The process is accelerated as anode current begins to flow. In the earlier stages of device turn-on, only the small spot near the gate lead is "turned on" and capable of conducting anode current. Since this spot, for a short time period, is of very small cross section compared to the total semiconductor chip cross section, its equivalent resistance to current flow is high. This situation results in high localized I^2R-losses which, in turn, may result in abnormal localized heat generation or "hot spots." Such conditions can cause device failure. For this reason, the rate at which anode current increases (determined at this time period by the external inductance between anode and cathode) must be kept within bounds, as specified on the device data sheet (its di/dt-rating, Figure 8.6). Turn-on times for SCRs vary from a few hundred nanoseconds to several microseconds. Turn-on time is one of the factors placing an upper bound on the switching frequency at which the device can be operated. During this turn-on period, the instantaneous power dissipated in the device may be very high and contributes to what is called the switching loss.

The general characteristics that are required for turn-on by the gate are shown in Figure 8.12, which depicts an acceptable region of gate voltage and current for device turn-on. The numerical values within this region vary somewhat with the device size and type. Device turn-on time can vary considerably

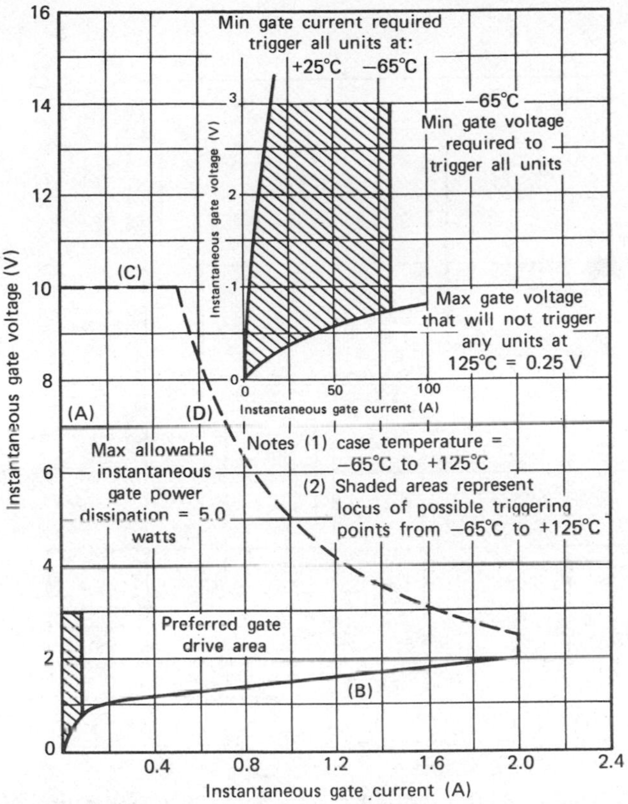

Figure 8.12. Gate turn-on characteristics. (Courtesy Westinghouse Electric Corp.)

for pulses of voltage and current within the designated region. Pulses of high current and short time duration generally reduce the device turn-on time. This is known as hard-firing or gate overdrive. A gate pulse shape that is widely used has an initial high value of current with very short rise time, which is quickly reduced to a lower current level and remains at this level for several microseconds.

8.5.2 GATE TRIGGERING CIRCUITS

Almost all types of pulse circuits are used for firing thyristors, and the gate triggering circuit design generally follows design methods used for other types of pulse circuits. Much will depend upon the nature of the input signals to the triggering circuitry (the control signals referred to above) of the requirements for operating the thyristor, and of the electromagnetic noise levels in the regions of the gate circuitry. Motor control systems generally operate under conditions of high electromagnetic noise.

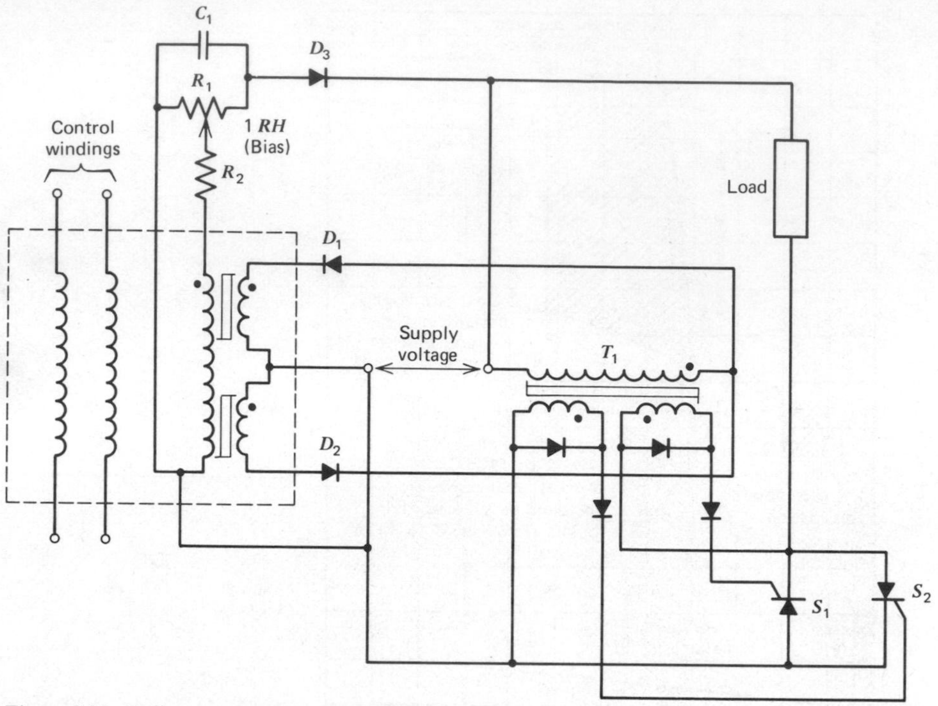

Figure 8.13. Full-wave magnetic amplifier firing circuit.

Components that have found widespread use in triggering circuits are of both semiconductor and magnetic types. Magnetic amplifiers have been frequently used for this purpose but have been displaced in recent years by semiconductor switching devices, which generally give faster switching times and lower weight triggering systems than magnetic devices. Figure 8.13 is an example

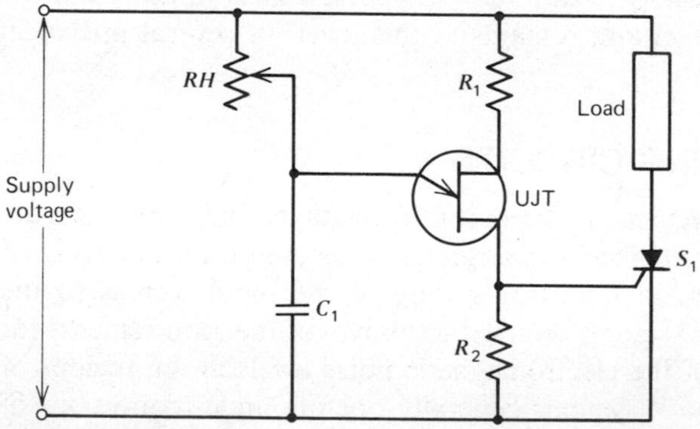

Figure 8.14. Half-wave phase-controlled semiconductor circuit.

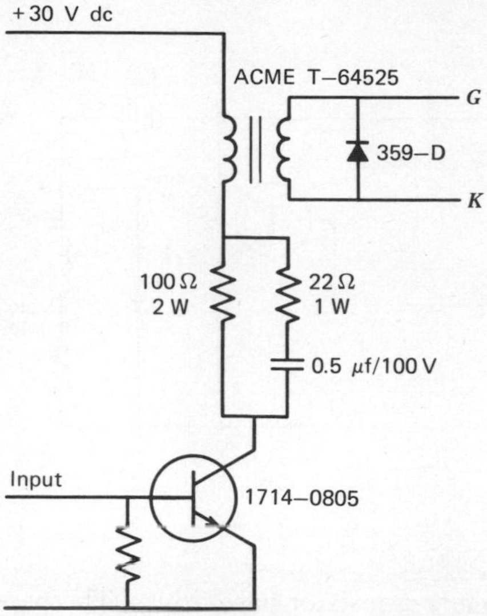

Figure 8.15. A firing module for SCR gate triggering circuit.

of a triggering circuit for a full-wave SCR bridge using magnetic amplifiers. Two types of semiconductor switches are useful in this application: the unijunction transistor (UJT) and the silicon controlled switch (SCS). Figure 8.14 illustrates a simple scheme using one UJT. The transistor is probably the most common device used in gate circuits. Figure 8.15 shows a transistor amplifier circuit with transformer coupling to the SCR gate. These examples do not even scratch the surface of the multitude of circuits and techniques for thyristor gate triggering. Many triggering systems are available fully packaged today and are widely used on standard applications, such as phase-controlled rectifiers.

Transformer coupling to the thyristor gate terminal, such as shown in Figure 8.15, serves a useful purpose in reducing noise signals at the gate which often cause false triggering of the thyristor. A false turn-on of a thyristor may be a very dangerous occurrence and often leads to catastrophic failure, since it can place the anode-cathode circuit in a direct electrical short across the power source. This will be apparent when we discuss motor controller circuits later in this chapter. Pulse transformers for use in gate triggering service have been discussed briefly in Chapter 3. To prevent the stray capacitances in a transformer from coupling noise signals into the gate, the transformer grounding and shielding must be carefully designed.

8.5.3 TRANSISTOR BASE DRIVES

Circuits for turning on power transistors are probably more familiar to the reader and are discussed in many textbooks. Most of the considerations dis-

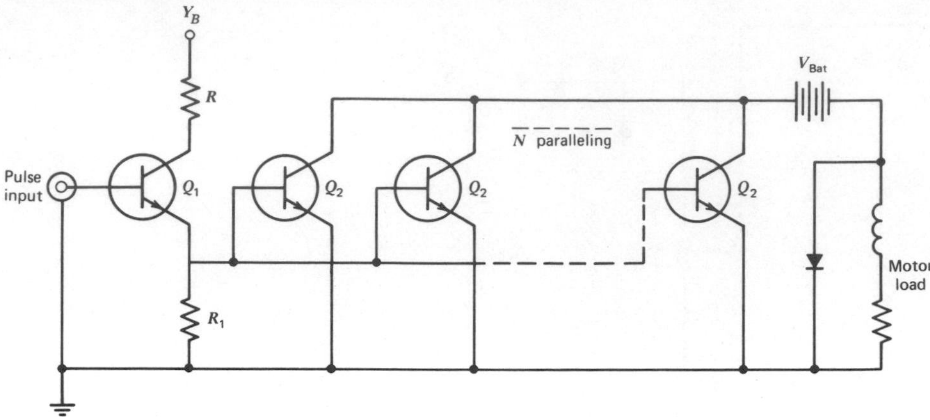

Figure 8.16. Common-emitter follower drive.

cussed for thyristor gate drives are applicable to transistor base drives. The base signal is, of course, maintained during the entire on-time of the device, which necessitates some differences in design as compared with the very short gate pulses required for thyristor turn-on. Some of the desired variations in base signal to obtain certain device turn-on and turn-off characteristics have been discussed in Section 8.3.5. Since transistors are frequently connected in parallel to achieve high current capability in motor control applications, proper synchronization of base signals is important. Figure 8.16 shows a common-emitter-follower drive for N parallel transistors where synchronization is obtained by direct electrical paralleling of the N base terminals. Figure 8.17 is a complete base drive for operating a transistor in a pulse-width-modulated mode, to be discussed later in this chapter.

8.6 POWER SEMICONDUCTOR PROTECTION

Power semiconductors are easily damaged by operation at excessive temperature, by overcurrents, by voltage spikes or transients which cause false triggering of the power device, and, in the case of thyristors, by an excessive rate of current rise at turn-on. The latter potential source of failure can be guarded against only by the proper design of the circuit at turn-on and by proper choice of the thyristor in the specific application.

Thermal protection of a power device must also be accomplished during the design stage of a motor control system. There is no true "protection" from excessive device temperatures, but, with sufficient analysis of the expected device loading, a device of sufficient current and voltage rating and an appropriate heat sink can be chosen for a specific application. The required analysis includes thermal analysis to determine the size and type of heat sink

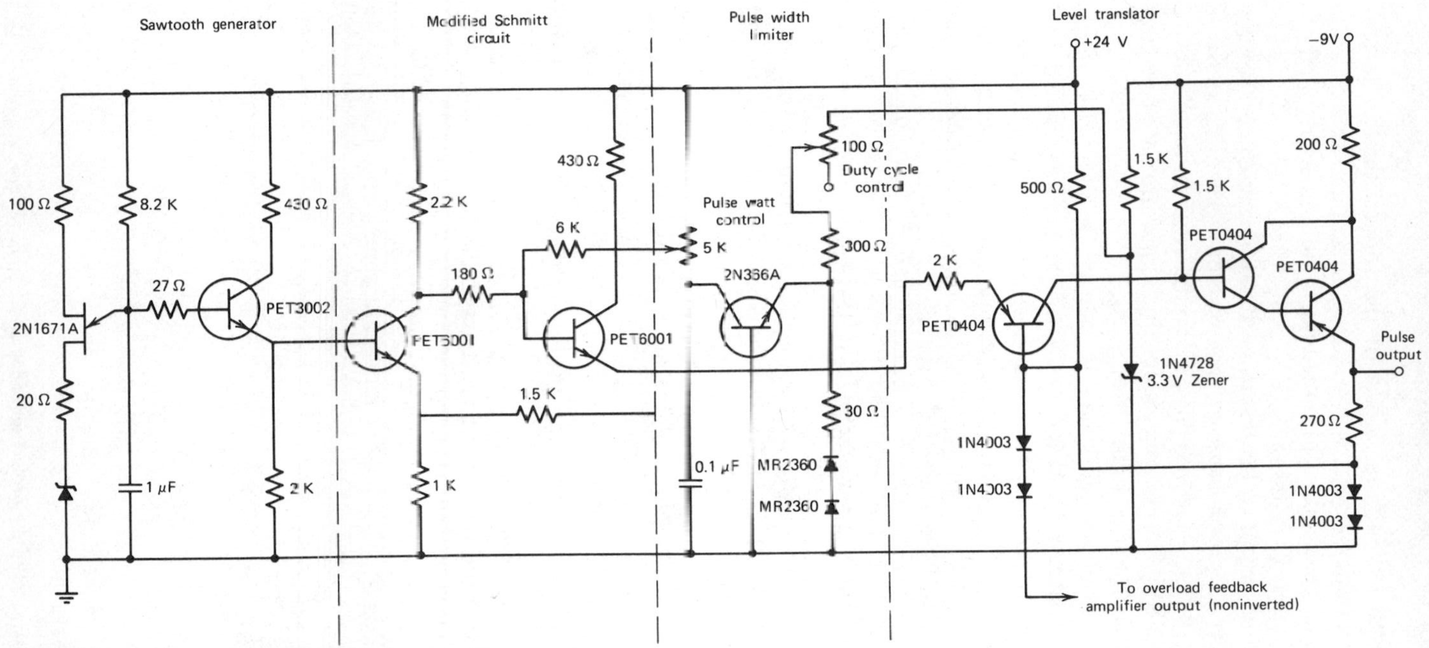

Figure 8.17. Pulse width modulator for pnp transistor.

required, based upon predicted internal losses of the semiconductor devices. Thermal analysis of semiconductor devices is beyond the scope of this text, but many guidelines in choosing heat sinks and forced-air or liquid cooling methods may be found in References 1, 2, 7, and 8.

Overcurrent protection requires fast-clearing fuses because of the low current overload capability of power semiconductors. The fuse rating must be well within the I^2t–rating of the semiconductor, a parameter given on the device data summary. (See Figure 8.6.) Fuses designed for this application are commercially available, and are illustrated in Figure 8.5.

Improper triggering due to voltage spikes is reduced primarily by proper design of the gate triggering circuit. However, to prevent triggering from external signals that may inadvertently be coupled to the device or its power circuitry, the anode-cathode terminals of thyristors are usually paralled by an *R-C* network, as shown in Figure 8.18. Methods for designing these networks, called snubber networks, are given in References 1, 2, and 16.

8.7 PHASE CONTROLLED RECTIFIERS

A lot of information about this circuit is contained in its name. The word *rectifier* implies that it is an ac to dc conversion device, which it is. The term, *controlled rectifier*, implies the use of silicon *controlled* rectifiers rather than common rectifiers, which also is true. And, finally *phase* implies control as a function of a signal related to the phase of the ac, also true. Phase controlled rectifiers refer to a collection of circuits which, in terms of their power circuitry, are identical to the well-known circuits used for ac/dc conversion at all power levels. Rectifier circuits have no inherent control properties but are conversion circuits with fixed ac-dc voltage ratios that vary only slightly when voltage drops are caused by load currents. By replacing the rectifying device with a controlled rectifier, specifically, an SCR, these circuits still have the ac/dc conversion capability. They can also effect a large measure of control over the magnitudes of the voltage and current on the dc side. Phase controlled rectifier circuits are widely used for the control of dc commutator motors at almost all power ratings

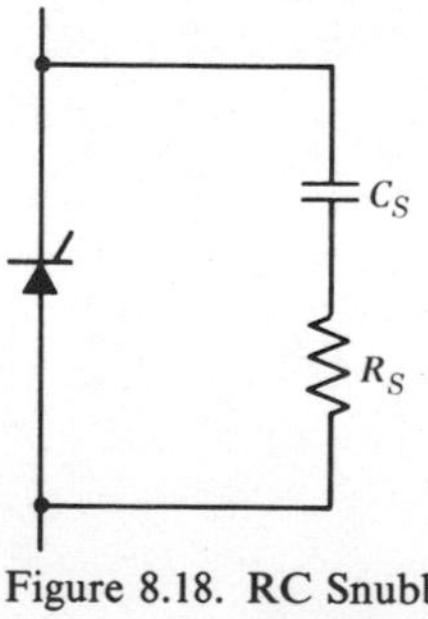

Figure 8.18. RC Snubber network, used in the protection of thyristors from voltage transients.

wherever the power supply is alternating current. They are the simplest form of electronic motor control of the four classes that we will be discussing. This is so in part because the SCRs are *line-commutated* from the ac source rather than requiring commutation by auxiliary circuits of the type discussed in Section 8.4.

In Chapter 5, the various techniques for dc commutator motor control were presented and it was noted that these fall into two general categories: armature control and field control. In the case of a series motor, the control voltage operates upon both the field and armature in series. These three general modes of dc commutator motor control have different implications concerning the performance, waveforms, and control techniques of the controller system. In the case of a separately excited motor in which only field control is to be used, the controller looks into a *static R-L* circuit, that is, the field circuit. In the case of armature control of a separately excited motor and the control of a series motor, the controller looks into and *active circuit* consisting of the motor back emf plus an equivalent series resistance and inductance. The distinction between *active* and *passive* loads will be of some importance in the analysis. Phase controlled rectifier circuits are used to control both field and armature (and series motor) circuits, so it is of interest to evaluate the performance of these circuits with both active and passive loads.

The simplest rectifier circuit is the half-wave rectifier, and we shall use this circuit to demonstrate the general principles of phase control. Figure 8.19 illustrates a simple half-wave phase controlled rectifier for controlling a *static* inductive load. The waveforms of several voltages and currents are also shown in Figure 8.19. Here, v_{in} is the input voltage, a sinusoidal quantity with frequency, $1/T_p$. The SCR is programmed to be turned on to a conductive state at a time interval, T_α, after each zero crossover with increasing voltage of each cycle. At the half-period time, T_o, the applied voltage becomes negative, applying a reverse bias to the SCR, S_1. However, because of inductive load, the current lags the voltage in phase and continues beyond the time, T_o. Reverse bias is applied to S_1 during the entire half-period, $T_p - T_o$, assuring adequate turn-off. The waveform of the load currents, i_l, is sinusoidal, and can be calculated from the conventional relationships for an R-L circuit energized from a sinusoidal voltage with zero time set equal to T_α. The rms and average values of the instantaneous voltage and currents shown in Figure 8.19 can be obtained with the aid of Table 8.1. It is seen that load voltage and current are controlled by the value of the time, T_α, which is determined by the timing of the gate pulse for S_1. Increasing T_α decreases the load rms and average current and voltage, according to the relationships given in item 7 of Table 8.1.

A full-wave bridge circuit is shown in Figure 8.20, along with selected waveforms for a static RL-load. It is seen that only two of the four semiconductor devices in the bridge circuit must be controlled rectifiers. This circuit is seldom practical as it is shown in Figure 8.20 for two reasons: The current in the load is of a pulsed nature with low ratio of average and rms to peak current and,

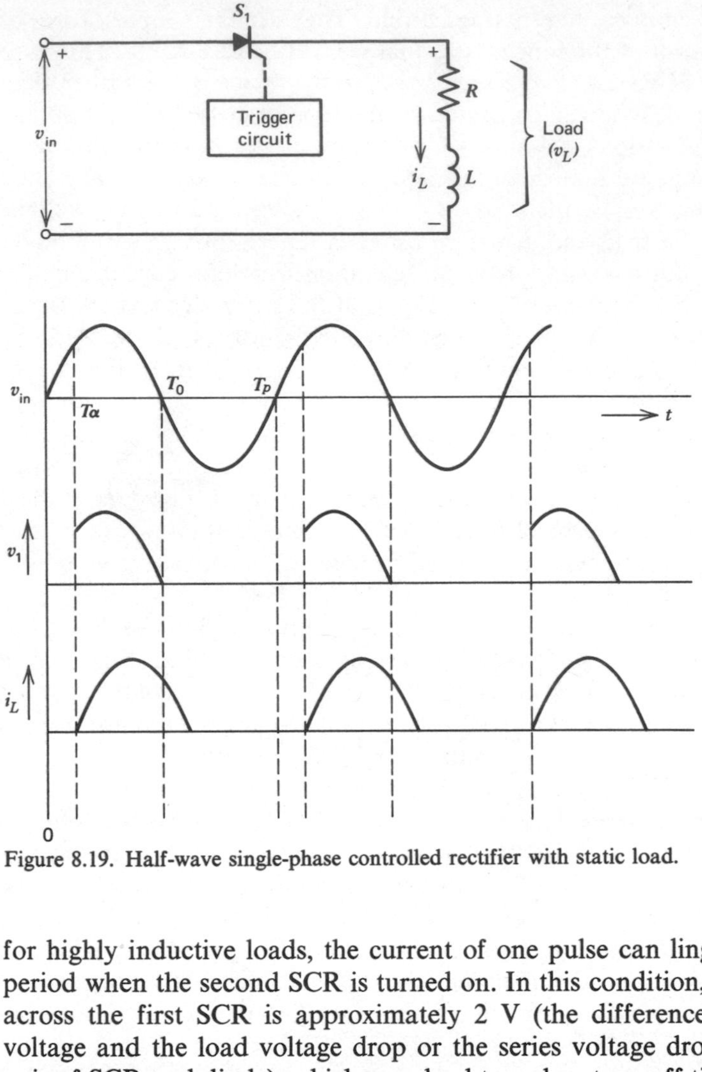

Figure 8.19. Half-wave single-phase controlled rectifier with static load.

for highly inductive loads, the current of one pulse can linger on into the time period when the second SCR is turned on. In this condition, the net reverse bias across the first SCR is approximately 2 V (the difference between the input voltage and the load voltage drop or the series voltage drop across the second pair of SCR and diode), which may lead to a slow turn-off time of the first SCR. In most inductive load applications, the load is in parallel with a freewheeling diode as shown in Figure 8.21. The freewheeling diode permits a more continuous load current or armature current and improves the turn-off capability of the first SCR.

The operation of the circuit in Figure 8.21 is as follows: Assume that the system is in a steady-state condition at $t = T_\alpha$ with a current, I_o, in the load and freewheeling diode and all semiconductors in the off-state. This analysis will begin with the turning on of S_1. When S_1 is turned on, it "sees" the parallel combination of the load and freewheeling diode plus the voltage drop across D_2.

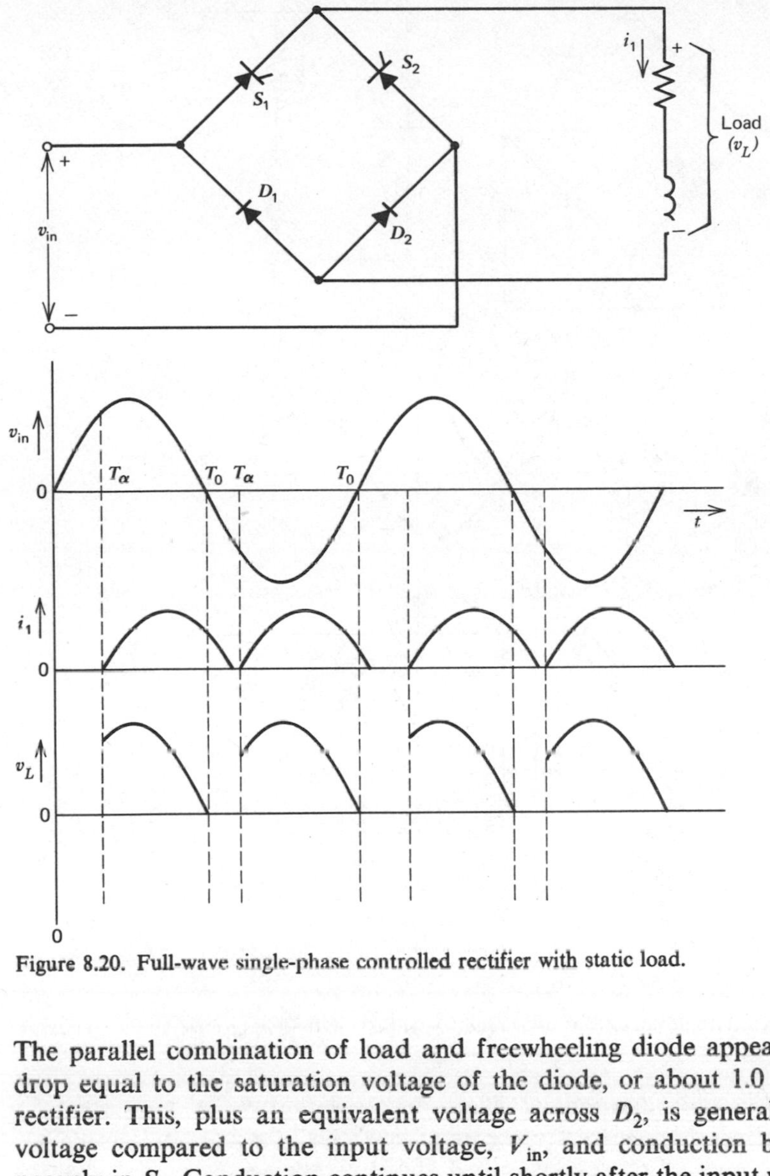

Figure 8.20. Full-wave single-phase controlled rectifier with static load.

The parallel combination of load and freewheeling diode appears as a voltage drop equal to the saturation voltage of the diode, or about 1.0 V for a silicon rectifier. This, plus an equivalent voltage across D_2, is generally a negligible voltage compared to the input voltage, V_{in}, and conduction begins instantaneously in S_1. Conduction continues until shortly after the input voltage reverses polarity at T_o—specifically, after the negative value of v_{in} exceeds the voltage drop of the load-freewheeling diode combination, or about -1.0 V. After this condition occurs, S_1 is reverse-biased and is commutated. Current continues to flow in the load-freewheeling diode combination until S_2 is turned on at an interval, T_α, after the first half-period. Conduction through S_2 is near instantaneous for the same reasons as given above when S_1 was turned on and continues

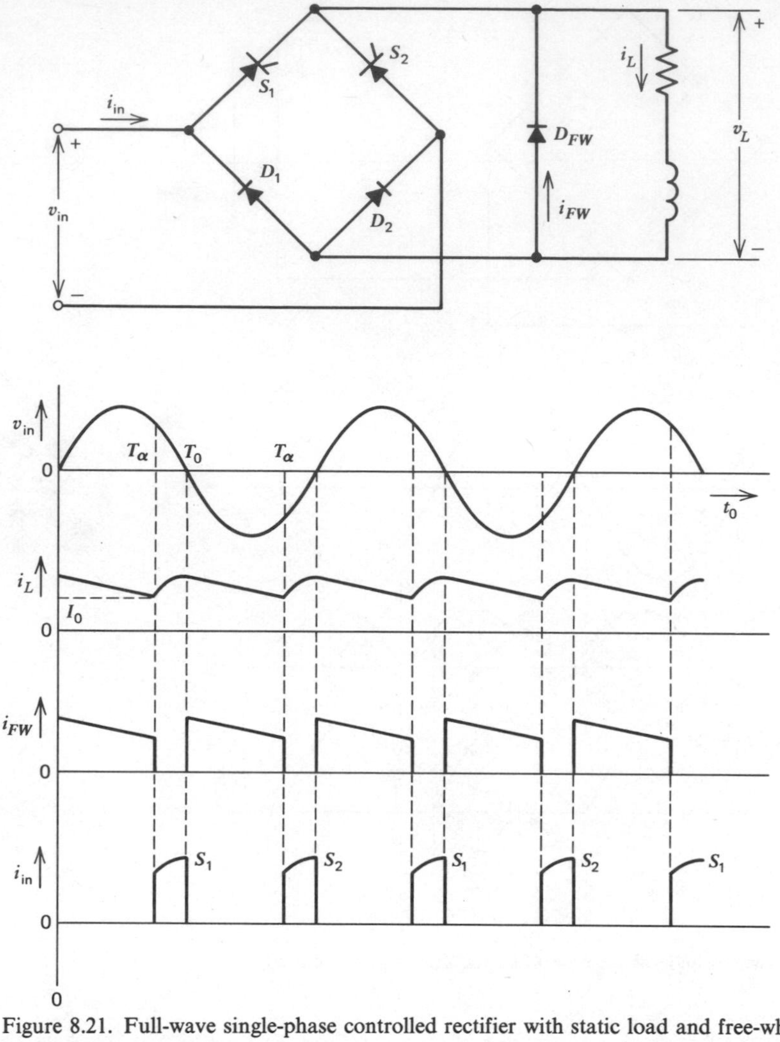

Figure 8.21. Full-wave single-phase controlled rectifier with static load and free-wheeling diode.

until shortly after the input voltage again becomes positive, commutating S_2. For a brief interval, the load current again "freewheels" through D_{FW} until S_1 is again turned on and the process described above is repeated.

During the period in which the freewheeling diode is conducting in the above operation of the circuit in Figure 8.21, the current can be calculated from the relationships for a transient current in an RL-circuit shorted across itself (which neglects the small voltage drop across the diode) with an initial ($t=0$) current equal to the current when the SCR ceases conduction. That is, the current during the freewheeling period has the waveform of transient current decay in an RL-circuit. The sinusoidal current pulse during the time when one of the SCRs is conducting is that of an RL-circuit, which is energized from a sinusoidal source. The beginning of this current ($t=0$) is carefully related to the time, T_α, in Figure 8.21. Thus, the load current shown in Figure 8.21 is the piecemeal combination of two types of transient currents, each of which can be readily calculated. This is characteristic of the current waveforms found in motor control systems.

With an active load, such as a motor armature, the waveforms are similar to those with a passive load and are sketched in Figure 8.22. Here, the armature emf, E_a, is shown as a constant magnitude with time. Actually, there is a slight ripple in the curve of E_a, since the motor speeds up slightly when energized from the bridge and slightly decreases in speed when the armature current is freewheeling. Except in motors with very light inertia, this ripple is generally too small to be observable. The principal difference between the operation of the circuit with active and passive loads is the instant at which the bridge SCRs turn off and freewheeling begins. With an active load represented by a voltage, E_a, this instant occurs when the line voltage, v_{in}, becomes less in magnitude than E_a. This instant is shown as time, T_1, in Figure 8.22. Average and rms values for the voltage and current waves shown for the phase controlled rectifier circuits can be calculated from the relationships given in Table 8.1.

Where the power source for the motor is polyphase, polyphase bridge circuits are used in phase control. Figure 8.23 illustrates the three-phase half-wave circuit and associated waveforms. Figure 8.24 shows the three-phase, full-wave bridge circuit which is the most common polyphase rectifier and inverter circuit.

8.7.1 FREEWHEELING DIODE CONSIDERATIONS

The freewheeling diode does require additional concerns in the sizing of the SCRs in the phase-controlled rectifier and in most other controllers. For example, when S_1 in Figure 8.22 is turned on, the freewheeling diode is reverse-biased and turns off, causing its current (which is also the armature current) to be transferred to the bridge circuit through S_1 and D_2. The buildup of current in S_1 is limited solely by the inductance in these two branches of the bridge circuit and the power source, since the armature inductance is already charged with this value of load current. The bridge and source inductance is often very low. For example, if the source is an "infinite bus," which is frequently the case in large power systems, it has no inductance. Inductances of 1 μH or less occurs, for example, which can allow very rapid buildup of current in S_1 that can lead to hot spots and device failure. This situation is frequently remedied by placing a small, saturable reactor in series with the anode of S_1 which will produce a time delay of 5 to 10 μs or so. Once the reactor is saturated, it results in a negligible voltage drop and power loss. The reverse

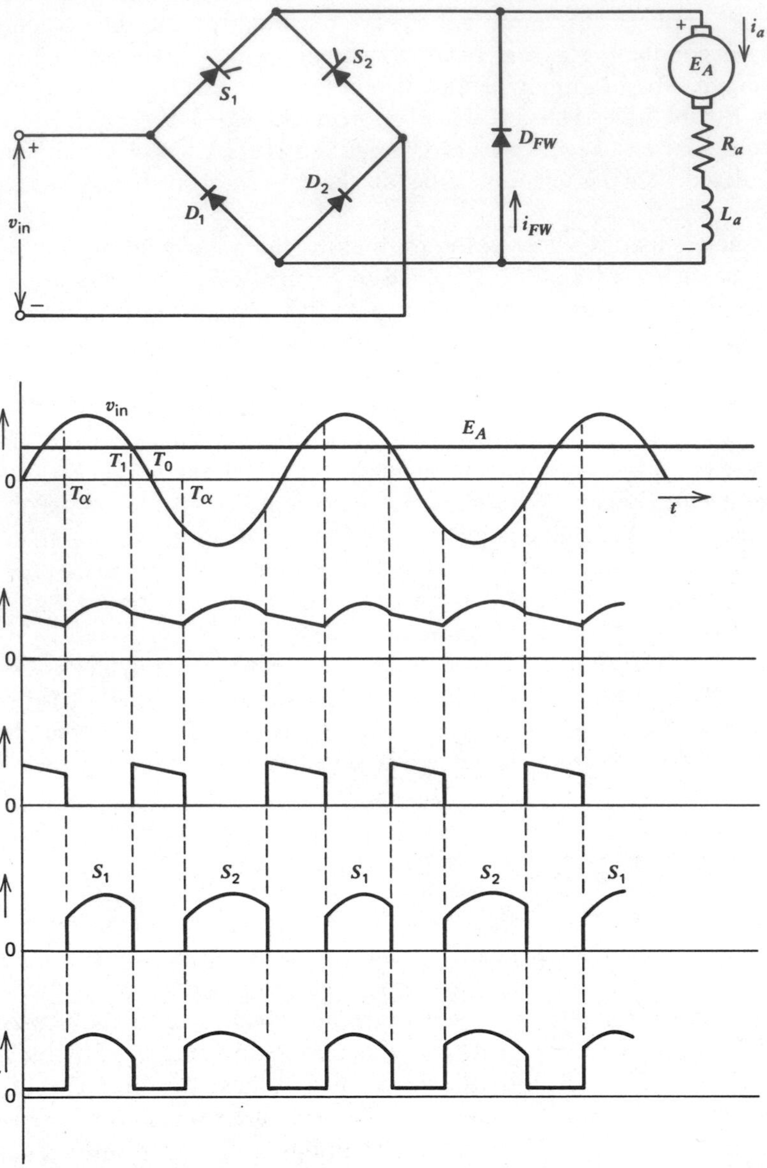

Figure 8.22. Full-wave single-phase controlled rectifier with active load and free-wheeling diode.

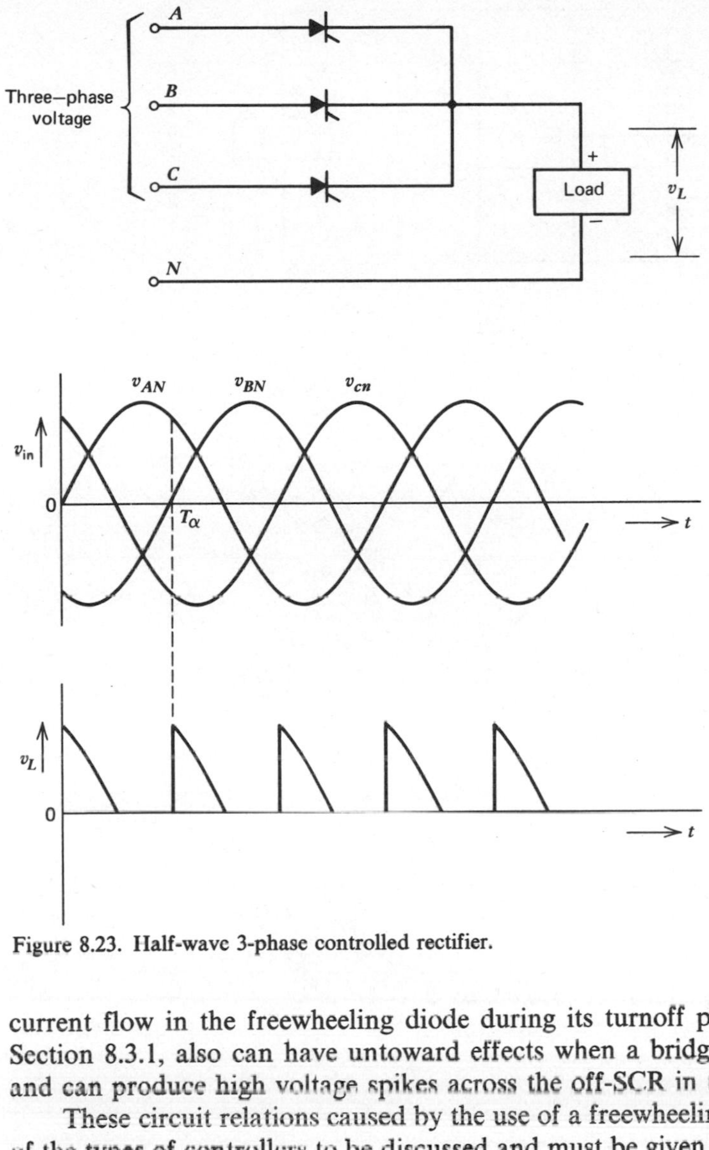

Figure 8.23. Half-wave 3-phase controlled rectifier.

current flow in the freewheeling diode during its turnoff process, discussed in Section 8.3.1, also can have untoward effects when a bridge SCR is turned on and can produce high voltage spikes across the off-SCR in the bridge.

These circuit relations caused by the use of a freewheeling diode occur in all of the types of controllers to be discussed and must be given consideration in the design of controller circuits.

Example 8.7

Determine the average voltage applied to the load for the half-wave phase controlled rectifier of Figure 8.19 if the input voltage, v_{in}, is 120 V and 60 Hz, and T_α is one-half of T_o. Repeat for $T_\alpha = 0.002$ s.

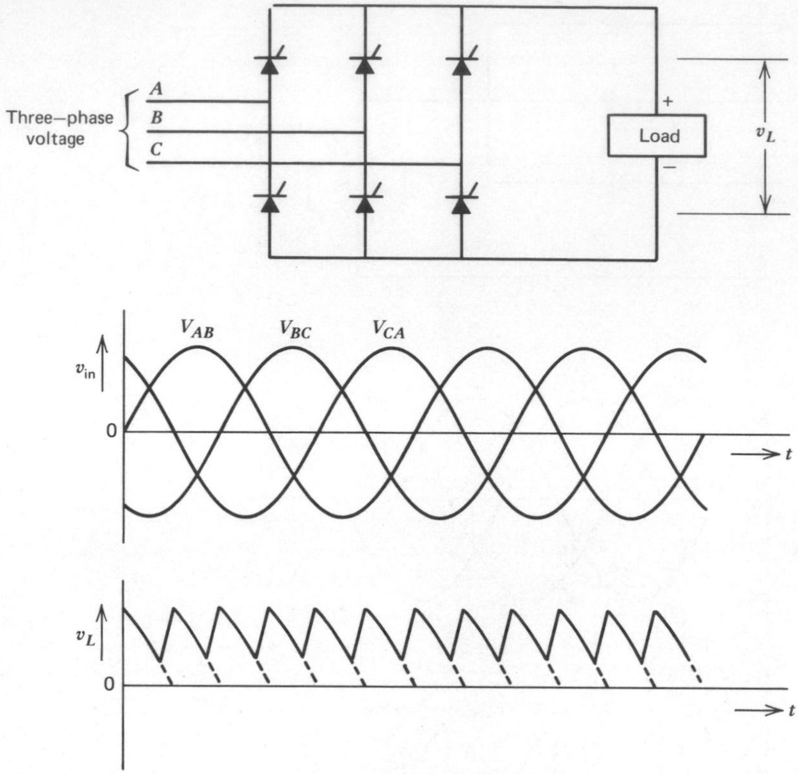

Figure 8.24. Full-wave three-phase controlled rectifier and voltage waveforms.

The average voltage can be obtained directly from Table 8.1, item 7. For 60 Hz, the full period, T_p, is 0.0167 s; T_o, the half-period is 0.0083 s. For T_α equal to half T_o,

$$V_{av} = \frac{1}{2\pi}(1 + \cos\frac{\pi}{2})\sqrt{2} \cdot 120 = 27 \text{ V}$$

For T_α equal to 0.002 s,

$$V_{av} = \frac{1}{2\pi}\left(1 + \text{cox}\,\pi\frac{0.002}{0.0083}\right)\sqrt{2} \cdot 120 = 46.5 \text{ V}$$

Example 8.8

Repeat Example 8.7 for the full-wave circuit of Figure 8.20.

For the purpose of calculating average values, the period of a full cycle of load voltage in the full-wave circuit is the half-period of the sine wave, shown as T_o in Figure 8.20. Therefore, using Table 8.1 again, for T_α equal to half T_o,

$$V_{av} = \frac{1}{\pi}\left(1+\cos\frac{\pi}{2}\right)\sqrt{2}\cdot 120 = 54 \text{ V}$$

which is twice the value found for the half-wave circuit, as we should expect. The average voltage with $T_\alpha = 0.002$ s will likewise be twice the half-wave circuits' value, or 93 V.

Example 8.9
If the load in the full-wave cirtcuit of Figure 8.20 consists of a resistance of 1.2 ohms and an inductance of 2 ohms, determine the average load current for the two conditions of firing angle used in Example 8.8.

The *average* voltage across the load inductance is zero over a complete period of applied voltage, since the integrated volt-seconds across the inductance during charging (i increasing) exactly equals the negative integrated voltseconds across the inductance during discharging (i decreasing). Hence, the average voltages calculated in the previous examples appear solely across the load resistance (neglecting the SCR and diode drops) and average load current can be found by dividing the average voltage by the load resistance. For T_α equal to half T_o, $I_{av} = 54/1.2 = 45$ A; for $T_\alpha = 0.002$ s, $I_{av} = 93/1.2 = 77.5$ A.

8.7.2 CONTROLLED RECTIFIER CURRENT RELATIONSHIPS

The previous examples have calculated several average voltages for phase controlled rectifier circuits, and it was found that the waveform parameters from Table 8.1 could be used for this purpose. The analysis of currents in controlled rectifier circuit is not as simple.

When a rectifier is turned on into an inductive load of R ohms and L henries at time T_α, the resulting current flow can be calculated from circuit theory. The voltage drops in this circuit are described by the differential equation,

$$E_m \sin \omega t = iR + L\frac{di}{dt} \tag{8.29}$$

The solution for current in (8.29) is

$$i = \frac{V_m}{Z}\left[\sin(\omega t-\phi) - e^{-[(t-T\alpha)/\tau]}.\sin(\omega T_\alpha - \phi)\right] + I(T_\alpha)\cdot e^{-[(t-T_\alpha)/\tau]} \tag{8.30}$$

where

$$Z=\sqrt{R^2+(\omega L)^2}\ ;\ \phi=\tan^{-1}(\omega L/R)$$

V_m = maximum value of the input voltage, v_{in}

$I(T_\alpha)$ = load current existing at time, $t=T_\alpha$

$\tau=L/R$, time constant of load

$\omega=2\pi f$ = angular frequency of v_{in}

This current consists of three components: the steady-state value, which is the first term in (8.30); a voltage-related transient component, the second term; and a second transient term resulting from an initial value of current in the inductance. The latter term is characteristic of systems with freewheeling diodes. In continuously energized circuits, these transient terms disappear after a few cycles of the input signal. In discontinuous circuits, these terms are usually present at the beginning of each switching interval and may have just as much influence on the shape of the current pulse and its period as the steady-state component. The shape of the current pulse described by (8.30), which is a combination of a sine pulse and exponential terms, is sometimes called quasi-sinusoidal because its appearance usually bears much resemblance to a sinusoidal shape. The pulse parameters, such as pulse period, maximum value, slopes, and so forth, cannot be calculated on the basis of sinusoidal assumptions. The analysis of this type of current pulse is further complicated by the fact that the first transient term in (8.30) varies with the instant at which the SCR is turned on; in this case, the time, T_α. If turn-on occurs such that $\omega T_\alpha=\phi$, the second term in (8.30) does not exist. If turn-on happens to come at the instant when $\omega T=\phi\pm\pi/2$, the second term is maximum. The reader may recall that this situation exists in the turning on or energizing of most inductive devices, such as transformers, from ac sources. We have already attempted to illustrate examples of these quasi-sine pulses: in Figures 8.19 and 8.20, the load current pulses are of this shape with $I(T_\alpha)$ zero. The SCR current pulses (and intervals of the load current) in Figures 8.21 and 8.22 are also of this shape but with a nonzero $I(T_\alpha)$.

Figure 8.25 is a more detailed picture of the type of pulse existing with zero initial current, $I(T_\alpha)$. The two components of the actual current pulse, the steadystate and transient components, are also shown. The pulse period is

$$T_o=T_\beta-T_\alpha \tag{8.31}$$

The value of T_β for a current pulse of this type can be found by setting $i=0$ and

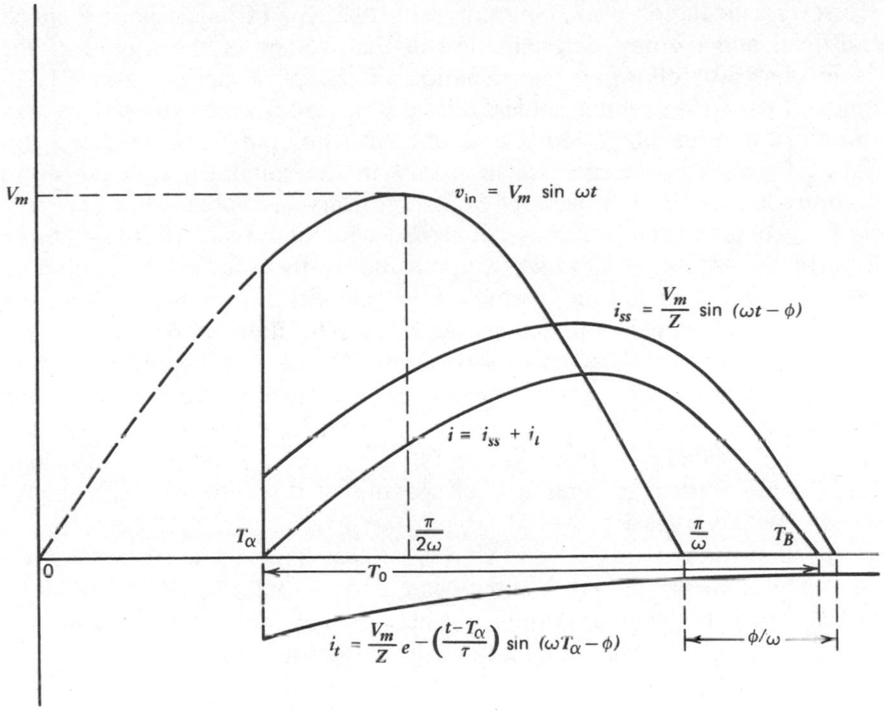

Figure 8.25. Load current pulse for Figure 8.19.

$t = T_\beta$ in (8.30), which gives the transcendental equation

$$\sin(\omega T_\beta - \phi) = e^{(-T_\beta - T_\alpha)/\tau} \sin(\omega T_\alpha - \phi) \tag{8.32}$$

This is of the same form as (8.18) developed for series commutation circuits. The use of hand calculations for obtaining numerical values from expressions of this form arc tedius and time-consuming, but computer routines are easily developed. Also, curves have been prepared giving the period, T_o, of this pulse shape as a function of T_α and ϕ.[17]

Fortunately, the calculation of the *time-average* value of current through the load resistance for current pulse forms of this type can be calculated by the method given in Example 8.9, since the time-average value of voltage across the load inductance is zero. Time-average of dc current is obtained by dividing the average value of applied voltage (found from Table 8.1, item 7) by the load resistance. This value is of use in determining system efficiencies, load power, and the average currents in the SCRs and diodes. Time-average values of current and voltage are the values measured by average-indicating instruments, such as D'Arsenvol instruments.

However, calculation of rms currents with this type of pulse shape is much more difficult and requires determination of the average of the square of the right side of (8.30) following determination of the pulse period from (8.32). Reference 17 has worked out graphical relationships from which rms values may be determined in terms of T_α and ϕ, and, of course, computer routines are again applicable. Often rms currents are necessary in determining ohmic losses in motor controllers and in the sizing of SCRs and electrical cables.

So far, we have been primarily concerned with analysis of (8.30) when the initial current, $I(T_\alpha)$ is zero, which is applicable to the load current pulses in Figures 8.19 and 8.20 and to many similar circuits. More generally, these relationships are applicable to *discontinuous* current flow in *RL*-circuits energized from sources with sinusoidal waveshape. The case of *continuous* current flow is also of concern and exists in many motor control systems in which the motor is paralleled with a freewheeling diode. Typical current waveshapes have already been presented in Figures 8.21 and 8.22. However, even though the load current is continuous, it is composed of sections of discontinuous currents of pulses. To use the correct technical jargon, these sections are called *piecewise continuous* currents. The load current shown in these figures is composed of two sections: the current through the source, load, and one set of SCR and diode in the bridge, shown as the input currents in these figures; and the current through the freewheeling diode, load combination, shown as i_{FW}.

The first of these piecewise continuous currents can be analyzed with the aid of (8.30) with $I(T_\alpha)$ at some nonzero value. The analysis is similar to that described in the previous paragraphs, except that it is further complicated by the addition of the second transient term involving $I(T_\alpha)$. Note that this term must be added to (8.32). The instant of time at which the pulse begins is T_α. The time instant at which the pulse ends is determined by the voltage relationships in the load circuit that result in the freewheeling diode becoming forward-biased, and in turning on. With a static load, this time is approximately at the zero crossing of the input voltage, shown as T_o in Figure 8.21. With an active load, this instant is approximately, when the input voltage becomes less in magnitude that the load voltage, shown as T_1 in Figure 8.22. These time instants are described as approximate instants, since they may be altered somewhat by the turn-on characteristics and voltage drop of the diode and by inductances in the diodes and SCRs and their leads. If these inductances are negligible (less than a μH), these time instants can be used in the analysis, as stated above.

The freewheeling diode current begins at the instant when the pulse through the bridge ends—again assuming that the secondary effects mentioned above are negligible. This current segment ends when the source is again applied to the load through the bridge, thus reverse-biasing the diode and turning it off. During conduction of the freewheeling diode when the load is static, the current in the load-diode loop can be described by the familiar inductive current decay expression

$$i_{FW} = I(T_o)e^{-t/\tau} \tag{8.33}$$

where $I(T_o)$ equals the load current at the zero crossing of the input voltage signal, T_o. The initial current, $I(T_o)$ can be calculated by means of (8.30). The instant, T_o, occurs shortly after the instant at which the current pulse is at its maximum value. When this current pulse is maximum, di/dt and the voltage across the load inductance are zero. Then the magnitude of the current pulse starts to decrease from this maximum and di/dt and the inductance voltage become negative. It is this action of the load inductance that forward-biases the freewheeling diode and causes conduction to begin when the inductive voltage becomes sufficiently negative to overcome the diode voltage drop (one volt or less). For systems with a dynamic load of voltage, E_a, current flow in the diode-load loop is given by

$$0 = Ri + L\frac{di}{dt} + E_a \tag{8.34}$$

The solution for current can be obtained by conventional methods from (8.34). Thus

$$i_{FW} = I(T_1)e^{-t/\tau} - \frac{E}{R}(1 - e^{-t/\tau}) \tag{8.35}$$

where $I(T_1)$ = the load current at the instant when $E_a = V_{\text{in}}$ and $\tau = L/R$. This time instant is shown as T_1 in Figure 8.22.

The shape of the freewheeling current is exponential, as shown by (8.33) and (8.35). These shapes are frequently approximated by straight-line segments with negative slopes. This is particularly true in the case of an active load, which is of interest in our discussion of dc commutator motor loads. In many circuits of this nature, the load resistance can be neglected and (8.34) becomes

$$0 = L\frac{di}{dt} + E_a \tag{8.36}$$

the solution of which is

$$i_{FW} = I(T_1) - \frac{E}{L}t \tag{8.37}$$

Thus, when the load resistance of an active load is negligible, the load current is truly a straight line with a negative slope. This assumption is frequently used in the analysis of dc commutator motors. The waveshapes in Figures 8.22 and 8.23

and item 6 of Table 8.1 make use of this assumption. It is also a common practice to approximate the rising portion of the load current (the segment when the source voltage is applied) with a straight line. This results in the zigzag waveform shown in item 5 of Table 8.1, which is commonly used for a motor load with a freewheeling diode. Approximations using striaight-line segments are especially convenient in calculating average and rms values of load currents and are sufficiently accurate for many purposes.

This discussion of current pulse shapes, periods, average, and rms values has been based upon the phase controlled rectifier system, but is applicable to many other situations in motor control. We shall refer to these relationships frequently in discussions of other types of motor control.

Example 8.10

A dc commutator motor is controlled from a single-phase, full-wave bridge controlled rectifier (Figure 8.22) energized from a 60-Hz, 120-V power source. The motor has an armature resistance of 0.1 ohm and an armature inductance of 1 mH. The armature is in parallel with a freewheeling diode. The motor is operating at a constant speed such that the armature back emf is 60 V. At a certain instant when one SCR is gated on ($T_\alpha = 0.006$ sec.) the armature current is 80 A. Determine

a. The value of armature current at the time T_1 when the freewheeling current begins and the SCR is turned off.

b. The current at the end of the freewheeling current segment.

V_m for a 120-V rms source is 169.7 V; $Z = \sqrt{0.1^2 + (377 \times .001)^2} = 0.39\angle 75.2°$. $\phi = 75.2° = 1.3125$ rad; $V_m \sin\omega T_1 = 60$; $\omega T_1 = \pi - 60/169.7 = 2.786 =$; $\omega \mathrm{T}_\alpha 377 \times 0.003 = 1.131$ rad; $T_0 = (1/2)(1/60) = 0.00833$ s; $\omega T_o = \pi$; $\tau = 0.001/0.1 = 0.01$ s. $i(T1)$

From (8.30)

$$i(T_1)$$

$$= \frac{169.7 - 60}{0.39}\left\{\sin(2.786 - 1.3125) - \exp\left[-\left(\frac{0.0083 - 0.003}{0.01}\right)\right]\sin(1.131 - 1.3125)\right\}$$

$$+ 80\exp\left[-\left(\frac{0.0083 - 0.003}{0.01}\right)\right]$$

$$= 281.3[0.995 - 0.586(-0.1805)] + 80 \times 0.586$$

$$= 356 \text{ A, at the end off the first segment} \qquad \text{(a)}$$

The freewheeling segment continues until the next SCR is gated on, which is T_α s after the zero crossover of the input voltage. Therefore, the free-wheeling current at T_α s is, from (8.35)

$$i_{FW}(T_\alpha)=356\exp\left(\frac{-0.003}{0.01}\right)-\frac{60}{0.1}\left[1-\exp\left(-\frac{0.003}{0.01}\right)\right]$$

$$=264-155.5=108.5 \text{ A} \tag{b}$$

Example 8.11

Figure 8.26 illustrates a steadystate load current in a motor armature in parallel with a freewheeling diode. Straight-line approximations of the current have been used, and maximum and minimum values are shown. Determine the average and rms values of the input, load, and freewheeling diode currents.

The ratio of minimum to maximum current in this waveform, ξ, is $100/150 = 2/3$. The input current can be calculated from the relationships of item 6 of Table 8.1:

$$I_{rms}=150\sqrt{\frac{1}{3}\left(\frac{0.002}{0.006}\right)\left(1+\frac{2}{3}+\frac{4}{9}\right)}=72.6 \text{ A}$$

$$I_{av}=150\left(\frac{1}{2}\right)\left(\frac{0.002}{0.006}\right)\left(1+\frac{2}{3}\right)=41.7 \text{ A}$$

The freewheeling diode current can be obtained from item 6 of Table 8.1:

$$I_{rms}=150\sqrt{\left(\frac{0.004}{0.006}\right)\left(1+\frac{2}{3}+\frac{4}{9}\right)\frac{1}{3}}=102.8 \text{ A}$$

$$I_{av}=150\left(\frac{1}{2}\right)\left(\frac{0.004}{0.006}\right)\left(1+\frac{2}{3}\right)=83.3 \text{ A}$$

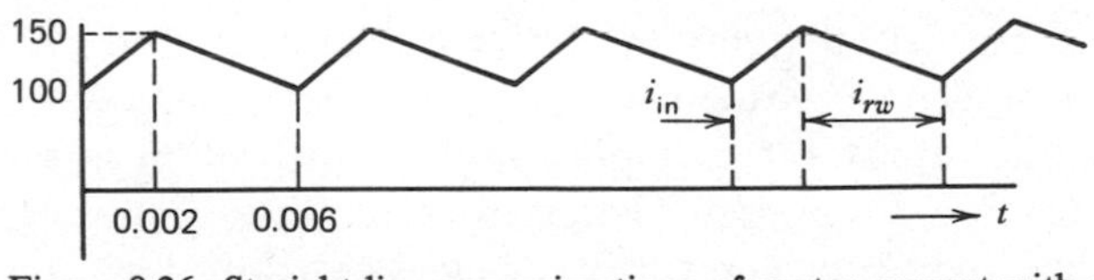

Figure 8.26. Straight-line approximations of motor current with chopper control and free-wheeling diode.

The load current is found from item 5 of Table 8.1:

$$I_{\text{rms}} = 150\sqrt{\frac{1}{3}\left(1 + \frac{2}{3} + \frac{4}{9}\right)} = 125.8 \text{ A}$$

$$I_{av} = 150\left(\frac{1}{2}\right)\left(1 + \frac{2}{3}\right) = 125 \text{ A}$$

8.8 CHOPPER CONTROLLERS

The principal dc commutator motor controller for control applications, in which the energy source is direct current, is known as a chopper controller or a voltage chopper. It is widely used in battery-energized systems such as the variable speed drives in golf carts, fork-lift trucks, and other electric vehicles. The term *chopper* is the generic name given to control devices in which a switch between the source voltage and load is periodically opened and closed in order to vary the load voltage. It originally served as a voltage regulator system and the switching device was frequently a mechanical switch, such as that in many automotive voltage regulators. The mechanism of voltage control is that of varying the on-time that a voltage of constant magnitude is applied to a load, rather than varying the magnitude of an input voltage continuously applied to the load. The on-time is varied by "chopping" the input voltage into discrete pulse, hence the name of this type of regulator. Motor control choppers differ from voltage regulators mainly in the range of load voltage over which control is maintained. For variable speed armature control, this range must be from zero to the maximum motor voltage. Motor control choppers are commercially available which use SCRs, GTOs, power transistors, and Darlingtons as the power switching device.

8.8.1 TIME RATIO CONTROL

Time ratio control refers to the control of a ratio of time parameters, usually expressed as the ratio of on-time to off-time, or the ratio of on-time to the time period of a pulse frequency, which is on-time plus off-time. It is because of the importance of this control concept that we have chosen to write the rms and average formulas given in Table 8.1 in terms of time parameters rather than angle parameters. In these formulas, T_o is the on-time of a pulse and T_p is the period of repetition and also the sum of off-time plus on-time. Inspection of the formulas in Table 8.1 indicates that, for an input voltage of fixed magnitude, V_m, the variation of rms or average voltage or current is achieved by controlling this ratio, T_o-T_p. This is the basis of time ratio control. Looking back at the discussions of phase-controlled rectifier controllers in the previous sections, we see that these also operate on the principle of time ratio control. In these controllers, the magnitude and frequency of the input voltage is fixed, and control is achieved by varying the ratio of on-time (by varying the SCR turnon time, T_α) to the repetition period which, in this case, is determined by the input signal frequency. Stated in another way, the average load voltage of a load is controlled by the variation of the SCR turn-on time over the time period of a half-cycle of the frequency of the input voltage.

There are several modes of time ratio control in common use in practical chopper controllers.

1. **Constant-frequency, variable pulsewidth.** In this control mode, the input voltage is gated on at a constant frequency determined by some fixed frequency signal, such as an oscillator, charging of an *R-C* network, line frequency, and so forth. The pulse width is determined by controlling the turn-off time of the power semiconductor or time at which SCR commutation is initiated. Or in some cases, the turn-off or commutation time is initiated at a fixed rate and turn-on time is varied to control the average load voltage.
2. **Variable frequency, constant pulsewidth.** In this mode, the turn-on time is controlled and varied as a function of required average load current or power. Once the power semiconductor is in the on-state, a fixed time delay is set into operation resulting in a fixed on-time period before commutation is initiated.
3. Combinations of the above two modes.

In any type of time-ratio control using SCRs, there is a fixed time period following the on-time pulse during which the SCR must be off before a subsequent on-time pulse can be initiated. This is the SCR parameter, t_{off}, described in Section 8.3.2, which determines the maximum value of average voltage that can be applied to the load. In many cases, this limitation is bypassed by directly applying the input voltage to the load, with no chopping. In the case of dc commutator motor loads, this can be done only at relatively high motor speeds that will provide sufficient back emf's to limit the armature current.

The waveforms of time ratio control are about as simple as possible. In Table 8-1, the load applied voltage can usually be represented by item 2; load current without a freewheeling diode can be described by item 4 (the slope of the parameter in this diagram—positive or negative—will not affect the average and rms formulas); item 5 is applicable to current in the load with a freewheeling diode; item 6 can be used when the straight-line approximations are applicable for current in the power semiconductor or the freewheeling diode, with the same comments on the slope of the function applicable as stated previously.

8.8.2 CHOPPER CIRCUITS

The variety of circuit configurations that come under the classification of choppers is far too numerous to describe here. Therefore, only a few of the basic circuits will be described. It is hoped that these descriptions will provide the reader with sufficient groundwork to permit the analysis of most types of chopper circuits. One basic criterion for classifying chopper circuits is their ability to accommodate *regeneration*, that is, classification in terms of unilateral or bilateral power flow. Most of the standard circuits fall into the former category. We shall illustrate several from each type of circuits.

A basic SCR chopper circuit is shown in Figure 8.10, which has already been discussed in relation to SCR commutation. Figure 8.10 has a static load, but this circuit is equally applicable to an active load, such as a dc commutator armature with back emf, E_a. The static load representation is, of course, applicable to an armature load at zero speed, where R and L represent the armature resistance and inductance, respectively. This chopper circuit is also frequently used to control field current in separately-excited dc commutator motors, in which case the load is a static load. This circuit is unilateral and will not accommodate regenerative braking of the motor. Figure 8.27 illustrates voltage and current waveforms for the circuit of Figure 8.10 applied to the control of a dc commutator motor armature with back emf E_a. Constant motor speed is assumed in this figure, which implies constant back emf, E_a for a constant field curcuit.

Figure 8.27 assumes that the system is in a steady-state condition with the load current equal to I_o at the beginning of each period, which starts with the gating on of the power SCR, S_1. Straight-line approximations have been used to describe the current segments in the main SCR and the freewheeling diode. These waveforms are representative of any of the time-ratio control modes described in Section 8.8.1. The minimum off-time for this circuit, that is, the period between T_o and T_p in Figure 8.27, is determined by the parameter, t_{off}, of S_1. Also, during this off period, the reversing pulse through S_3 (Figure 8.10) must be completed which, in most circuits, requires a longer time period than the commutating time of S_1. To summarize the operation of this basic chopper circuit, the following description of the sequence of events during steady-state operation is given.

1. At $t=0$ with load current equal to I_o, a gate pulse is applied to the gate terminal of S_1, causing it to turn on after a microsecond or so; because of the voltage, $V_{in}-E_a$, applied to the load, current in S_1 and the load increases exponentially; this current is approximated by a straignt line with positive slope as shown in Figure 8.27. The capacitor has been previously charged to a potential somewhat greater than V_{in} with the upper plate negative.
2. At some subsequent time, T_c (which is determined by the desired average

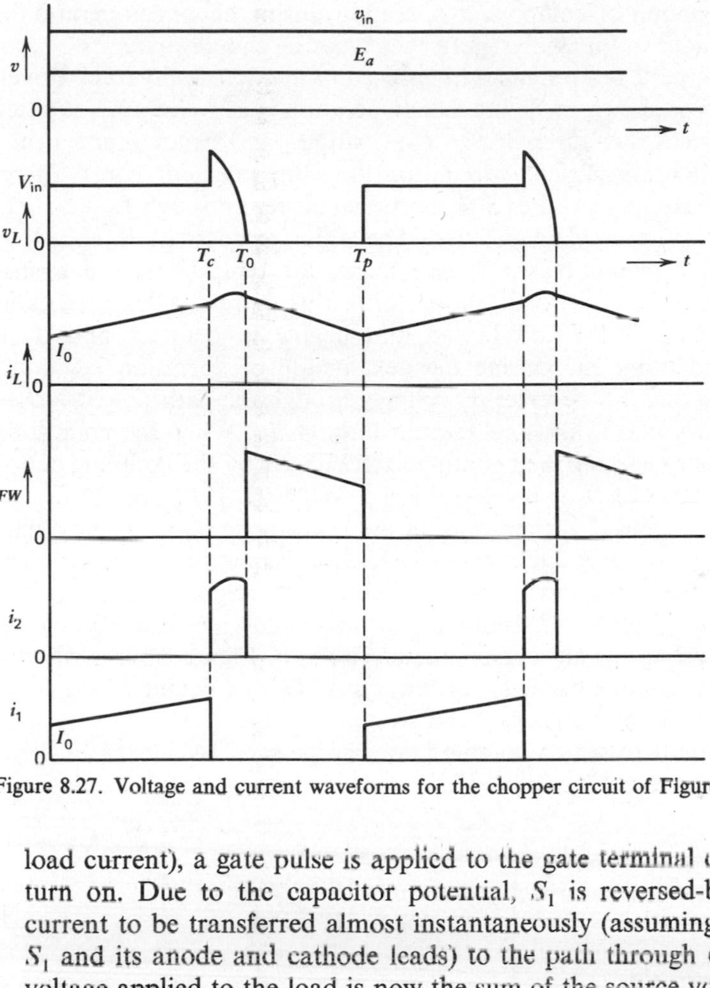

Figure 8.27. Voltage and current waveforms for the chopper circuit of Figure 8.10.

load current), a gate pulse is applied to the gate terminal of S_2, causing it to turn on. Due to the capacitor potential, S_1 is reversed-biased, causing its current to be transferred almost instantaneously (assuming no inductance in S_1 and its anode and cathode leads) to the path through C and S_2. The net voltage applied to the load is now the sum of the source voltage, V_{in}, and the capacitor voltage, causing a sinusoidal current pulse through the circuit $E_b - C - S_2 - L - R$, which can be described by the equations presented in Section 8.4 for series commutation.

3. Shortly after this current pulse, i_2, reaches its maximum value, at which time the voltage across the load inductance is approximately zero, S_2 is reversed-biased and begins its commutation period. This time is shown as T_o in Figure 8.27. Current begins to flow in the freewheeling diode at approximately this same time instant; C is charged with the upper terminal positive.

4. From T_o to T_p, load current is continued through the freewheeling diode. Also, the charge on C must be reversed through S_3 during this time interval (current i_3 in Fig. 8.10).

This detailed description of component operation during chopper operation is given to aid the reader in the analysis of other types of chopper circuits.

The Jones chopper[18] is a popular circuit used in many fork-lift truck drives. The power circuitry is shown in Figure 8.28. Operation is as follows: S_1 is gated on, resulting in current flow through $S_1 - L_2 -$load; L_2 and L_1 are magnetically coupled (both windings are usually wound on the same magnetic core). Therefore, current flow through L_2 causes a proportional current through $L_1 - C - D_1$, charging C with the lower plate positive. The pulse through S_1 is ended by turning on S_2, which reverse-biases S_1 and causes the typical sinusoidal pulse through the path $C - S_2 - L_2 -$load. Shortly after this current pulse reaches its maximum value, D_{FW} is forward biased and begins to conduct and S_2 is reversed-biased and turns off. During the next period of operation beginning with the turning on of S_1, C reverses its voltage through the path $C - S_1 - D_1 - L_1$, which also contributes to the load current through L_2. When the voltage on C is reversed, further charging may continue through D_1 by the coupling action of load current in L_2 until D_1 is reversed-biased. When D_1 is turned off by this reverse-bias condition and C is charged with the lower plate positive, the circuit is in the condition to repeat the above-described sequence of events. The disadvantages of the Jones circuit are the size and weight of the coupled inductances, L_1 and L_2. However, until recently, this circuit has resulted in lower manufacturing cost than the all-semiconductor circuit of Figure 8.10, partly due to the use of a commutating capacitor of lower size. Design details of the Jones chopper may be found in References 1 and 18.

A chopper circuit in which commutation is achieved by means of series commutation is shown in Figure 8.29. This circuit uses a minimum number of

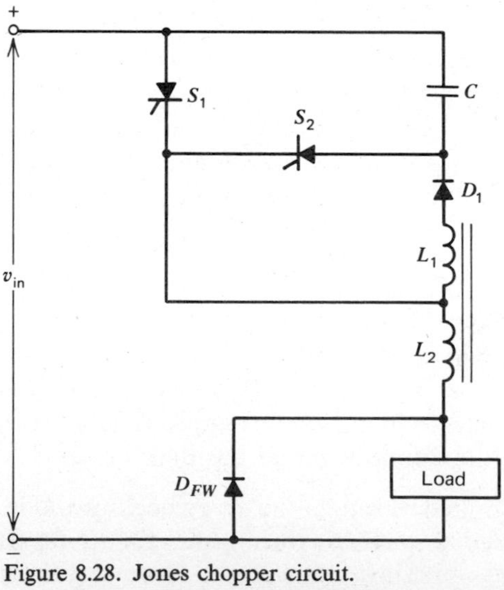

Figure 8.28. Jones chopper circuit.

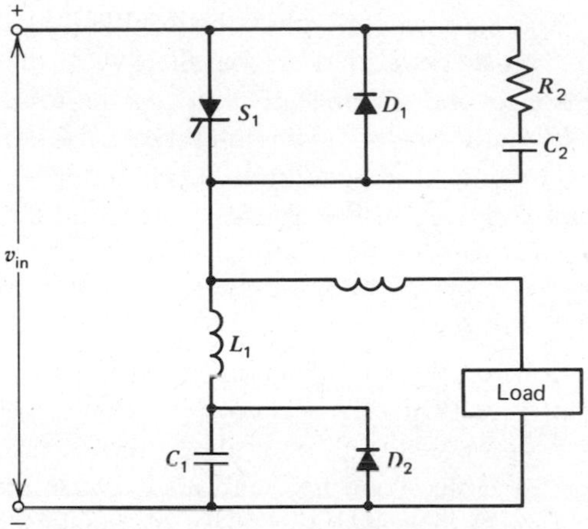

Figure 8.29. Chopper using series commutation.

power semiconductor devices.[19] Commutation of the only SCR required in this circuit, S_1, is by means of the resonant charging of C_1.

In reference to the basic SCR chopper of Figure 8.10, several design criteria have been developed.[20] By assuming the turn-off time of S_1 and S_2 to be identical and equal to t_{off}, the commutating capacitor is related to circuit parameters by

$$C \geqslant \frac{1}{xR}\left[\frac{4.5}{\pi}t_{\text{off}}\right] \tag{8.38}$$

The reversing inductor, under the same assumption, is

$$L_c \geqslant xR\left[\frac{4.5}{\pi}t_{\text{off}}\right] \tag{8.39}$$

where

$$x = \frac{1}{V_{\text{in}}\sqrt{C/L_c}} \approx 0.2 - 1.0$$

V_{in} = voltage applied to chopper

R = load resistance, ohms

t_{off} = turn-off time of SCRs

When bilateral power flow is required, as in the case of traction-type dc commutator motors using regenerative braking, the power semiconductor circuitry as shown in Figures 8.10, 8.28, and 8.29 generally must be doubled; i.e., another chopper with reverse polarity must be added in parallel with the chopper used for motor control. There are several schemes of reducing the total number of power devices required for bilateral power flow. Reference 21 is an example of such circuits. Further discussions of regeneration using chopper circuits to control dc commutator motors are given in References 17, 22, and 23.

8.9 AC MOTOR CONTROL

Our discussion of motor control in previous sections has focused mainly on the control of dc commutator motors. We now turn our attention to electronic control of induction and synchronous motors. The dc commutator motor has traditionally been more widely used for variable speed applications because of the simplicity and flexibility of control techniques for this class of machine. With the advent of the electronic controllers discussed in this chapter, it is still more widely used in variable speed and in many variable torque control applications, for the same reasons. This is particularly true at the higher motor power ratings. Many useful and interesting electronic circuits have been developed for variable-speed control of ac motors and many have been tested and found to have good technical performance characteristics. The goal of these efforts has been the replacement of dc commutator machines by "brushless" ac machines because of the operational and maintenance problems associated with the brush/commutator system. However, variable speed controllers for ac motors are more complex and costly and use more power semiconductor devices than their dc counterparts. In general, the substitution of ac machines with electronic control in the traditional variable speed applications has not materialized, despite many excellent technical achievements in this area. One exception to this statement is in the textile industry, where both induction and reluctance synchronous motors with electronic control are in use today.

Before discussing the major types of ac motor speed control, inverters and cycloconverters, it is worthwhile noting some of the schemes used in the control of smaller machines, mainly in voltage control of single-phase induction motors and universal (series ac) motors. In induction motors, some measure of speed control under loaded conditions can be achieved by variation of the input voltage. (See Chapter 6) The range of speed variation that can be obtained in this manner without exceeding the current rating of the motor is seldom more than 10% below synchronous speed. Also, voltage control is often desirable when the power source has poor voltage regulation and, of course, for induction motor starting. The speed of universal motors is controlled primarily by means of the input voltage, much as in the dc commutator series motor. The simplest means of electronic voltage control is a variation of the phase-controlled rectifier in which the voltage output of the controller is alternating. All of the three-terminal devices listed in Table 8.2 are used in these applications. Figure 8.30

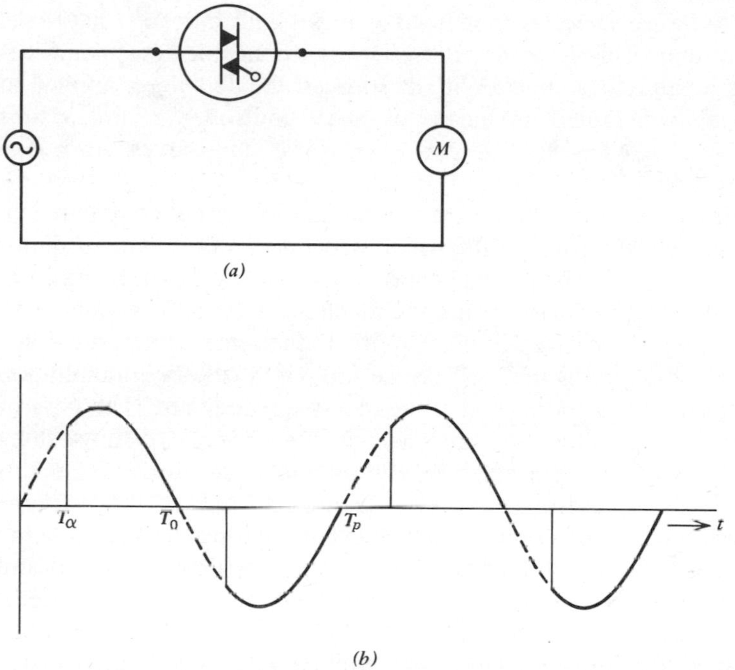

Figure 8.30. (*a*) Triac control of a single-phase motor. (*b*) Voltage applied to motor.

illustrates the use of a TRIAC for phase control of a single-phase ac machine and the associated motor input voltage. Note that the formula for rms voltage given in item 7 of Table 8.1 is equally applicable to the ac motor waveform. The formula for average voltage is also applicable if interpreted as the *half-wave* average.

The principal means of speed control of the cage-type induction and synchronous motors is by varying the frequency of the applied voltage. We are excluding from this discussion speed control of wound-rotor or slip-ring induction motors for which many electromechanical and electronic techniques for varying the frequency in the rotor to control speed are in use. The control of stator frequency in induction motors also implies control of the stator voltage in order to operate the motor at a constant level of magnetic excitation, since the magnetic flux is determined by the ratio of voltage to frequency as shown in Chapter 6. This is accomplished by several techniques in induction motor controllers which may be placed in two general classifications: pulse-width modulation and control of the dc voltage applied to the inverter. Pulse-width modulation is a form of time-ratio control discussed in Section 8.8.1 in which the magnitude of the motor applied voltage is varied by changing the ratio of on-time to off-time of the input voltage pulses. Voltage-controlled inverters are often termed *square-wave inverters*, and load voltages and currents may have high harmonic content. This class of controllers is applicable to systems energized from both ac and dc sources. With ac sources, the dc voltage applied to the inverter is usually obtained by means of phase-controlled rectifiers; this system is sometimes called a *dc link controller*. With dc sources, a voltage chopper is usually used.

There is an important difference in the commutation circuitry required in these types of ac controllers using SCRs, which depend upon the type of motor being controlled. With synchronous machines having dc field windings, commutation can frequently be achieved using the machines internal generated emf, requiring relatively little auxiliary circuitry. With induction motors and singly excited synchronous motors, such as reluctance motors, auxiliary commutation circuits somewhat similar to those used in choppers are required. The type of commutation circuits used in these controllers is the principal distinguishing feature among the many inverter circuits used in ac motor control.

In the variable-speed control of induction motors, there is a third variable that must be considered in the control scheme along with input frequency and voltage. This is rotor slip, the difference between rotor mechanical speed and synchronous speed. It is possible to leave slip relatively uncontrolled and let it vary with load as in fixed applied-frequency cases. However, this leaves open the likelihood that at some operating point slip will be relatively large, causing excessive rotor I^2R loss and at other points, slip will be unduly small, resulting in saturation of the magnetic circuit of the motor. The control of *slip frequency* in variable-speed drives evolved in Europe many years ago[24] and has been used in most systems since. It can be shown that controlling slip-frequency results in higher motor efficiency than allowing slip to be uncontrolled.[25] In this method, the signal that controls motor input frequency—which is determined by the triggering timing of the power semiconductors—is made up of a signal proportional to mechanical speed plus a small increment which represents slip frequency (or rpm). This increment is actually the torque control of the motor, since it determines the operating point on the motor's speed-torque characteristic (see Chapter 6).

Squirrel-cage induction motors for variable speed applications are generally constructed to have very low rotor resistance and thus operate at relatively low values of slip. The shape of the speed-torque characteristic changes relatively little over a wide range of frequencies, permitting a wide range of speed control. At very low input frequencies—frequencies approaching zero or direct current—the rotor and stator resistances become the dominant impedances in the motor equivalent circuit. This requires modification of the fixed V/f ratio required at other frequencies; the voltage must be increased over this ratio to overcome these resistive voltage drops. The use of controlled slip frequency also permits a simple, smooth transition between motor and generator operation of the induction machine. When the voltage increment in the control signal is reduced and becomes zero, the machine is at a no-load condition with its mechanical losses being supplied by the mechanical load. If the increment is further reduced and becomes negative, this calls for a mechanical speed larger than the synchronous speed. Then the machine becomes regenerative and supplies power to the dc bus at the inverter input terminals. Most inverter circuits are capable of accepting this reverse power flow due the circuitry required to handle the reactive power of induction motor loads.

Only a few of the many interesting circuit configurations used in ac motor control can be included in this chapter. For further study of these circuits, consult References 13, 17, 22, and 24 to 31.

8.9.1 SERIES INVERTERS

The series inverter is one of the simplest inverter circuits and uses a minimum number of SCRs where these are used as the power semiconductor. Series inverters are primarily used in fixed frequency applications,[17] but have been adapted for use as variable-speed controllers with induction and reluctance machines in certain types of load applications. Figure 8.31 illustrates one configuration of series inverter that has a measure of both voltage and frequency control.[13] Current and capacitor waveforms are also shown; it is seen that the series inverter operates with discontinuous current pulses, which is a result of the use of series commutation. The minimum off-time is determined by the t_{off} specification of the SCRs. This determines the maximum rms current at a fixed voltage. The normal operation of Figure 8.31 is with alternate firing of S_1 and S_2, which produces the discontinuous sine pulses shown. This may result in the buildup of an excessive voltage on the capacitor at light loads, as discussed in Section 8.5. This can be reduced by alternate firings of S_1 and S_3. When S_3 is turned on (with the proper capacitor polarity, of course), there is a current pulse through the battery causing energy to be returned to the battery and a reduction in the energy stored on the capacitor.

The maximum frequency of operation of this circuit is determined by the sum of the off-time required for safe SCR operation and the pulse period as determined by (8.12). The latter is a function of the parameters of the series *RLC* circuit, where *R* includes the load equivalent resistance. Operation at lower frequencies than this maximum is possible by increasing the off-time. This also reduces the rms current, as can be seen from item 1 of Table 8.1, and the load capability of the motor. For certain types of loads, namely, windage-type loads such as fans, blowers, and certain types of pumps, this characteristic of declining power capability with speed is acceptable. This type of inverter has been successfully used for automotive blower applications.

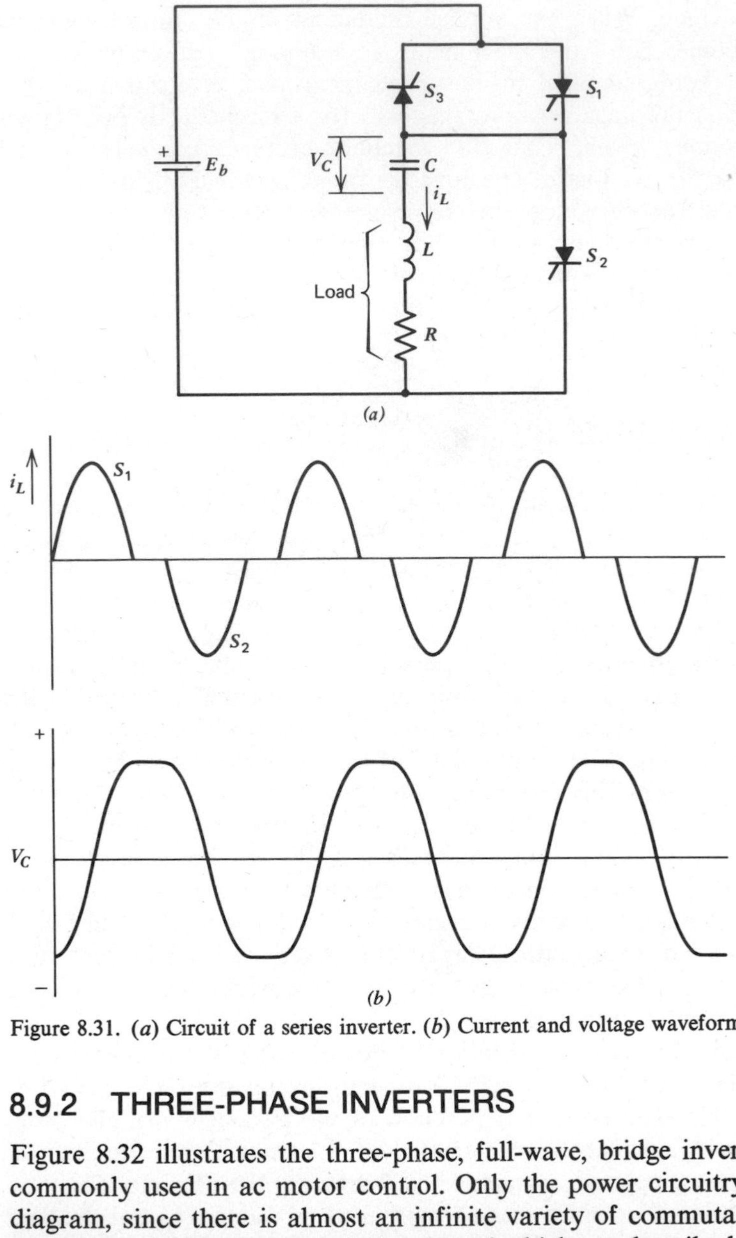

Figure 8.31. (*a*) Circuit of a series inverter. (*b*) Current and voltage waveforms.

8.9.2 THREE-PHASE INVERTERS

Figure 8.32 illustrates the three-phase, full-wave, bridge inverter circuit that is commonly used in ac motor control. Only the power circuitry is shown in this diagram, since there is almost an infinite variety of commutation schemes that have been used with this circuit, many of which are described in the references. The waveforms of various voltage and current components in the inverter, assuming input voltage control, are illustrated in Figure 8.33. Inverter waveforms are generally more complex than those found in the dc motor controllers; the use of Fourier series methods is more common. The magnitude of the motor

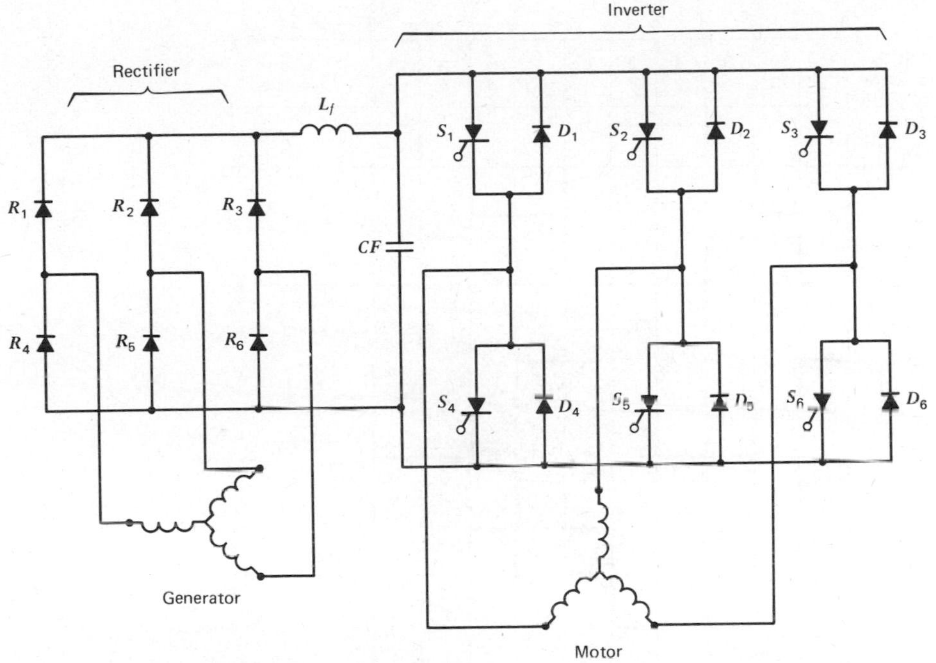

Figure 8.32. Bridge inverter with voltage input control.

current is controlled by controlling the input dc voltage to the inverter which, in this case, is done by the generator. The frequency of the motor current is determined by the timing of the SCR gate signals. As noted above, this timing is often a function of a signal that is proportional to speed plus rotor (slip) frequency. The diodes in antiparallel with the SCRs are required to provide a path for reactive current flow after the SCR has been turned off; they also aid in providing paths for reverse power flow when the motor is operating as a generator. There is a very rigid sequence of SCR gate pulses in the bridge inverter; incorrect firing of an SCR, which might be caused by noise input at its gate terminal, may result in opposite SCRs being on at the same time, causing a direct short across the dc input terminals.

Detailed examination of Figure 8.33 will aid in the understanding of the operation of a voltage-controlled inverter. At the top of this figure is shown the intervals of angular frequency, in radians, when SCR gating occurs. The next line shows the SCRs, as numbered in Figure 8.32, which are gated on in each interval. In voltage-controlled inverters, the SCRs must be gated on during the entire interval of SCR on-time because anode current may not be flowing during the interval to maintain the SCR in a conductive state. This is always the case with inductive loads such as induction motors, for, during a portion of each SCR on-time, the current will be flowing through one of the antiparallel diodes

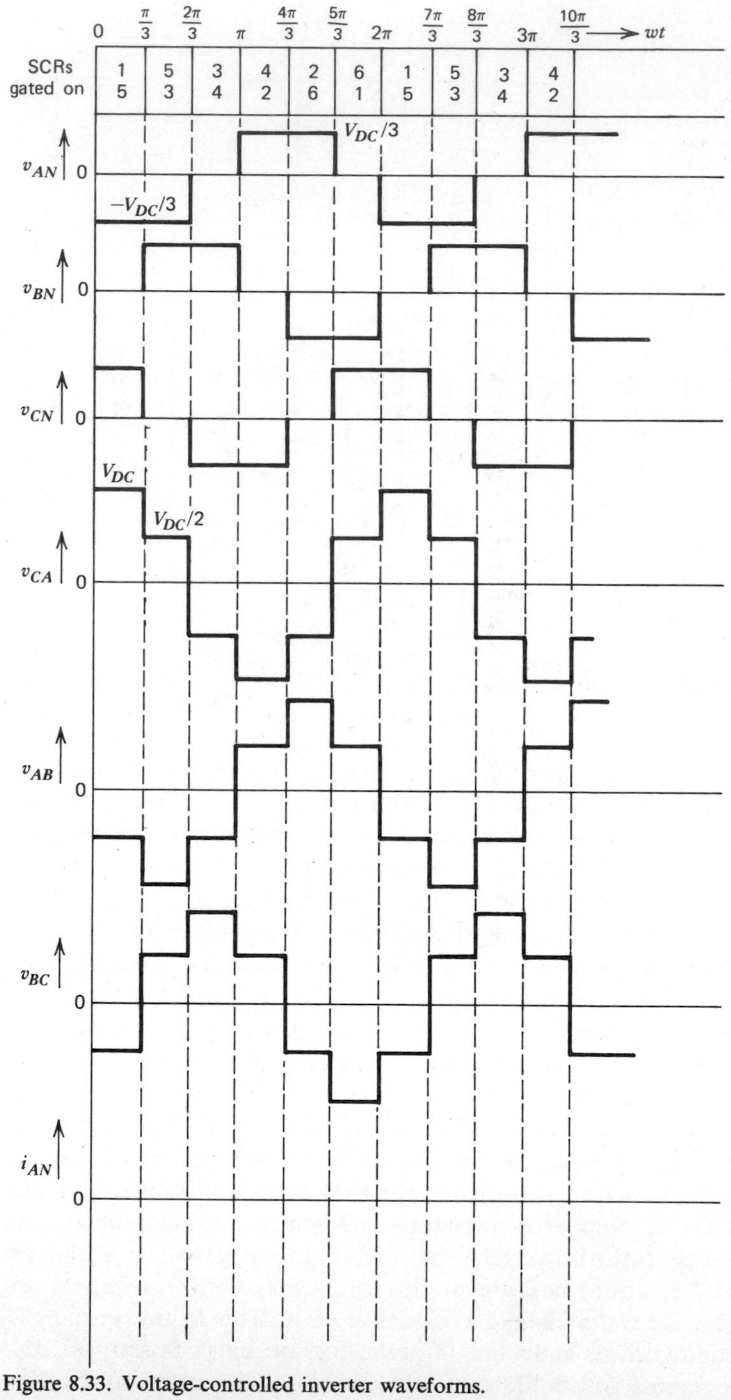

Figure 8.33. Voltage-controlled inverter waveforms.

rather than the SCR. In the gating sequence shown here, only two SCRs are on at any given instant. It is seen that the interval during which an SCR is gated on is $2\pi/3$ radians. The next three diagrams in Figure 8.33 illustrate the load line-neutral or phase voltages; the following three diagrams illustrate the line-to-line voltages. These diagrams represent idealized waveshapes that are strictly valid only for a pure resistive, balanced load. This is because, during each interval, one of the terminals of the load is "floating," i.e., it is not connected to a gated SCR. Its potential will be determined by the nature of the load current and is not easily predicted.

The voltage waveforms in Figure 8.33 are based upon the assumption that the floating terminal is at a potential of $\pm V_{\text{dc}}/2$ with respect to the other load terminals, which is true only for the balanced resistive load. It is a reasonable approximation for other *balanced* loads, also. The magnitudes of the voltage levels are indicated in Figure 8.33 on the diagrams for v_{AN} and v_{CA}. From item 1 of Table 8.1 (which is applicable for either alternating or unilateral square pulses), it is seen that the rms value of the phase voltages is

$$V_p = \frac{V_{\text{dc}}}{3}\sqrt{\left(\frac{2}{3}\right)} = 0.272\ \text{V}_{\text{dc}} \tag{8.40}$$

The line-to-line voltages can be found in a similar manner as

$$V_L = \frac{V_{\text{dc}}}{\sqrt{2}} \tag{8.41}$$

The reader should observe that the phase and line voltages of this inverter are a set of balanced three-phase voltage, i.e., both the individual phase voltages and the individual line voltages are displaced from each other by $2\pi/3$ rad, and the instantaneous sum of the individual phase and line voltages is zero. The fundamental component of current in phase A is shown in Figure 8.33 for an inductive load.

The SCR gating sequence shown in Figure 8.33, which has only two SCRs on in each interval, is desirable from the standpoint of SCR commutation times. It is seen that there is an interval of $\pi/3$ rad following the commutation of an SCR before the opposite SCR connected in series across the dc input voltage is gated on. For example, S_5 is commutated at $2\pi/3$ rad and the opposite SCR, S_2, is not gated on until π rad. However, this sequence results in a low ratio of rms phase voltage to dc input voltage, as seen from (8.40). This voltage ratio can be increased, but at the expense of reduced commutation time, by a gating sequence such that three SCRs are on in each interval. In this sequence, in an idealized gating sequence, the one SCR would be gated on at the instant its opposite SCR is commutated. However, in practice there must be a time delay between these two events to permit safe commutation of the first SCR. This sequence tends to increase the likelihood of a direct short through the two SCRs due to inadequate commutation. The load voltage waveforms are just the reverse of those shown in Figure 8.33 when three SCRs are gated on; the phase voltage is 0.471 V_{dc} and the line voltage is 0.815 V_{dc}.

The power circuit configuration of an inverter with *pulse modulation control* is essentially the same as those with variable voltage control, and the three-phase bridge circuit shown in Figure 8.32 is commonly used. However, the SCR gating schemes are quite different. In phase-modulated inverters, the phase voltage is composed of a series of square pulses of short duration compared to the fundamental frequency. These pulses are of constant magnitude and variable pulse widths. The output voltage waveform is a pulse train whose polarity reverses periodically to provide the fundamental frequency. The repetition rate of the output pulse train is referred to as the carrier frequency, f_c. The ratio of carrier frequency to fundamental frequency should be as high as possible to minimize load harmonics. Pulse-width modulation eliminates the need for an additional voltage controller such as phase-controlled rectifier or the alternator-rectifier system shown in Figure 8.32.

Figure 8.34 illustrates a portion of inverter output voltage controlled by pulse-width modulation. The desired sinusoidal waveform is also shown. This period of voltage should be compared with the waveforms shown in Figure 8.33 for the voltage-controlled inverter. Variation of the width of the pulses that make up each half-period—sometimes called "pedistals"—controls the rms voltage magnitude. This variation or modulation is achieved in practice by many techniques. A common technique is known as the triangulation of subharmonic method, in which the off-time between pulses is determined by the crossover points of two reference signals: a high-frequency triangular voltage wave and a sinusoidal voltage varying at fundamental frequency. Other techniques are discussed in the references. The minimum value of off-time between pulses, which determines the carrier frequency, is limited by SCR turn-off times.

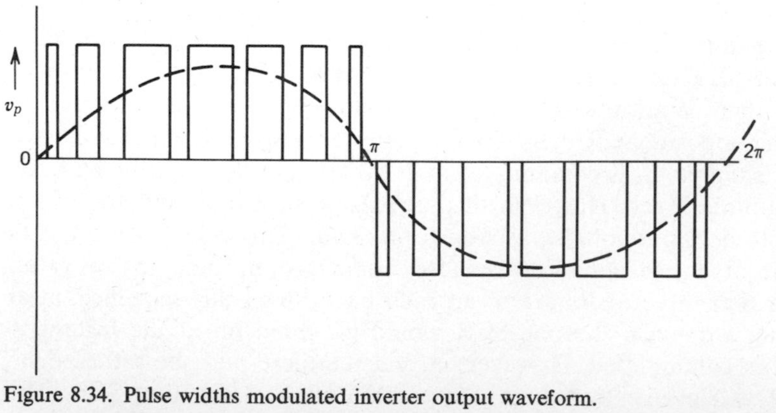

Figure 8.34. Pulse widths modulated inverter output waveform.

8.9.3 CYCLOCONVERTERS

The cycloconverter is a control device used in variable-speed motor applications, where the input power source is alternating current. It is a means of converting a source at fixed voltage and fixed frequency to an output with variable voltage and variable frequency. The source frequency must be at least three to four times the maximum frequency of the output. Figure 8.35 illustrates the circuit configuration and output voltage waveform of one phase of a three-phase cycloconverter applicable to polyphase motor control. The mechanism of voltage and frequency control of the cycloconverter is a combination of the principles applied in the phase-controlled rectifier and the pulse-modulated inverter. The fundamental frequency of the output voltage of the cycloconverter is determined by the number of pulses of input frequency used in the pulse train that makes up a half-period of the output frequency. The magnitude of the

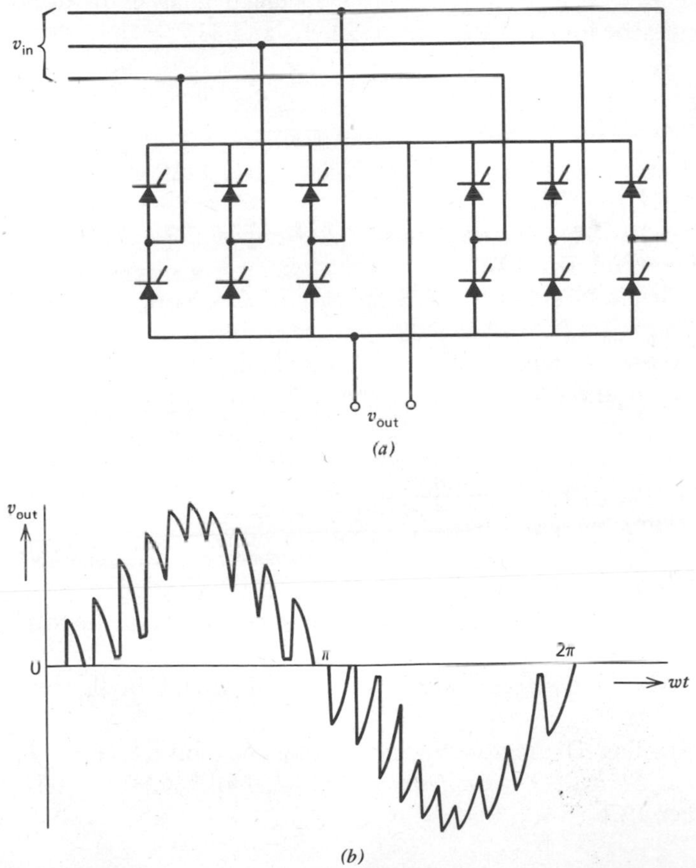

Figure 8.35. (*a*) Cycloconverter single-phase section. (*b*) Output voltage waveform.

output voltage is determined by the portion of each half-cycle of input frequency that exists during the on-time of the SCR, as in phase controlled rectifier control. Both of these parameters, output frequency and output voltage, are determined by the timing of the gate pulses of the SCRs in Figure 8.35. The SCR commutation is by means of line commutation, that is, reverse bias resulting from the input voltage reversing polarity, as in a phase controlled rectifier.

From Figure 8.35 it would seem that a very large number of power semiconductor devices are required for the control of a polyphase motor, since this figure shows only one phase of such a control system. The cycloconverter does require a larger number of power semiconductors than inverters of equivalent power rating. However, the ratings of individual devices in a cycloconverter are lower than in those in an inverter. In comparing cycloconverters with inverters, it is a matter of comparing the relative cost, size, and weight of the use of many power semiconductors of relatively low current rating with the use of a few power semiconductors of high current rating. Detailed analysis of cycloconverter controllers may be found in References 22, 32, and 33.

REFERENCES

1. General Electric Co., *General Electric SCR manual*, Fifth Edition, Electronics Park, Syracuse, N.Y., 1972.
2. Westinghouse Electric Corp., *Westinghouse SCR Designers Handbook*, Second Edition, Youngwood, Pa., 1970.
3. R. Allan, "Power Semiconductors," *IEEE Spectrum*, New York, November 1975.
4. F. E. Gentry, R. I. Scace, and J. F. Flowers, "Bidirectional Triode *p-n-p-n* Switches," *Proceedings of the IEEE*, Vol. 53, No. 4, New York, April 1965, pp. 355–369.
5. "Electric Motors and Controls," *Machine Design Reference Issue*, Vol. 48, No. 10, Cleveland, Ohio, April 29, 1976.
6. T. H. Szypulski, "Some Studies on Parallel and Series Operations of Triacs," *IEEE Industrial Applications Society* 1976 *Conference Record*, No. 76CH1122-1-1A, New York, October 1976.
7. R. S. Ramshaw, *Power Electronics*, London, Chapman, and Hall, London, 1973.
8. D. G. Fink, ed., *Electric Engineers Handbook*, McGraw-Hill Book Company, New York, 1975.
9. A. B. Roby, "Designing Darlington Variable Frequency Invert *IEEE Industrial Applications Society* 1976 *Conference Record*,, No. 76CH1122-1-1A, New York, October 1976.

10. C. D. Todd, *Zener and Avalanche Diodes*, Wiley Interscience, New York, 1970.
11. J. B. Rice and L. E. Nickels, "Commutation *dv/dt* Effects in Thyristors Three-Phase Bridge Converters, *IEEE Transactions on Industry and General Applications*, New York, November 1968.
12. R. F. Dyer, "Concurrent Characteristics of SCR Switching Parameters for Inverter Applications, *SCR and Solid State Technology*, Syracuse, N.Y., April 1965.
13. L. E. Unnewehr, "Series-Commutated SCR Controllers for Variable-Speed Reluctance Motor Drives," *IEEE Power Electronics Specialists Conference Record*, No. 73CH0787-2 AES, Pasadena, Calif., June 1973.
14. Y. Takeda, et al., "Analysis and Design of the Morgan Circuit with Resistor Loads, *Electrical Engineering in Japan*, October 1967.
15. R. G. Schieman, "Power Line Pollution by 3-Phase Thyristor Motor Drives, *IEEE Industrial Applications Society Conference Record*, No. 76CH1122-1-1A, New York, October 1976.
16. J. K. Hall and D. S. Palmer, "Electrical Noise Generated by Thyristor Control, *Proceedings of the IEEE*, Vol. 123, No. 8, New York, August 1976.
17. S. B. Dewan and A. Straughen, *Power Semiconductor Circuits*, Wiley Interscience, New York, 1975.
18. T. Malarkey, "SCR Controller for a Series Field DC Motor," *Motorola Applications Note AN*-734, Phoenix, Ariz., 1974.
19. R. G. Hoft, H. S. Patel, and Y. Dote, "Thyristor Series Resonant DC-DC Chopper," *IEEE Transactions on Magnetics*, New York, September 1972.
20. G. N. Revanker and P. K. Palsetia, "Design Criteria of a Commutating Circuit of a DC Chopper," *IEEE Electronics and Control*, Vol. IEC 1-19, No. 3, New York, August 1972.
21. B. Berman, "Battery Powered Regenerative SCR Drive" *IEEE Transactions on Industrial Applications*, Vol. IA-8, No. 2, New York, March 1972.
22. J. D. Harnden and F. B. Golden, "Power Semiconductor Applications," Vol. I, IEEE Press, 1972.
23. A. Kusko, *Solid-State Motor Drives*, MIT Press, Cambridge, Mass., 1969.
24. D. A. Bradley, et al., "Adjustable-Frequency Inverters and Their Application to Variable-Speed Drivers," *Proceedings of the IEEE*, Vol. 111, No. 11, New York, November 1964.
25. P. D. Agarwal, "The GM High Performance Induction Motor Drive Systems," *IEEE Transaction Paper*, No. 68TP107-PWR, New York, January 1968.
26. B. D. Bedford and R. G. Hoft, *Principles of Inverter Circuits*, John Wiley & Sons, Inc., New York, 1964.
27. J. J. Pollack, "Advanced Pulsewidth Modulated Inverter Techniques," *IEEE*

Transactions on Industry Applications, Vol. IA-8, No. 2, New York, March 1972.

28. T. A. Lipo, "Performance Calculations of a Reluctance Motor Drive by D. Harmonics," *IEEE Industrial Application* 1976 *Conference Record*, No. 76CH1122-1-1A, New York, October 1976.
29. K. Y. G. Li, "New 3-Phase Inverter Circuit," *Proceedings of the IEEE*, Vol. 115, No. 11, New York, November 1968.
30. P. H. Nayak and R. G. Hoft, "Optimizing the PWM Waveform of a Thyristor Inverter," *IEEE Industrial Applications Society* 1974 *Conference Record*, No. 74CH0833-41A, New York, October 1974.
31. I. L. Kosow, *Control of Electric Machines*, Prentice-Hall, Inc., Englewood Cliffs, New Jersey, 1973.
32. C. J. Amato, "Variable Speed with Controlled Slip Induction Motor," *IEEE Conference Record*, Industrial Static Power Conversion Conference, Philadelphia, Pa., November 1965.
33. L. J. Jacovides, "Analysis of a Cycloconverter-Induction Motor Drive System Allowing for Stator Current Discontinuities," *IEEE Transactions on Industry Applications*, Vol. IA-9, No. 2, New York, March 1973.

PROBLEMS

8.1. For the first three waveforms of Table 8.1, determine the *form factor*, that is, the ratio of rms to average values if $T_o = 1/2T_p$.

8.2. A true square wave of current is seldom realizable because of circuit inductances. Rather, current pulses are trapezoidal in shape, although they are often approximated by square pulses. Assume that the waveform of Table 8.1 item 2 is composed of trapezoidal pulses rather than the square pulses shown. The wave is depicted in Figure 8.36. Determine the expressions for

(a) The rms current, I_{rms}.

(b) The time-average current, I_{av}.

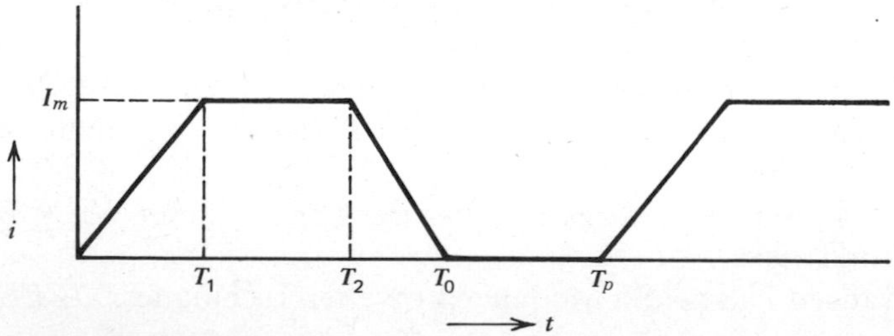

Figure 8.36. Problem 8.2.

8.3. Determine the magnitude and phase of the first five terms of the Fourier series representation of Table 8.1, item 4.

8.4. Assume that a current of peak magnitude, $I_m = 10$ A, and waveform of Table 8.1, item 4, flows in a 1-ohm resistor. Determine the ohmic loss in the resistor by means of the equations in Table 8.1, and compare the value calculated with that calculated by means of the Fourier components derived in Problem 8.3. ($T_o = T_p/3$).

8.5. A saturable reactor is used in a circuit in which the voltage is the unidirectional square pulses of the form shown in item 2 of Table 8.1, with $V_m = 50$ V, $T_o = 100$ μs, and $f = 1/T_p = 3200$ Hz. The saturable reactor has a toroidal core (zero air gap) of the material 48NI (Figure 2.4). The core is saturated in one polarity by each pulse of the applied square wave.

Assume that a 50 V source is available for resetting the core to the opposite polarity. Design a coil that is capable of resetting (to the opposite polarity) the core in the allowable time $(T_p - T_o)$ for the applied waveform of item 2 in Table 8.1 from a source of 50 V.

8.6. In Figure 8.9, let $E_b = 100$ V, $C = 120$ μF, $L = 200$ μH, $R = 0.05$ ohm. Determine Z_o, α, and β, assuming that $V_{co} = I_{1o} = 0$. Under these conditions, determine the charge on the capacitor following the sine current pulse after S_1 is triggered. Determine the peak current of the sine pulse and the pulse width.

8.7. Determine the maximum rate of rise of current in S_1 in Problem 8.6 if S_1 is closed on the static R-L circuit.

8.8. The load in Figure 8.9 is to be operated from the 100 V battery of Problem 8.6, with a pulse frequency of 1000 Hz.

(a) Determine the value of L_C for reversing the capacitor in the required time ($C = 120$ μF); neglect the resistance of the reversing circuit.

(b) Determine the steady-state voltage on the capacitor, V_{co}, for continuous operation of the circuit (Equation 8.18a).

8.9. In the circuit of Figure 8.19, $R = 1.0$ ohm and $L = 2.0$ mH. A voltage of 120 V, 400 Hz is applied. Determine the rms and time-average load voltage and time-average load current if $T_\alpha = 20\%$ of T_o.

8.10. Repeat Problem 8.9 for the circuit of Figure 8.20.

8.11. Repeat Problem 8.9 for the circuit of Figure 8.21, neglecting the impedances of the free-wheeling diode.

8.12. A 120 V, 50-A dc motor is to be controlled from a 120-V, 60-Hz source. Determine the rating of the SCRs to be used in a single-phase bridge phase-controlled rectifier for this application. The rating should include rms, average current, and PRV.

8.13. A 10-hp, 100-V separately excited dc commutator motor is to be operated at variable speed and variable torque. At speeds below 3000 rpm, the torque is to be controlled by means of an armature chopper from a 120 V battery. The battery has an internal resistance of 0.09 ohm; armature resistance is 0.025 ohm; armature inductance is 0.5 mH. The motor constant (see Equations 5.49 and 5.50) is 0.028 V-s/A; the maximum field current is 10 A. The maximum motor torque (one-minute rating) is 71 N-*m* with $I_f = 10$ A.

A chopper of the type shown in Figure 8.10 is to be used to control torque between 0 and 71 N-*m* in the 0 to 3000 rpm speed range.

(a) Assume that S_1 is to be turned on at a constant rate and that the average armature current is to be controlled by controlling the on-time of the current pulse. A free-wheeling diode will be used. Choose a pulse frequency (500 Hz is a good starting point) and determine the range of on-time pulse width to vary the armature current from 10 to 250 A at 3000 rpm.

(b) Repeat (a) at 1000 rpm.

(c) Repeat (a) at zero rpm; are there any problems maintaining the current within 250 A average with your original choice of pulse frequency? Would added armature-circuit inductance be helpful at standstill?

8.14. Specify an SCR for use as S_1 in Problem 8.13. Specification should include I_{rms}, I_{av}, forward- and reverse-voltage rating, di/dt, and reapplied dv/dt.

8.15. Design the commutation circuit for the chopper of Problems 8.13 and 8.14.

8.16. If you have access to analog or digital computer modeling facilities, prepare a model of the chopper motor control system of Problems 8.13 to 8.15 and evaluate the system to see if S_1 is always operated within your specifications of Problem 8.14. In your model, S_1, S_2, S_3, and D_{FW} can be assumed to be open circuits when off; when conducting, all four devices should be modeled by an inductance of about 1 μH and a resistive voltage drop of about 1.5 V for the SCRs and 1 V for the diode.

8.17. A single-phase induction motor is rated 1/4 hp, 60 Hz, 1750 rpm, 120 V. When the motor is operated at its rated condition, the input current is 4.5 A; during starting, a peak current of 8 A is drawn when the motor is connected directly across a 120-V, 60-Hz source.

This motor is to be used in an application requiring some measure of speed control. A triac controller, Figure 8.30, is to be used for this purpose, with speed control being obtained by varying the applied motor voltage in the phase-controlled manner shown in Figure 8.30b.

(a) Assume that the maximum (pull-out) torque of this motor is 1.5 times its rated torque with current of 1.2 times rated, and that the triac controller must be capable of supplying this maximum torque with rated voltage applied to the motor (minimum T_α in Figure 8.30). Specify the triac PRV, rms current, average current, and t_{off}.

(b) What minimum delay time, T_α in Figure 8.30, would you suggest for the maximum torque condition in (a)?

(c) What is the rms motor current at this condition?
(d) What are the advantages of the triac control during motor start-up and stall conditions?

8.18. A portable grinding tool uses a 24-V battery and a single-phase induction motor rated 1/8 hp. The power output required by the grinding operation is relatively constant but a small amount of voltage control on the induction motor is considered necessary. The motor is to be controlled by the single-phase SCR inverter circuit of Figure 8.31. The equivalent impedance of the motor at its normal running condition, as seen at the motor terminals, is $R=0.62$ ohm and $L=0.36$ mH. The input frequency is to be 400 Hz. At 1/8 hp, the motor requires an rms input current of 22 A and a rotor current of 15 A.
(a) Determine the required size and ratings of S_1, S_2, and C for this application, neglecting battery, lead, and device resistances and inductances.
(b) Will the voltage on C be excessive at 1/8 hp? (See Equation 8.18a.) Is the use of S_3 desirable to limit the circuit voltage as well as to obtain speed control?

8.19. As an alternate to the use of the inverter-induction motor in the portable grinding tool application, described in Problem 8.18, a chopper-series (universal) motor system is proposed. The 1/8 hp series motor to be evaluated runs at 5000 rpm, 8.6 A, from a 24-V dc or 60 Hz source of voltage.
(a) This motor impedance was evaluated by a blocked rotor test, giving the following test data: 8.7 A, 40 W, 16.7 V, 60 Hz; the voltage across the armature terminals$=2.7$ V, across field terminals $=14$ V; field winding resistance at 20°C$=-0.232$ ohm$-$armature widing resistance$=0.196$ at 20°C. Determine the blodked rotor inductances of the armature and field.
(b) Design a power transistor chopper to drive this motor at its rated conditions from a 24-V battery, neglecting circuit resistances and inductances as in Problem 8.18. Specify transistor I_C, h_{FE} (min), V_{CE} (sat), P_r and f_R (typ).
(c) Compare the chopper-series motor with the inverter-induction motor system in terms of relative ease of speed control and speed regulation.

Chapter 9
General Theory of Electric Machines

In the preceding chapters we have considered each machine as a separate entity in a piece meal fashion and emphasized mostly the steady-state behavior of rotating machines. We recall, however, that for electromechanical systems (Chapter 4), we formulated and solved the dynamical equations of motion under certain constraints. A similar approach is applicable to rotating machines, also, and we shall discuss this in subsequent sections. Here, we shall attempt to present various general methods which could be applied to a variety of rotating machines. In fact, generalized machine theory has been developed to a point that it could be used to analyze almost two dozen different types of rotating machines. We restrict ourselves to machines having the following three basic magnetic structures:

1. Machine with saliency on stator.
2. Machine with no saliency either on stator or on rotor.
3. Machine with saliency on rotor.

Examples of 1, 2, and 3, respectively, are dc commutator machines, induction machines, and salient pole synchronous machines. We are already familiar with these machines.

Generalized machine theory can be used as a powerful line of attack on new problems in addition to providing a systematic approach to solving routine problems. Before we discuss the details, let us review the nature of the problem and the objectives of developing a general theory.

9.1 GOALS AND THE NATURE OF THE PROBLEM

As mentioned above, our goal in developing a general theory is to obtain an analytical method which is applicable to most machines operating under steady-state or transient conditions. In principle, this goal can be achieved if (1) we know the parameters of the equations of motion and if (2) we solve these equations for given operating conditions. We recall from Chapter 4 that the general equations of motion of an electromechanical system are, in matrix notation

$$\mathbf{v}=\frac{d}{dt}(\mathbf{Li})+\mathbf{ri} \tag{9.1}$$

$$T_e=-\frac{1}{2}\tilde{\mathbf{i}}\,\frac{\partial}{\partial\theta}[\mathbf{L}]\mathbf{i} \tag{9.2}$$

These equations are applicable to electric machines also. The problem then resolves to:

1. Determining the inductance and resistance matrices, **L** and **r** respectively, of the machine.
2. Specifying the terminal conditions, such as either the voltages **v** or the currents **i**.
3. Solving (9.1) and (9.2), indicating whether dynamic or steady state solutions are desired.

In the following sections we shall consider the details of carrying out these steps. Here we merely wish to indicate the methods in outline. First of all, to find the resistances, the conductor dimensions (length and cross section) and resistivity must be known. If necessary temperature corrections may be applied.

The inductances may be determined from the defining equation for inductance as flux linkage per ampere, that is

$$L_{pq}=\frac{\lambda_{pq}}{i_q} \tag{9.3}$$

where λ_{pq} is the flux linkage with a coil (say p) attributable to a current i_q in another coil (say q) and L_{pq} is the inductance between the coils p and q.

Alternatively, we may obtain the inductances by the energy-stored method where the energy stored in inductive circuits (mutually coupled, for generality) is equated to the energy stored in the magnetic field; that is,

$$\frac{1}{2}\sum_{k=1}^{n}\sum_{m=1}^{n} L_{km} i_k i_m = \frac{1}{2}\int_{\text{vol}} \mathbf{B}\cdot\mathbf{H}\,dv \tag{9.4}$$

We shall illustrate the applications of both methods, (9.3) and (9.4), in the next section. After having determined the parameters, the next most important, but difficult, step is to obtain solutions to the equations of motion. In carrying through this step the usefulness of the general theory becomes evident. One approach to solving these equations is the "brute force" method namely, by numerical integration. We term this method the time-domain formulation or state-variables method and we shall discuss it in a later section. The second and more common method is the transformation method, where the equations are solved by introducing a change of variables. For this purpose, we rewrite (9.1) as

$$\mathbf{v} = \mathbf{Z}\mathbf{i} \tag{9.5}$$

where $\mathbf{Z} = Z(p) = p\mathbf{L} + \mathbf{r}$ and is often called the operational impedance. Now, we introduce a change of variables such that

$$\mathbf{v} = \mathbf{S}\mathbf{v}' \tag{9.6}$$

and

$$\mathbf{i} = \mathbf{S}\mathbf{i}'$$

where the primed quantities are the new variables and are related to the unprimed (old) variables by a transformation matrix $\mathbf{S}$. Substituting (9.6) in (9.5) yields

$$\mathbf{S}\mathbf{v}' = \mathbf{Z}\mathbf{S}\mathbf{i}' \tag{9.7}$$

We may choose $\mathbf{S}$ such that it is nonsingular (that is, its inverse, $\mathbf{S}^{-1}$, exists) and then premultiply both sides of (9.7) by $\mathbf{S}^{-1}$ to obtain

$$\mathbf{v}' = (\mathbf{S}^{-1}\mathbf{Z}\mathbf{S})\mathbf{i}' = \mathbf{Z}'\mathbf{i}' \tag{9.8}$$

Notice that (9.8) is an alternate form of (9.5). We recall that it is difficult to solve (9.5) because it contains time- (or θ-) dependent coefficients. To overcome this difficulty we may choose the transformation matrix $\mathbf{S}$, such that θ-dependence is eliminated from the equations of motion. The equation can then be solved in a straightforward manner. We shall have more to say about this method also later. For the present we wish to point out that there are a number of transformations which are applicable to the equations for various electric machines. The three commonly used transformations are the following

1. symmetrical–component (or $\pm$) transformation.
2. forward–backward (or *fb*) transformation.
3. *dq*–transformation.

The last-mentioned transformation shows an equivalence between an actual electric machine and a hypothetical (or transformed) machine which has windings only along two mutually perpendicular axes, called the direct-(d-) and the quadrature-(q-) axes. That is, under certain constraints, most machines may be "expressed" as transformed machines having windings along the d- and the q-axes. Conversely, if we start with a machine which has windings only along the above two axes, we should expect to be able to "construct" (or simulate) many types of conventional machines. This thought process has been the key to the generalized machine theory. A machine having coils along the d- and q-axes, (with appropriate brushes and commutator), has been termed the "primitive machine." Thus from the primitive machine, many types of actual machines can be derived. We shall discuss this in a later section but first we shall determine the machine inductances, as shown in the next section.

9.2 DETERMINATION OF MACHINE INDUCTANCES

From the previous chapters we observe that, considering the topology of the magnetic circuit, an electric machine may either be (1) a cylindrical rotor machine, such as the induction machine and round-rotor synchronous machine; or (2) a salient pole machine, where the saliency may be on the stator, as in a dc machine, or on the rotor as in a salient pole synchronous machine. In determining the machine inductances, we shall first consider a round rotor machine and then a salient pole machine. Assume that the machine has sinusoidally distributed windings and flux densities. This assumption is made to simplify the algebraic details and to illustrate the principles. Nonsinusoidal winding and flux density distributions, however, can be taken into account by including the harmonic terms of the corresponding Fourier-series representations.[1] Because the concentrated and distributed windings differ only in physical layout, rather than in principle of operation we shall consider only the concentrated winding as a representative model. This is especially so because a distributed winding is magnetically equivalent to a concentrated winding having N-turns, if $N=2Z$, where $Z=$ number of series conductors/radian of a distributed winding.

9.2.1 INDUCTANCES OF A ROUND-ROTOR MACHINE

A uniform air-gap (round-rotor) machine is shown in Figure 9.1 a. We assume that the stator and the rotor each has one concentrated winding with N^s and N^r turns, respectively. Also, we assume $\mu_{\text{iron}} \gg \mu_o$, $r \gg g$ (Figure 9.1a), and $H_r \gg H_\theta$. Furthermore, the fundamental components of the stator and rotor mmf's are given, respectively, by

$$\begin{aligned} F^s &= k^s N^s i^s \sin\theta \\ F^r &= k^r N^r i^r \sin(\theta + \theta_0) \end{aligned} \tag{9.9}$$

as shown in Figure 9.1a and b where k^s and k^r are winding factors. (See Chapter 6.) Notice that θ is any arbitrary position around the periphery in the

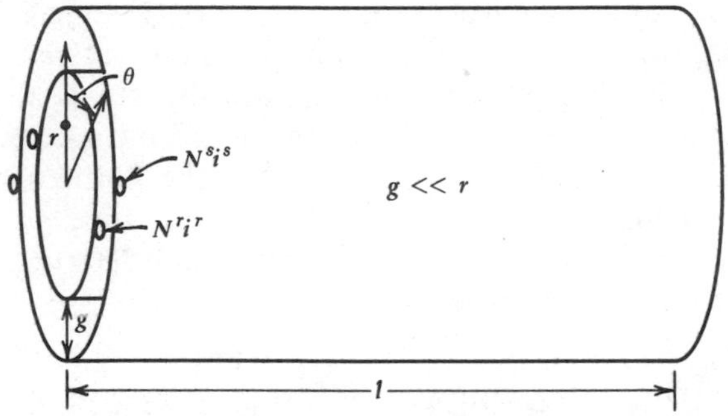

(a)

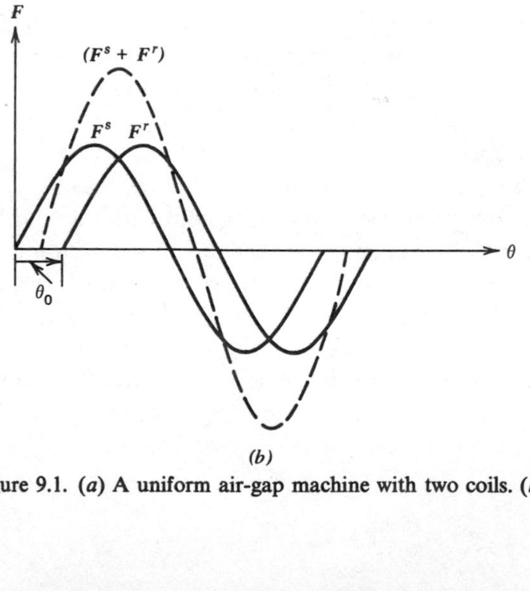

(b)

Figure 9.1. (a) A uniform air-gap machine with two coils. (b) The mmfs of the coils.

air gap and θ_o is the displacement between F^s and F^r. We determine the self- and mutual inductance by the energy storage method as follows.

Consider an elementary surface area Δs through which a flux $\Delta\phi$ passes (in the r-direction). Then

$$\Delta\phi = B_r \Delta s = \mu_o H_r \Delta_s \tag{9.10}$$

The permeance of this flux path is

$$\Delta P = \mu_o \frac{\Delta s}{g} \tag{9.11}$$

If Ni is the resultant magnetomotive force, then the flux $\Delta\phi$ is also given by

$$\Delta\phi = Ni\Delta P = \mu_o \frac{Ni}{g} \Delta s \tag{9.12}$$

Comparing (9.10) and (9.12), we have

$$H_r = \frac{Ni}{g} \tag{9.13}$$

We could also verify (9.13) from the fact that the mmf drop, Ni, is only across the air gap, g, μ_{iron} being much greater than μ_0. Now, Ni is found from the phasor addition of the stator and rotor mmf's. That is, the maximum value of the resultant mmf is

$$(Ni)^2 = (k^s N^s i^s)^2 + (k^r N^r i^r)^2 + 2k^s k^r N^s N^r i^s i^r \sin\theta_0 \tag{9.14}$$

From (9.13) and (9.14) we notice that the maximum value of H_r, when the rotor winding is displaced from the stator winding by θ_o, is such that

$$(H_r)^2_{\text{max}} = \frac{1}{g^2}\left[(k^s N^s i^s)^2 + (k^r N^r i^r)^2 + 2k^s k^r N^s N^r i^s i^r \sin\theta_o\right] \tag{9.15}$$

and H_r varies sinusoidally with θ. The average value of $(H_r)^2$ is simply $[(H_r)^2_{\text{max}}]/2$ and the magnetic energy stored in the air gap, W_m, is

$$W_m = \frac{1}{4}\mu_0 (H_r)^2_{\text{max}} (\text{volume of air gap}) = \frac{\mu_0 \pi r l g}{2} (H_r)^2_{\text{max}} \tag{9.16}$$

In terms of the inductances, the same energy may be expressed as

$$W_m = \frac{1}{2} L^{ss} (i^s)^2 + \frac{1}{2} L^{rr} (i^r)^2 + L^{sr} i^s i^r \tag{9.17}$$

But (9.15) and (9.16) yield

$$W_m = \frac{\mu_0 \pi r l}{2g}\left[(k^s N^s)^2 (i^s)^2 + (k^r N^r)^2 (i^r)^2 + 2k^s k^r N^s N^r \sin\theta_0 i^s i^r\right] \quad (9.18)$$

Comparing (9.17) and (9.18), we obtain the desired inductances as

$$L^{ss} = \frac{\mu_0 \pi r l}{g}(k^s N^s)^2 = L^s \quad (9.19a)$$

$$L^{rr} = \frac{\mu_0 \pi r l}{g}(k^r N^r)^2 = L^r \quad (9.19b)$$

$$L^{sr} = \frac{\mu_0 \pi r l}{g} k^s k^r N^s N^r \sin\theta_o = L^{sr} \sin\theta_o \quad (9.19c)$$

So far we have considered only one winding on the stator and one on the rotor. In a two-phase machine we have two windings on each member. The two windings on the stator are identical, are mutually perpendicular and thus magnetically decoupled. Those on the rotor are also, as shown in Figure 9.2. In this case, stator and rotor inductances, expressed as matrices are

$$\mathbf{L}^{ss} = \begin{bmatrix} L^s & 0 \\ 0 & L^s \end{bmatrix} \quad \text{and} \quad \mathbf{L}^{rr} = \begin{bmatrix} L^r & 0 \\ 0 & L^r \end{bmatrix} \quad (9.20)$$

The stator-to-rotor mutual inductances can be obtained from (9.19). Thus

$$\mathbf{L}^{sr} = \begin{bmatrix} L_{aa}^{sr} & L_{ab}^{sr} \\ L_{ba}^{sr} & L_{bb}^{sr} \end{bmatrix} = \begin{bmatrix} L^{sr}\cos\theta_0 & -L^{sr}\sin\theta_0 \\ L^{sr}\sin\theta_0 & L^{sr}\cos\theta_0 \end{bmatrix} \quad (9.21)$$

In (9.21) the subscripts on L's correspond to the phases (a and b) and the superscripts are used to denote stator or rotor (s or r). Thus, for instance, L_{ab}^{sr} = mutual inductance between the phase a on the stator and phase b on the rotor. The elements of the matrix in (9.21) are obtained from (9.19c) by setting $\theta_0 = (90 - \theta)$ for L_{aa}^{sr}, $\theta_o = (90 + \theta)$ for L_{ab}^{sr}, and so on. The entire **L**-matrix for the two-phase machine is, therefore,

$$\mathbf{L} = \left[\begin{array}{c|c} \mathbf{L}^{ss} & \mathbf{L}^{sr} \\ \hline \mathbf{L}^{rs} & \mathbf{L}^{rr} \end{array}\right] \quad (9.22)$$

where the submatrices are given by (9.20) and (9.21) and $\mathbf{L}^{rs}$ is the transpose of $\mathbf{L}^{sr}$.

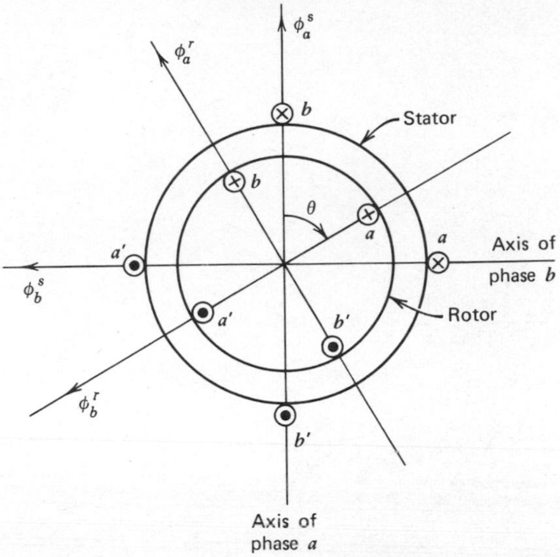

Figure 9.2. A two-phase round rotor machine.

9.2.2. INDUCTANCES OF MACHINES WITH SALIENCY

A two-phase machine, with saliency on the rotor, is shown in Figure 9.3. Assuming sinusoidal flux density distribution and considering only the fundamental components of the mmf's, we use the flux linkage method to find the machine inductances. The rotor has only one winding and the L-matrix of the machine can be written as

$$\mathbf{L}=\left[\begin{array}{cc|c} \mathbf{L}_{aa}^{ss} & \mathbf{L}_{ab}^{ss} & \mathbf{L}_{ar}^{sr} \\ \mathbf{L}_{ba}^{ss} & \mathbf{L}_{bb}^{ss} & \mathbf{L}_{br}^{sr} \\ \hline \mathbf{L}_{ra}^{rs} & \mathbf{L}_{rb}^{rs} & \mathbf{L}_{rr}^{rr} \end{array}\right] \tag{9.23}$$

where the nomenclature has been explained in the last section. We first consider the rotor self-inductance, which is independent of the rotor position. Therefore,

$$L_{rr}^{rr}=L_f=\frac{\mu_0 N_f^2 A_f}{g_d}=\text{a constant} \tag{9.24}$$

where N_f = number of turns of the rotor winding; A_f = rotor-pole surface area; and g_d = air gap at the direct axis. The mutual inductances between the stator and rotor windings can be obtained by the method discussed in the last section.

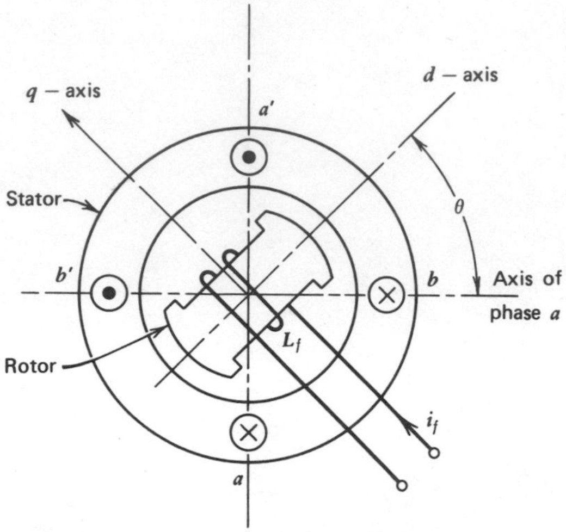

Figure 9.3. A two-phase salient rotor machine.

Without repeating the details, we may write the following:

$$L_{ar}^{sr} = L_{ra}^{rs} = L^{sr}\cos\theta$$
$$L_{br}^{sr} = L_{rb}^{rs} = -L^{sr}\sin\theta \tag{9.25}$$

To find the stator self- and mutual-inductances, consider the mmf $F_a^s = k^sN^si_a^s$ arising from phase a and resolve it along the d- and q-axes (Figure 9.3) to obtain

$$F_{da}^s = F_a^s\cos\theta = k^sN^si_a^s\cos\theta$$
$$F_{qa}^s = F_a^s\sin\theta = -k^sN^si_a^s\sin\theta \tag{9.26}$$

If we define P_d and P_q as permeances along the d- and q-axes, respectively, the air-gap fluxes can be expressed as

$$\phi_{da}^s = P_dF_{da}^s$$
$$\phi_{qa}^s = P_qF_{qa}^s \tag{9.27}$$

The flux linking phase a, resulting from the current i_a^s is

$$\lambda_{aa}^{ss} = k^sN^s\phi_a^s = k^sN^s(\phi_{da}^s\cos\theta - \phi_{qa}^s\sin\theta) \tag{9.28}$$

Combining (9.26) to (9.28) yields

$$\lambda_{aa}^{ss}=\frac{1}{2}(k_sN^s)^2 i_a^s\left[(P_d+P_q)+(P_d-P_q)\cos 2\theta\right] \tag{9.29}$$

or

$$L_{aa}^{ss}=\frac{\lambda_{aa}^{ss}}{i_a^s}=L^s+L_0^s\cos 2\theta \tag{9.30}$$

where $L^s=\frac{1}{2}(k^sN^s)^2(P_d+P_q)$ and $L_0^s=\frac{1}{2}(k^sN^s)^2(P_d-P_q)$. Substituting $\theta=\theta+\frac{\pi}{2}$ in (9.30) gives

$$L_{bb}^{ss}=L^s-L_0^s\cos 2\theta \tag{9.31}$$

To determine the mutual inductances, we recall that

$$L_{ab}^{ss}=L_{ba}^{ss}=\frac{\lambda_{ab}^{ss}}{i_a^s} \tag{9.32}$$

The flux linking phase b, λ_{ab}^{ss}, resulting from the current in phase a, i_a^s, is found from (9.27) and (9.28) by setting $\theta=\theta+\pi/2$ so that

$$\lambda_{ab}^{ss}=k^sN^s\left[\phi_{da}^s\cos\left(\theta+\frac{\pi}{2}\right)-\phi_{qa}^s\sin\left(\theta+\frac{\pi}{2}\right)\right] \tag{9.33}$$

Substituting (9.26) and (9.27) in (9.32) yields

$$\lambda_{ab}^{ss}=(k^sN^s)^2 i_a^s\left[P_d\cos\theta\cos\left(\theta+\frac{\pi}{2}\right)+P_q\sin\theta\sin\left(\theta+\frac{\pi}{2}\right)\right]$$

which, when substituted in (9.32) and simplified, results in

$$L_{ab}^{ss}=L_{ba}^{ss}=-L_0^s\sin 2\theta \tag{9.34}$$

where L_0^s has been defined in conjunction with (9.30).

Expressed in matrix notation, L-matrix of a two-phase salient rotor machine is

$$L=\left[\begin{array}{cc|c} L^s+L_0^s\cos 2\theta & -L_0^s\sin 2\theta & L^{sr}\cos\theta \\ -L_0^s\sin 2\theta & L^s-L_0^s\cos 2\theta & -L^{sr}\sin\theta \\ \hline L^{sr}\cos\theta & -L^{sr}\sin\theta & L_f \end{array}\right] \tag{9.35}$$

The third kind of magnetic topology in electric machines is the saliency on the stator, such as in a dc machine. One such two-pole machine, with a full-pitch coil on the rotor, is shown in Figure 9.4*a*. Figure 9.4*b* shows the stator mmf distribution and the fundamental component of the flux density distribution in the air gap. By Fourier analysis of the flux density distribution for its fundamental component we obtain (relating B to H to mmf)

$$B(\theta) = \frac{4\mu_o}{\pi g} N^s i^s \cos\alpha \cos\theta \tag{9.36}$$

The flux linking the rotor coil is, therefore,

$$\lambda^r = N^r l r \int_{\theta_o - \pi/2}^{\theta_o + \pi/2} B(\theta)\, d\theta \tag{9.37}$$

Substituting (9.36) in (9.37) and simplifying gives

$$\lambda^r = 2N^r l r i^s \left(\frac{4\mu_o}{\pi g} N^s \cos\alpha \right) \cos\theta_o \tag{9.38}$$

from which the stator-to-rotor mutual inductance is

$$L^{sr} = \frac{\lambda^r}{i^s} = \frac{8\mu_o}{\pi g} l r N^s N^r \cos\alpha \cos\theta_o = L_m \cos\theta_o \tag{9.39}$$

where $L_m = \dfrac{8\mu_o}{\pi g} l r N^s N^r \cos\alpha$; α and r are defined in Figure 9.4; l is axial length of the machine; g is the air gap and N^s and N^r are the number of turns on the stator and on the rotor, respectively.

Notice that the self-inductance of the stator is a constant and may be denoted by L_f.

The self-inductance of the rotor is a θ_o-dependent quantity and may be determined by a procedure similar to that for the salient rotor machine discussed previously in this section. The details are left as an exercise. As an end result, the rotor self-inductance is

$$L^{rr} = L^r + L_o \cos 2\theta_o \tag{9.40}$$

Finally, the self-inductance of the stator is a constant (independent of the rotor position); that is,

$$L^{ss} = (N^s)^2 \mu_o \frac{A_s}{g_d} = L^s \tag{9.41}$$

where A_s = surface area of the stator pole face and g_d = total air gap in the d-axis.

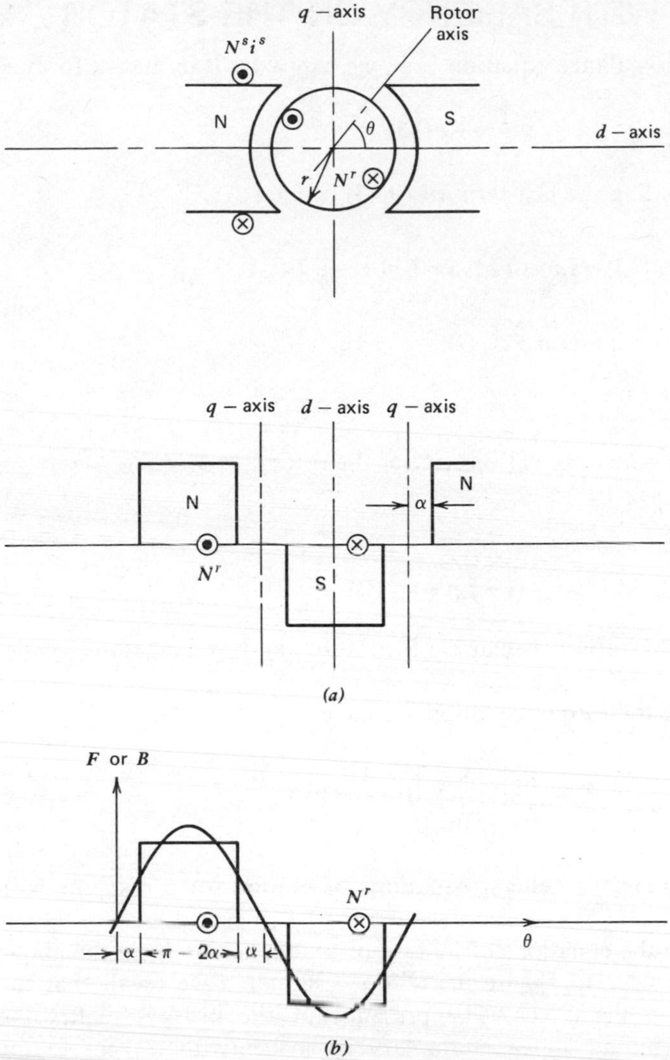

(b)

Figure 9.4. (*a*) A two-pole salient rotor machine. (*b*) Fundamental component of $B(\theta)$.

The **L**-matrix for the machine, with saliency on the stator, is then given by

$$\left[\begin{array}{c|c} \mathbf{L}^{ss} & \mathbf{L}^{sr} \\ \hline \mathbf{L}^{rs} & \mathbf{L}^{rr} \end{array}\right] = \left[\begin{array}{c|c} L^s & L_m \cos\theta_o \\ \hline L_m \cos\theta_o & L^r + L_o \cos 2\theta_o \end{array}\right] \tag{9.42}$$

We now have the **L**-matrices of the three basic forms of electric machines. These inductances may be substituted in the equations of motion to obtain the machine characteristics, as discussed in the following sections.

9.3 MACHINE WITH SALIENCY ON THE STATOR

Considering the voltage-balance equation first, we can write it in matrix form as

$$\mathbf{v}=\mathbf{ri}+p(\mathbf{Li}) \tag{9.43}$$

where $p=d/dt$. Expanding the last term of (9.43), we get

$$\begin{aligned} p(\mathbf{Li}) &= \mathbf{L}p\mathbf{i}+(p\mathbf{L})\mathbf{i}=\mathbf{L}p\mathbf{i}+\frac{d}{d\theta_o}\mathbf{L}\omega_m\mathbf{i} \\ &=(\mathbf{L}p+\omega_m\mathbf{G})\mathbf{i} \end{aligned} \tag{9.44}$$

where

$$\omega_m=d\theta_o/dt=\text{mechanical velocity of the rotor } \mathbf{G}=d\mathbf{L}/d\theta_o.$$

Substituting (9.44) in (9.43) yields

$$\mathbf{v}=(\mathbf{r}+\mathbf{L}p+\omega_m\mathbf{G})\mathbf{i} \tag{9.45}$$

The last term in (9.45) arises because of rotation and is sometimes called *rotational voltage*.

Next, considering the torque equation we have

$$T_e=\frac{1}{2}\tilde{\mathbf{i}}\left[\frac{\partial}{\partial\theta_0}\mathbf{L}\right]\mathbf{i}=\frac{1}{2}\tilde{\mathbf{i}}\mathbf{G}\mathbf{i} \tag{9.46}$$

Thus (9.45) and (9.46) are the general equations of motion for a machine with saliency on the stator.

In order to obtain the equations of motion of a dc machine from the above general equations, we refer to Figure 9.5. From Chapter 5 we recall that the brushes are located on the q-axis. The presence of the brushes makes the armature (or rotor) winding appear stationary in space (with respect to the stator) as far as the q-axis flux is concerned, even though in reality the armature conductors are rotating. Because the rotor flux is stationary, it generates no motional voltage in the stator. Consequently, the **G**-matrix for the dc machine becomes, from (9.44) and (9.42)

$$\mathbf{G}=\begin{bmatrix} 0 & 0 \\ -L_m\sin\theta_o & -2L_o\sin 2\theta_o \end{bmatrix} \tag{9.47}$$

And the desired general equations of motion of a dc machine become, with

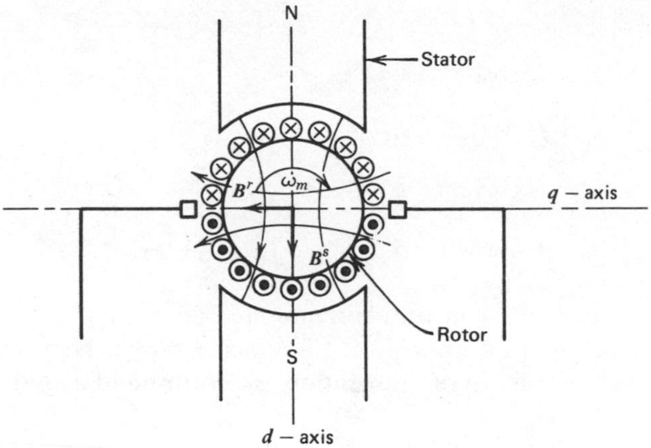

Figure 9.5. A dc machine showing stator- and rotor-flux axes.

$\theta_o = -\pi/2$ (the brushes being along the q-axis)

$$v^s = r^s i^s + L^s p i^s$$

$$v^r = r^r i^r + L^r p i^r + \omega_m L_m i^s \tag{9.48}$$

$$T_e = L_m i^s i^r$$

Various constraints may be applied to (9.48) to obtain the characteristics of a given dc machine under a specified operating condition.

9.4 CYLINDRICAL ROTOR MACHINE

Knowing the parameters, we can obtain the characteristics of an induction motor from the general equations (9.45) and (9.46). For this purpose, consider a two-phase machine with two independent windings on the stator and two on the rotor. The L-matrix is given by (9.18) and the constraints for an induction motor are (1) the rotor windings are short circuited, that is, $v_a^r = v_b^r = 0$ and (2) as shown in Chapter 6, the rotor current frequency $\omega^r = \omega^s - \omega_m$, where ω_m = rotor speed = $\dot{\theta}_0$ and ω^s = stator frequency. Next, we assume balanced two-phase excitation on the stator such that

$$v_a^s = V^s \cos \omega^s t$$
$$v_b^s = V^s \sin \omega^s t \tag{9.49}$$

and

$$
\begin{aligned}
i_a^s &= I^s \cos(\omega^s t - \phi^s) \\
i_b^s &= I^s \sin(\omega^s t - \phi^s) \\
i_a^r &= I^r \cos(\omega^r t - \phi^r) \\
i_b^r &= I^r \sin(\omega^r t - \phi^r)
\end{aligned}
\tag{9.50}
$$

Substituting the parameter values, the constraints mentioned above, and (9.49) and (9.50) in (9.45), we obtain (9.51), with $\theta_o = \omega_m + \delta$ (since $\dot{\theta}_o = \omega_m$). Here we observe that, although δ is a constant of integration, it corresponds to the power-angle.

$$
\begin{aligned}
V^s \cos\omega^s t + \omega^s L^{sr} I^r \sin(\omega^s t + \delta - \phi^r) &= (r^s + L^s p) i_a^s \\
V^s \sin\omega^s t - \omega^s L^{sr} I^r \cos(\omega^s t + \delta - \phi^r) &= (r^s + L^s p) i_b^s \\
\omega^r L^{sr} I^s \sin(\omega^s t + \delta - \phi^s) &= (r^r + L^r p) i_a^r \\
-\omega^r L^{sr} I^s \cos(\omega^s t + \delta - \phi^s) &= (r^r + L^r p) i_b^r
\end{aligned}
\tag{9.51}
$$

Because the currents and voltages arc sinusoidally-varying quantities, we can express (9.51) in terms of complex exponentials by assuming

$$\mathbf{V}^s = V^s e^{j0}$$

$$\mathbf{I}^s = I^s e^{-j\phi^s}$$

and

$$\mathbf{I}^r = I^r e^{-j(\phi^r - \delta)}$$

in which case (9.51) reduce to the phasor equations

$$\mathbf{V}^s - j\omega^s L^{sr} \mathbf{I}^r = (r^s + j\omega^s L^s) \mathbf{I}^s \tag{9.52}$$

and

$$-j\omega^r L^{sr} \mathbf{I}^s = (r^r + j\omega^r L^r) \mathbf{I}^r \tag{9.53}$$

Recalling the definition of slip $s = (\omega^s - \omega^r)/\omega^s$, we rewrite (9.53)

$$-j\omega^s L^{sr} \mathbf{I}^s = \left(\frac{r^r}{s} + j\omega^s L^r \right) \mathbf{I}^r \tag{9.54}$$

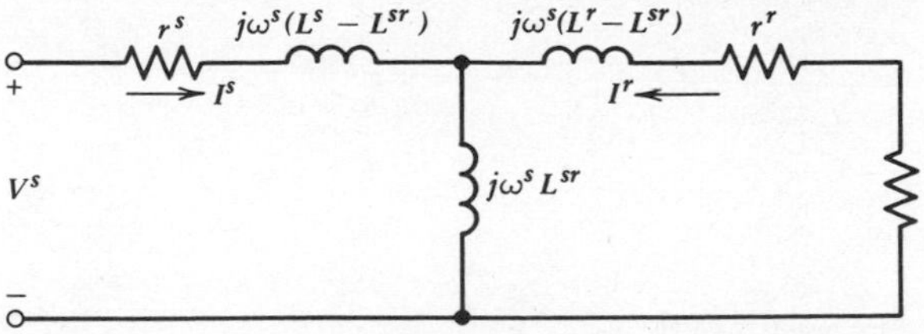

Figure 9.6. A per phase equivalent circuit of a two-phase induction motor.

We can represent (9.52) and (9.54) by the circuit shown in Figure 9.6 which is the per-phase steady-state equivalent circuit of an induction motor. Here, $\omega^s(L^s - L^{sr})$ is the stator leakage reactance and $\omega^s L^{sr}$ is the magnetizing reactance.

Having derived the equivalent circuit we can obtain the complete characteristics of the induction motor, as has been demonstrated in Chapter 6. In a manner similar to the above, the characteristics of a round-rotor synchronous machine can also be obtained. We pose this as an exercise (see Problem 9.4).

9.5 MACHINE WITH SALIENCY ON THE ROTOR

The **L**-matrix of a two-phase salient rotor machine is given by (9.35). When this is substituted in the general equations of motion, such as (9.1) and (9.2), we get differential equations with time-varying coefficients. In order to obtain the machine characteristics, these equations must be solved under specified constraints. However, analytical solutions are not forthcoming. Therefore, we solve the equations either numerically, which we will discuss in a later section, or by a *transformation* technique discussed in the following section.

9.5.1 THE *dq*-TRANSFORMATION

We observe that the difficulty in solving equations of motion arises from the relative motion between the rotor and the stator leading to the time-varying coefficients. Thus, it seems reasonable that if the stator coils (or windings) are replaced by fictitious coils mounted on the rotor, the problem arising from relative motion will be eliminated. For equivalence between the original stator coils and new fictitious coils on the rotor, we must have an invariance of the air-gap fields. Stated more simply, the mmf produced by the new coils on the rotor must be the same as the mmf of the original stator coils. With this idea in mind, we refer to Figure 9.7, which shows a salient pole machine with the original stator and rotor coils as well as new *d* and *q* coils, respectively, placed along the *d*- and *q*-axis. The currents in the various coils and their numbers of

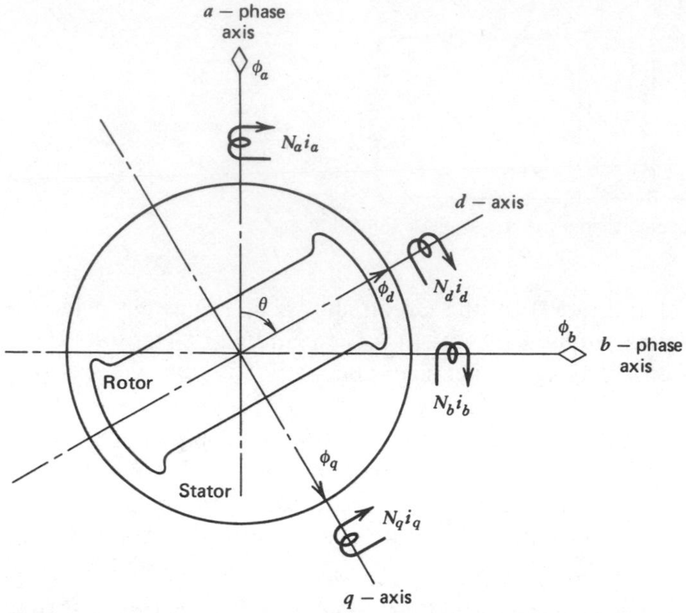

Figure 9.7. The *ab*-to-*dq* transformation in a salient rotor machine.

turns are like those marked on Figure 9.7. We assume the rotor to be unexcited because we want to find the conditions for equivalence between the original *ab*-coils of the stator and the new *dq*-coils on the rotor. Resolving the mmf's along the *d*- and *q*-axes, we get

$$\begin{aligned} F_d &= N_d i_d = N(i_a \cos\theta + i_b \sin\theta) \\ F_q &= N_q i_q = N(-i_a \sin\theta + i_b \cos\theta) \end{aligned} \tag{9.55}$$

Because the coils N_d and N_q are fictitious, we may choose $N_d = N_q = N$ and express (9.55) in matrix form as

$$\begin{bmatrix} i_d \\ i_q \end{bmatrix} = \begin{bmatrix} \cos\theta & \sin\theta \\ -\sin\theta & \cos\theta \end{bmatrix} \begin{bmatrix} i_a \\ i_b \end{bmatrix} \tag{9.56}$$

Thus, the original phase currents i_a and i_b are transformed to the new currents i_d and i_q, as given by (9.56). Because the rotor (or field) winding remains on the rotor, the field current, i_f, does not undergo any transformation. Including i_f, we may rewrite (9.56) as

$$\begin{bmatrix} i_d \\ i_q \\ \hline i_f \end{bmatrix} = \left[\begin{array}{cc|c} \cos\theta & \sin\theta & 0 \\ -\sin\theta & \cos\theta & 0 \\ \hline 0 & 0 & 1 \end{array}\right] \begin{bmatrix} i_a \\ i_b \\ \hline i_f \end{bmatrix} \tag{9.57}$$

Or by inverting we obtain

$$\begin{bmatrix} i_a \\ i_b \\ \hline i_f \end{bmatrix} = \left[\begin{array}{cc|c} \cos\theta & -\sin\theta & 0 \\ \sin\theta & \cos\theta & 0 \\ \hline 0 & 0 & 1 \end{array}\right] \begin{bmatrix} i_d \\ i_q \\ \hline i_f \end{bmatrix} \tag{9.58}$$

In general (9.58) can be written as

$$\mathbf{i} = \mathbf{S}_{dq}\mathbf{i}' \tag{9.59}$$

where $\mathbf{i}'$ denotes the new currents and $\mathbf{S}_{dq}$ is the transformation matrix relating $\mathbf{i}'$ to the old currents $\mathbf{i}$. The $\mathbf{S}_{dq}$ in (9.59) is known as the *dq-transformation*. Also, we observe that

$$\mathbf{S}_{dq}^{-1} = \tilde{\mathbf{S}}_{dq} \tag{9.60}$$

where $\sim$ denotes the transpose implying that the transformation is an orthogonal transformation.

In summary, the *ab*-variables are related to the *dq*-variables by (9.59). For instance

$$\mathbf{i}_{ab} = \mathbf{S}_{dq}\mathbf{i}_{dq}$$

and

$$\mathbf{v}_{ab} = \mathbf{S}_{dq}\mathbf{v}_{dq} \tag{9.61}$$

with

$$\mathbf{S}_{dq} = \left[\begin{array}{cc|c} \cos\theta & -\sin\theta & 0 \\ \sin\theta & \cos\theta & 0 \\ \hline 0 & 0\theta & 1 \end{array}\right] \tag{9.62}$$

and

$$\mathbf{S}_{dq}^{-1} = \tilde{\mathbf{S}}_{dq}$$

Substituting (9.61) in (9.1) and (9.2) gives the equations of motion in terms of transformed variables as

$$\mathbf{v}_{dq}=\mathbf{r}\mathbf{i}_{dq}+\mathbf{S}_{dq}^{-1}p(\mathbf{L}\mathbf{S}_{dq}\mathbf{i}_{dq}) \tag{9.63a}$$

$$T_e=-\frac{1}{2}[\tilde{\mathbf{i}}_{dq}][\tilde{\mathbf{S}}_{dq}]\frac{\partial}{\partial\theta}[\mathbf{L}]\mathbf{S}_{dq}\mathbf{i}_{dq} \tag{9.63b}$$

In order to expand these equations, consider the second term

$$\begin{aligned}\mathbf{S}_{dq}^{-1}p(\mathbf{L}\mathbf{S}_{dq}\mathbf{i}_{dq})&=\mathbf{S}_{dq}^{-1}\left[\left(\frac{\partial\mathbf{L}}{\partial\theta}\dot{\theta}\right)\mathbf{S}_{dq}\mathbf{i}_{dq}+\mathbf{L}\frac{\partial}{\partial\theta}(\mathbf{S}_{dq})\dot{\theta}\mathbf{i}_{dq}+\mathbf{L}\mathbf{S}_{dq}(p\mathbf{i}_{dq})\right]\\&=\left[\mathbf{S}_{dq}^{-1}\frac{\partial\mathbf{L}}{\partial\theta}\mathbf{S}_{dq}+\mathbf{S}_{dq}^{-1}\mathbf{L}\frac{\partial}{\partial\theta}(\mathbf{S}_{dq})\right]\dot{\theta}\mathbf{i}_{dq}+(\mathbf{S}_{dq}^{-1}\mathbf{L}\mathbf{S}_{dq})p\mathbf{i}_{dq}\\&=\left\{\mathbf{S}_{dq}^{-1}\left[\frac{\partial}{\partial\theta}(\mathbf{L}\mathbf{S}_{dq})\right]\dot{\theta}+(\mathbf{S}_{dq}^{-1}\mathbf{L}\mathbf{S}_{dq})p\right\}\mathbf{i}_{dq}\end{aligned} \tag{9.64}$$

Substituting (9.35) and (9.62) we get, after considerable simplification

$$\mathbf{S}_{dq}^{-1}p(\mathbf{L}\mathbf{S}_{dq}\mathbf{i}_{dq})=\left[\begin{array}{cc|c}L_dp & \dot{\theta}L_q & L^{sr}p\\-\dot{\theta}L_d & L_qp & -\dot{\theta}L^{sr}\\\hline -L^{sr}p & 0 & \theta L_fp\end{array}\right]\left[\begin{array}{c}i_d^s\\i_q^s\\\hline i_f\end{array}\right] \tag{9.65}$$

where $L_d=L^s+L_0^s$ and $L_q=L^s-L_0^s$. From (9.63*a*), (9.63*b*) and (9.65), therefore,

$$\left[\begin{array}{c}v_d^s\\v_q^s\\\hline v_f\end{array}\right]=\left[\begin{array}{cc|c}r^s+L_dp & \dot{\theta}L_q & L^{sr}p\\-\dot{\theta}L_d & r^s+L_qp & -\dot{\theta}L^{sr}\\\hline L^{sr}p & 0 & r^r+L_fp\end{array}\right]\left[\begin{array}{c}i_d^s\\i_q^s\\\hline i_f\end{array}\right] \tag{9.66}$$

and

$$T_e=i_d^si_q^s(L_d-L_q)+i_q^si_fL^{sr} \tag{9.67}$$

Notice that the angular dependence has now been eliminated from the equations of motion, and for constant-speed operation, (9.60) is a set of linear differential equations with constant coefficients. From these equations the characteristics of a salient rotor machine, such as a salient pole synchronous generator or motor can be obtained.

We also wish to point out that, because a round-rotor machine is a special case of a salient pole machine (for which we set saliency effects to zero), the *dq*-transformation is equally applicable to the round-rotor machine. For instance, the two-phase induction motor could be analyzed by the *dq*-transformation. (See Problem 9.5.)

9.6 CERTAIN OTHER TRANSFORMATIONS

From the preceding discussions we noticed that analytical solutions to the equations of motion were not forthcoming. Explicit solutions could only be obtained by use of the *dq*-transformation to eliminate the *dq*-dependence from the equations. Although we were able to obtain the equivalent circuit of the induction motor under steady-state, direct solutions to the equations of motion are not readily obtainable for transient conditions. An alternative approach to the *dq*-transformation to obtain the dynamical characteristics of a round-rotor machine is through the use of the symmetrical component $(+\,-)$ transformation and forward-backward (*fb*) transformation. Whereas a mathematical basis exists for these transformations,[1] here we take them as defined. The matrix defining the $+\,-$ transformation for a two-phase machine is

$$\mathbf{S}_{+-}=\frac{1}{\sqrt{2}}\begin{bmatrix} 1 & 1 \\ -j & j \end{bmatrix} \tag{9.68}$$

and

$$\mathbf{S}_{+-}^{-1}=\tilde{\mathbf{S}}_{+-}^{*}=\frac{1}{\sqrt{2}}\begin{bmatrix} 1 & j \\ 1 & -j \end{bmatrix} \tag{9.69}$$

where $\sim$ denotes the transpose, $*$ the complex conjugate, and -1 the inverse of a matrix. The relationships between the original (unprimed) and transformed (primed) quantities are

$$\mathbf{i}=\mathbf{S}_{+-}\mathbf{i}'$$

and (9.70)

$$\mathbf{v}=\mathbf{S}_{+-}\mathbf{v}'$$

To apply this transformation to a two-phase machine having the **L**-matrix given by (9.20) and (9.21), we substitute these and (9.68) to (9.70) in (9.1) and

(9.2) to obtain the transformed equations of motion in expanded form as

$$\begin{bmatrix} v_+^s \\ v_-^s \\ \hline v_+^r \\ v_-^r \end{bmatrix} = \left[\begin{array}{cc|cc} r^s+L^s p & 0 & L^{sr}e^{j\theta}(p+j\dot{\theta}) & 0 \\ 0 & r^s+L^s p & 0 & L^{sr}e^{-j\theta}(p-j\dot{\theta}) \\ \hline L^{sr}e^{-j\theta}(p-j\dot{\theta}) & 0 & r^r+L^r p & 0 \\ 0 & L^{sr}e^{j\theta}(p+j\dot{\theta}) & 0 & r^r+L^r p \end{array}\right] \begin{bmatrix} i_+^s \\ i_-^s \\ \hline i_+^r \\ i_-^r \end{bmatrix} \tag{9.71}$$

$$T_e = -jL^{sr}\left[(i_+^{s*}i_+^r)e^{j\theta} - (i_+^s i_+^{r*})e^{-j\theta}\right] \tag{9.72}$$

For a given speed $\omega_m = \dot{\theta}$, (9.71) is a set of linear differential equations with time-varying coefficients. However, (9.72) is still nonlinear. To overcome this difficulty, the rotor quantities are referred to the stator by a further transformation, the *fb* transformation. The *fb* components are related to the $\pm$ components by

$$\mathbf{v}_{+-}^r = \mathbf{S}_{fb}\mathbf{v}_{fb}^r \tag{9.73}$$

where

$$\mathbf{S}_{fb} = \begin{bmatrix} e^{-j\theta} & 0 \\ 0 & e^{j\theta} \end{bmatrix} \tag{9.74}$$

When this transformation is carried out on the rotor voltages and currents, (9.71) and (9.72) become

$$\begin{bmatrix} v_+^s \\ v_-^s \\ \hline v_f^r \\ v_b^r \end{bmatrix} = \left[\begin{array}{cc|cc} r^s+L^s p & 0 & L^{sr}p & 0 \\ 0 & r^s+L^s p & 0 & L^{sr}p \\ \hline L^{sr}(p-j\omega_m) & 0 & r^r+L^r(p-j\omega_m) & 0 \\ 0 & L^{wr}(p+j\omega_m) & 0 & r^r+L^r(p+j\omega_m) \end{array}\right] \begin{bmatrix} i_+^s \\ i_-^s \\ \hline i_f^r \\ i_b^r \end{bmatrix} \tag{9.75}$$

$$T_e = jL^{sr}(i_+^{s*}i_f^r - i_+^s i_f^{r*}) \tag{9.76}$$

An equivalent circuit representing (9.75) is shown in Figure 9.8, from which the machine characteristics can be obtained for given constraints.

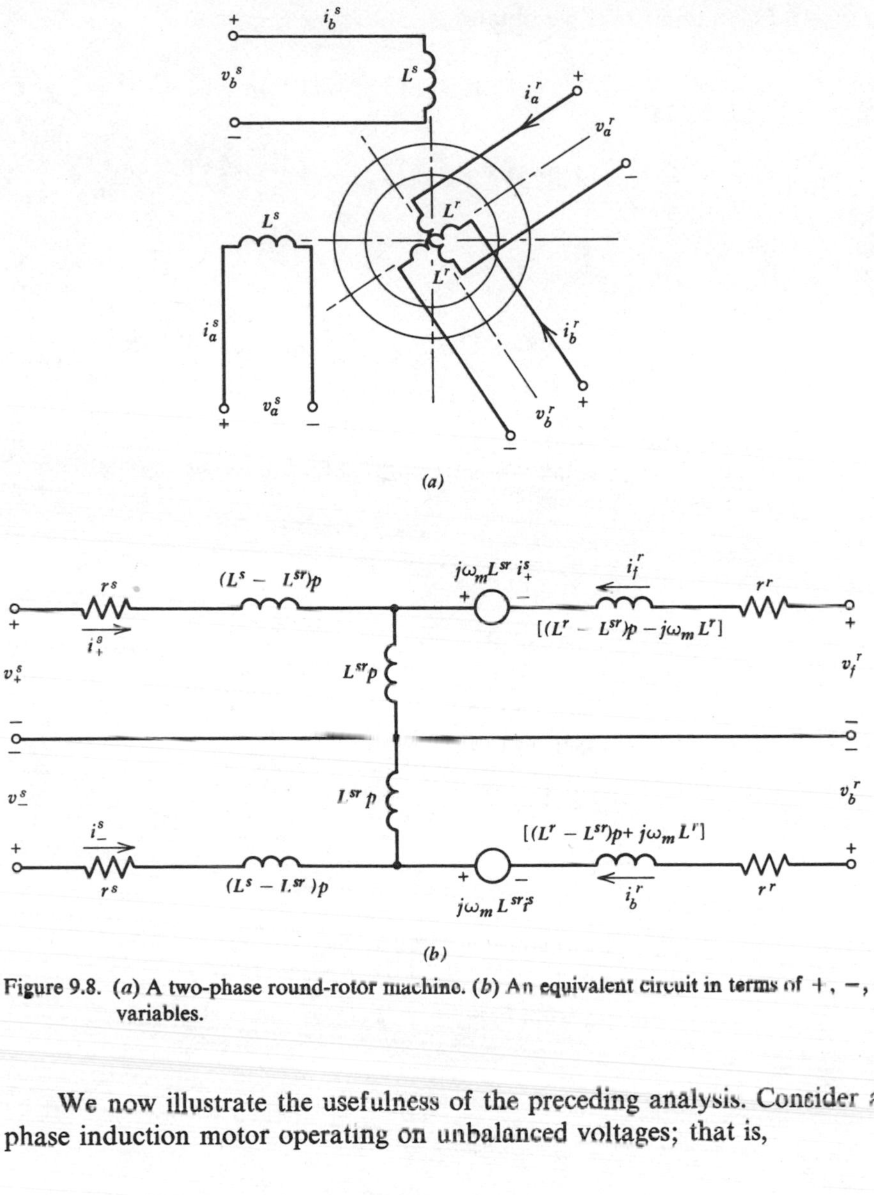

Figure 9.8. (*a*) A two-phase round-rotor machine. (*b*) An equivalent circuit in terms of +, −, and *fb* variables.

We now illustrate the usefulness of the preceding analysis. Consider a two phase induction motor operating on unbalanced voltages; that is,

$$v_a^s = V_a^s \cos \omega t$$

$$v_b^s = V_b^s \sin(\omega t + \phi) \tag{9.77}$$

$$v_a^r = v_b^r = v_f^r = v_b^r = 0 \tag{9.78}$$

From (9.68), (9.70) and (9.77) we obtain

$$
\begin{aligned}
v_+^s &= \frac{1}{2\sqrt{2}}\left[(V_a^s + jV_b^s)e^{j\omega t} + (V_a^{s*} + jV_b^{s*})e^{-j\omega t} \right] \\
v_-^s &= \frac{1}{2\sqrt{2}}\left[(V_a^s - jV_b^s)e^{j\omega t} + (V_a^{s*} - jV_b^{s*})e^{-j\omega t} \right]
\end{aligned}
\tag{9.79}
$$

where

$$
\begin{aligned}
V_a^s &= V_a^s \\
V_b^s &= -jV_b^s e^{j\phi}
\end{aligned}
\tag{9.80}
$$

and the superscript * denotes the complex conjugate. Under steady-state conditions

$$
\begin{aligned}
V_+^s &= \frac{1}{\sqrt{2}}(V_a^s + jV_b^s) \\
V_-^s &= \frac{1}{\sqrt{2}}(V_a^s - jV_b^s)
\end{aligned}
\tag{9.81}
$$

Using (9.81), (9.79) can be expressed as

$$
\begin{aligned}
v_+^s &= \frac{1}{2}\left(V_+^s e^{j\omega t} + V_-^{s*} e^{-j\omega t} \right) \\
v_-^s &= v_+^{s*}
\end{aligned}
\tag{9.82}
$$

Similarly, for the steady-state currents we can write

$$
\begin{aligned}
i_+^s &= \frac{1}{2}\left(I_+^s e^{j\omega t} + I_-^{s*} e^{-j\omega t} \right) \\
i_-^s &= i_+^{s*} \\
i_f^r &= \frac{1}{2}\left(I_f^r e^{j\omega t} + I_b^{r*} e^{-j\omega t} \right) \\
i_b^r &= i_f^{r*}
\end{aligned}
\tag{9.83}
$$

From (9.79) we observe that the voltage can be resolved into two components: $V_+^s e^{j\omega t}$ and $V_-^{s*} e^{-j\omega t}$. Consequently, (9.79) to (9.83), when substituted in (9.71), yield for a rotor speed ω_m and for an excitation $V_+^s e^{j\omega t}$

$$\begin{aligned} V_+^s &= (r^s + j\omega L^s) I_+^s + j\omega L^{sr} I_f^r \\ 0 &= j(\omega - \omega_m) L^{sr} I_f^s + [r^r + j(\omega - \omega_m) L^r] I_f^r \end{aligned} \tag{9.84}$$

For an excitation $V_-^{s*} e^{-j\omega t}$ we have

$$\begin{aligned} V_-^{s*} &= (r^s - j\omega L^s) I_-^{s*} - j\omega L^{sr} I_b^{r*} \\ 0 &= -j(\omega + \omega_m) L^{sr} I_-^{s*} + [r^r - j(\omega + \omega_m) L^r] I_b^{r*} \end{aligned} \tag{9.85}$$

But $(\omega - \omega_m) = s\omega$ and $(\omega + \omega_m) = (2 - s)\omega$, where s = slip. Therefore, we can rewrite (9.84) and (9.85) in the final form in terms of unconjugated voltages and currents as follows:

$$\begin{aligned} V_+^s &= (r^s + j\omega L^s) I_+^s + j\omega L^{sr} I_f^r \\ 0 &= j\omega L^{sr} I_+^s + \left(\frac{r^s}{s} + j\omega L^r \right) I_f^r \end{aligned} \tag{9.86}$$

and

$$\begin{aligned} V_-^s &= (r^s + j\omega L^s) I_-^s + j\omega L^{sr} I_b^r \\ 0 &= j\omega L^{sr} I_-^s + \left(\frac{r^r}{2 - s} + j\omega L^r \right) I_b^r \end{aligned} \tag{9.87}$$

These two sets of equations give the volt-ampere characteristics of the motor. Similarly, we can express the torque equation, (9.76), by using (9.83), as

$$\begin{aligned} T_e = \frac{1}{4} jL^{sr} \Big[&\left(I_+^{s*} I_f^r - I_+^s I_f^{r*} \right) + \left(I_-^s I_b^{r*} - I_-^{s*} I_b^r \right) \\ &+ \left(I_-^s I_f^r - I_+^s I_b^r \right) e^{j2\omega t} + \left(I_+^{s*} I_b^{r*} - I_-^{s*} I_f^{r*} \right) e^{-j2\omega t} \Big] \end{aligned} \tag{9.88}$$

Solving for the currents from (9.86) and (9.87), substituting in (9.89), and noting that the average value of $e^{\pm j2\omega t}$ is zero, we obtain the expression for the average torque as

$$T_{av}=\frac{\left(\frac{1}{2}\right)\omega(L^{sr})^2\frac{r^r}{s}(V_+^s)^2}{\left\{\frac{r^r r^s}{s}-\omega^2\left[L^r L^s-(L^{sr})^2\right]\right\}^2+\omega^2\left(r^s L^r+\frac{r^r L^r}{s}\right)^2}$$

$$-\frac{\left(\frac{1}{2}\right)\omega(L^{sr})^2\frac{r^r}{2-s}(V_-^s)^2}{\left\{\frac{r^r r^s}{2-s}-\omega^2\left[L^r L^s-(L^{sr})^2\right]\right\}^2+\omega^2\left(r^s L^r+\frac{r^r}{2-s}L^r\right)^2} \tag{9.89}$$

A summary of the two-phase transformation discussed so far is given in Figure 9.9, where the superscripts -1, $*$, and $\sim$ denote the inverse, complex conjugate, and transpose of the matrix respectively.

Consider the following example to illustrate an application of the preceding discussions.

Example 9.1

A two-phase two-pole 400 Hz servomotor has the following (per-phase circuit constants:

$$r^s=50 \text{ ohm} \qquad \omega L^s=610 \text{ ohm}$$

$$r^r=430 \text{ ohm} \qquad \omega L^r=460 \text{ ohm}$$

$$\text{and } \omega L^{sr}=380 \text{ ohm}$$

For a speed of 6000 rpm, calculate the steady-state torque, if the two phases are supplied by the voltages $v_a=230\angle 0°$ and $v_b=115\angle -90°$.

From the given data we have

$$n_s=\frac{120\times 400}{2}=24{,}000 \text{ rpm}$$

$$s=\frac{24{,}000-6000}{24{,}000}=0.75$$

$$\omega=2\pi\times 400=2513 \text{ r/s}$$

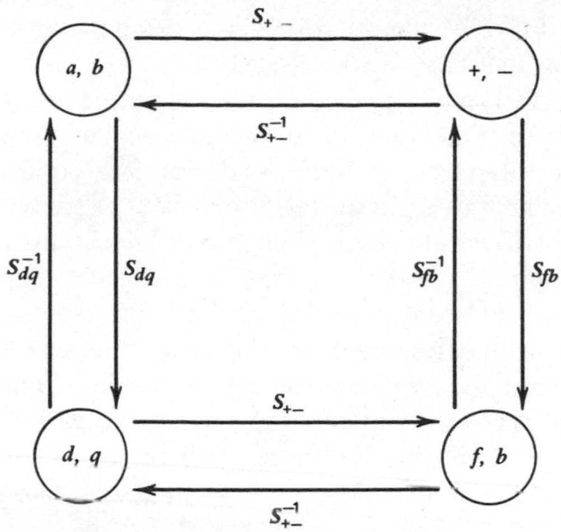

$$S_{+-} = \frac{1}{\sqrt{2}}\begin{bmatrix} 1 & 1 \\ -j & j \end{bmatrix}, \qquad S_{+-}^{-1} = \tilde{S}_{+-}^{*}$$

$$S_{fb} = \begin{bmatrix} e^{-j\theta} & 0 \\ 0 & e^{j\theta} \end{bmatrix}, \qquad S_{fb}^{-1} = \tilde{S}_{fb}^{*}$$

$$S_{dq} = \begin{bmatrix} \cos\theta & \sin\theta \\ -\sin\theta & \cos\theta \end{bmatrix}, \qquad S_{dq}^{-1} = \tilde{S}_{dq}$$

Figure 9.9. Summary of two-phase transformations.

And from (9.81) we get

$$V_{+}^{s} = \frac{1}{\sqrt{2}}\left[230 + j(-j115)\right] = 244 \text{ V}$$

$$V_{-}^{s} = \frac{1}{\sqrt{2}}\left[230 - j(-j115)\right] = 81.3 \text{ V}$$

Substituting the above numerical values and given data in (9.89) yields

$$T_{\text{av}} = 1.04\times 10^{-2} - 0.197\times 10^{-2} = 0.843\times 10^{-2}\text{N}-\text{m}$$

9.7 NOTE ON THREE-PHASE TRANSFORMATIONS

In our discussions thus far we have generally considered two-phase machines. From the preceding chapters, however, we know that most industrial ac machines are three-phase machines. Naturally, the question arises if the methods presented so far are applicable to such machines also.

The underlying principles of the analyses in preceding sections are equally applicable to three-phase, or multiphase, machines as to two-phase machines. For instance, it can be shown that a polyphase machine, under certain conditions can be reduced to an equivalent two-phase machine. We restricted ourselves to the two-phase machine to illustrate basic principles while keeping algebraic manipulations to a minimum. Although we shall not consider the details of three-phase transformations, we should point out that the three-phase transformations are expressed as (3 by 3) matrices. Because there are three phase variables (a, b, c), there are correspondingly three transformed variables. Thus the three components of the symmetrical components are the positive-, negative- and zero-sequence ($+ -0$) components. Similarly, the three components of the dq-transformation are the d, q, o components. Also, transformations from three-phases to two-phases exist. Because details are available elsewhere[1–3] on three phase machines, there is no need for further comment here. Since the usefulness of transformation techniques is restricted to somewhat idealized cases, we shall briefly review the primitive, or generalized, machine in the next section and will present digital solutions to a few realistic examples thereafter.

9.8 THE PRIMITIVE MACHINE

In Section 9.1 we indicated that a number of machines may be derived from a 'primitive' machine having stationary windings along the d- and q-axis, as shown in Figure 9.10. The rotor windings "appear" to be stationary because of the presence of the brushes. We assume saliency on the stator. Thus, we have a four-brush dc machine in which the coils along the d-axis are not electromagnetically coupled with these on the q-axis. The voltages induced in the various coils are as follows:

1. Voltages from self-inductances.
2. Transformer-type voltages (from mutual inductances) in mutually coupled coils.
3. Rotational (or speed) voltages in the rotor d-coil because of the stator q-coil field and in the rotor q-coil because of the stator d-coil field. (See also Section 9.3.)
4. Rotational voltages in the rotor d-coil because of the rotor q-coil field and in the rotor q-coil because of the rotor d-coil field.

In writing the voltage-balance equations we must account for all the above-mentioned voltages. Upon referring to Figure 9.10, the voltage equation

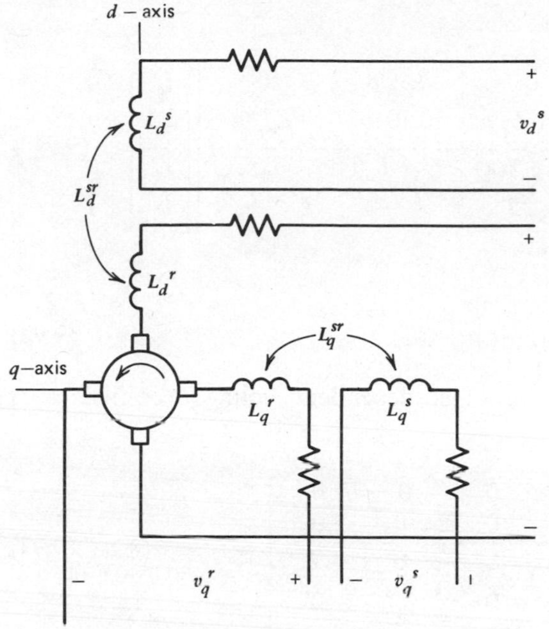

Figure 9.10. A generalized primitive machine.

in matrix notation becomes

$$\begin{bmatrix} v_d^s \\ v_q^s \\ \hline v_d^r \\ v_q^r \end{bmatrix} = \left[\begin{array}{cc|cc} r_d^s + L_d^s p & 0 & L_{dd}^{sr} p & 0 \\ 0 & r_q^s + L_q^s p & 0 & L_{qq}^{sr} p \\ \hline L_{dd}^{sr} p & G_{dq}^{rs} \omega_m & r^r + L_d^r p & G_{dq}^{rr} \omega_m \\ G_{qd}^{rs} \omega_m & L_{qq}^{rs} p & G_{qd}^{rr} \omega_m & r^r + L_q^r p \end{array}\right] \begin{bmatrix} i_d^s \\ i_q^s \\ \hline i_d^r \\ i_q^r \end{bmatrix} \tag{9.90}$$

where the G's are coefficients for speed-voltages; the subscripts correspond to the d- and q-axis, and the superscripts imply stator or rotor. For instance G_{dq}^{rs} represents the speed-voltage coefficient for the voltage induced in the rotor (superscript r) direct-axis coil (subscript d) because of rotation in the stator (superscript s) quadrature-axis (subscript q) field. Other symbols are denoted in Figure 9.10. Notice that $r_d^r = r_q^r = r^r$. We now recall the significance of the G's from Section 9.3, where we have $G = (\partial L^{sr}/\partial\theta)_{\theta=\theta_o}$. Consequently, we can substitute the following in (9.71)

$$G_{dq}^{rs} = -L_{qq}^{rs}, G_{qd}^{rs} = L_{dd}^{sr}$$

$$G_{dq}^{rr} = -L_q^r \quad \text{and} \quad G_{dq}^{rr} = L_d^r$$

to obtain

$$\begin{bmatrix} v_d^s \\ v_q^s \\ \hline v_d^r \\ v_q^r \end{bmatrix} = \left[\begin{array}{cc|cc} r_d^s + L_d^s p & 0 & L_{dd}^{sr} p & 0 \\ 0 & r_q^s + L_q^s p & 0 & L_{qq}^{sr} p \\ \hline L_{dd}^{sr} & -L_{qq}^{rs}\omega_m & r^r + L_d^r p & -L_q^r \omega_m \\ L_{dd}^{sr}\omega_m & L_{qq}^{rs} p & L_d^r \omega_m & r^r + L_q^r p \end{array}\right] \begin{bmatrix} i_d^s \\ i_q^s \\ \hline i_d^r \\ i_q^r \end{bmatrix} \tag{9.91}$$

This equation can also be written as

$$\mathbf{v} = (\mathbf{r} + \mathbf{L}p + \mathbf{G}\omega_m)\mathbf{i} \tag{9.92}$$

where the **G**-matrix containing the speed-dependent voltage coefficients is simply

$$\mathbf{G} = \left[\begin{array}{cc|cc} 0 & 0 & 0 & 0 \\ 0 & 0 & 0 & 0 \\ \hline 0 & -L_{qq}^{rs} & 0 & -L_q^r \\ L_{dd}^{sr} & 0 & L_d^r & 0 \end{array}\right] \tag{9.93}$$

The torque equation then becomes

$$T_e = \tilde{\mathbf{i}}\mathbf{G}\mathbf{i} \tag{9.94}$$

These equations are similar to those for machines studied earlier, as may be seen from (9.3). Thus we conclude that the primitive machine is a general machine from which the various other machines can be derived by applying proper constraints. Because the primitive machine is now of almost insignificant practical importance, we shall not pursue this discussion further. Details are available in References 2 to 6. We now turn to the applications of digital computers to obtain the solutions to the equations of motion directly.

9.9 THE TIME-DOMAIN FORMULATION

We recall that the voltage and torque equations of an electric machine may be routinely written as (9.1) and (9.2) (repeated below for convenience)

$$\mathbf{v} = \mathbf{r}\mathbf{i} + \frac{d}{dt}(\mathbf{L}\mathbf{i}) \tag{9.95}$$

$$T_e = -\frac{1}{2}\tilde{\mathbf{i}}\frac{\partial}{\partial\theta}(\mathbf{L}\mathbf{i}) \tag{9.96}$$

The performance characteristics of the machine are obtained by solving these equations. In this connection, we notice from the discussions of the last few sections that generalized machine theory aims at solving these equations by a transformation method. Whereas the use of a transformation method (such as the *dq* or ± transformation) facilitates the solutions of the equations of motion, this approach is applicable only when the machine is idealized and its inductance matrices satisfy certain properties. There are many instances where the inductances do not vary in an ideal (sinusoidal) fashion or the coefficients of sinusoid may not display a certain symmetry. In such cases transformations do not yield solutions to the equations of motion, because the position- dependent terms are not eliminated. However, a formulation in the time-domain offers one way of solving the equations. With such a formulation, we plan to solve (9.95) and (9.96) by a numerical method on a digital computer. A standard method of numerical integration, such as the Runge-Kutta-Gauss-Seidel (RKGS) method, and the Continuous System Modeling Program (CSMP) for certain cases, are applicable to (9.95) and (9.96). Thus, we obtain the solutions to these equations directly. We shall illustrate the details by considering a number of examples.

Example 9.2. A synchronous generator
For generator operation, it is advantageous to express (9.95) in the state-variables form. For this purpose, we rewrite (9.95) as

$$\mathbf{ri}+\mathbf{L}\frac{d}{dt}\mathbf{i}+\left(\frac{d}{dt}\mathbf{L}\right)\mathbf{i}=v \tag{9.97}$$

If the synchronous generator operates at a speed $\omega_m=\dot{\theta}$, (9.97) can be expressed as

$$\omega_m\mathbf{L}\frac{d}{d\theta}\mathbf{i}=-\left(\mathbf{r}+\omega_m\frac{d}{d\theta}\mathbf{L}\right)\mathbf{i}+v \tag{9.98}$$

Premultiplying both sides of (9.98) by $\omega_m\mathbf{L}^{-1}$ finally yields

$$\frac{d}{d\theta}\mathbf{i}=\mathbf{Zi}+\mathbf{Bv} \tag{9.99}$$

where $\mathbf{Z}=-\omega_m\mathbf{L}^{-1}\left(\mathbf{r}+\omega_m\frac{d}{d\theta}\mathbf{L}\right)$ and $\mathbf{B}=\omega_m\mathbf{L}^{-1}$. Notice that (9.99) is the state equation in the standard form. We now apply this formulation to a salient pole synchronous generator supplying a purely resistive load. Because of an inherent dissymmetry in the machine, the inductance coefficient of L_{ab}^{ss}, (9.34), is not L_o^s. Consequently, the *dq*-transformation shall not eliminate the θ-dependent terms. We must, therefore, use the time-domain formulation and solve the equations by numerical integration.

For this example, we consider a 10-kW, 220-V, two-phase, 12-pole, 50-Hz salient pole machine having the following parameters: $L_{aa}^{ss}=(15.93+2.96\cos 2\theta)$, mH; $L_{ab}^{ss}={}_{ab}9.56\sin 2\theta$, mH; $L_{bb}^{ss}=(15.93-2.96\cos 2\theta)$, mH; $L_{ar}^{sr}=L_{ra}^{rs}=0.197\sin\theta$, H; $L_{br}^{sr}=L_{rb}^{rs}=0.197\cos\theta$, H; $L_{rr}^{rr}=L_f=2.52$ H; $r^s=r_a+r_{\text{load}}=(0.1+9.68)=9.78$ ohm; $r^r=r_f=7.5$ ohm; $v_a=v_b=0$; and $v_f=3.75$ V, where the symbols are defined in (9.23) and (9.35). With these numerical values given, the problem is to compute the steady-state armature currents. One possible method of solving (9.97), when this is reduced to the form of (9.99), is by using the RKGS, a numerical integration subroutine from the IBM-360 Scientific Subroutine Package. This subroutine requires two user-supplied subroutines: an output subroutine and a subroutine to evaluate the right-hand side of (9.99) for a given i-θ pair. A typical i-θ pair may be the initial current, i_o, and the initial angular position, θ_o, all of which are assumed to be zero, except for $i^r=I_f=5$ A., for the present example. The flow chart for the use of the RKGS subroutine is shown in Figure 9.11*a*, and a flow chart for the evaluation of $FCT=d\mathbf{i}/d\theta$ is given in Figure 9.11*b*. The $d/d\theta$ terms in $\mathbf{Z}$ in (9.99) can be evaluated numerically for the given **Z**-functions. The procedure is as follows.

Considering the **Z** and **B** matrices of (9.99), we see that these matrices involve arithmetic operations such as matrix inversion, multiplication by a constant, and so forth. It is not necessary to carry out these operations *a priori*. Rather, appropriate data could be fed into the computer and the desired operation performed by the computer. For this purpose, (9.98) is rewritten as

$$\mathbf{L}\frac{d\mathbf{i}}{d\theta}=\mathbf{Z}_1\mathbf{i}+\mathbf{v}_e$$

where **L** is the original **L**-matrix, $\mathbf{Z}_1=-(\mathbf{r}/\omega_m)+(d\mathbf{L}/d\theta)$, and $\mathbf{v}_e=(\mathbf{v}/\omega_m)$.

For the example under consideration, with the **Z**-matrix known and the numerical values given above, the computed armature currents for the two phases supplying purely resistive loads are shown in Figure 9.12. It is evident that the output currents, as a consequence of the dissymmetry in the machine, depart substantially from being sinusoidal.

It is to be noted that steady-state values of the currents are desired for this example; that is $i(\theta)=i(\theta+2\pi)$. In principle, $i(0)=0$ could be taken as the initial condition. However, in addition to yielding the steady-state solution, this procedure would give the transient solution at the expense of computation time. On the other hand, if $i(\theta)$ could be known at some angle, $\theta=n\pi$, large enough that the transients have settled down, the steady-state solution could be computed in only one cycle. Although the current, $i(n\pi)$, is not known, it can be estimated, thus saving the necessity of computing much of the transient solution. This estimate is obtained by taking $i(0)$ at the standstill steady-state currents are: $i_f=5$ A.; $i_a=0$ A.; $i_b=0$ A. The iteration was carried out at 25 points for each interval of π for six cycles. Then $i(12\pi)$ was taken as the estimate for a steady state. The final calculations were done for six cycles using 200 iteration points for each interval of π. Figure 9.12 is a plot of the current for the sixth cycle.

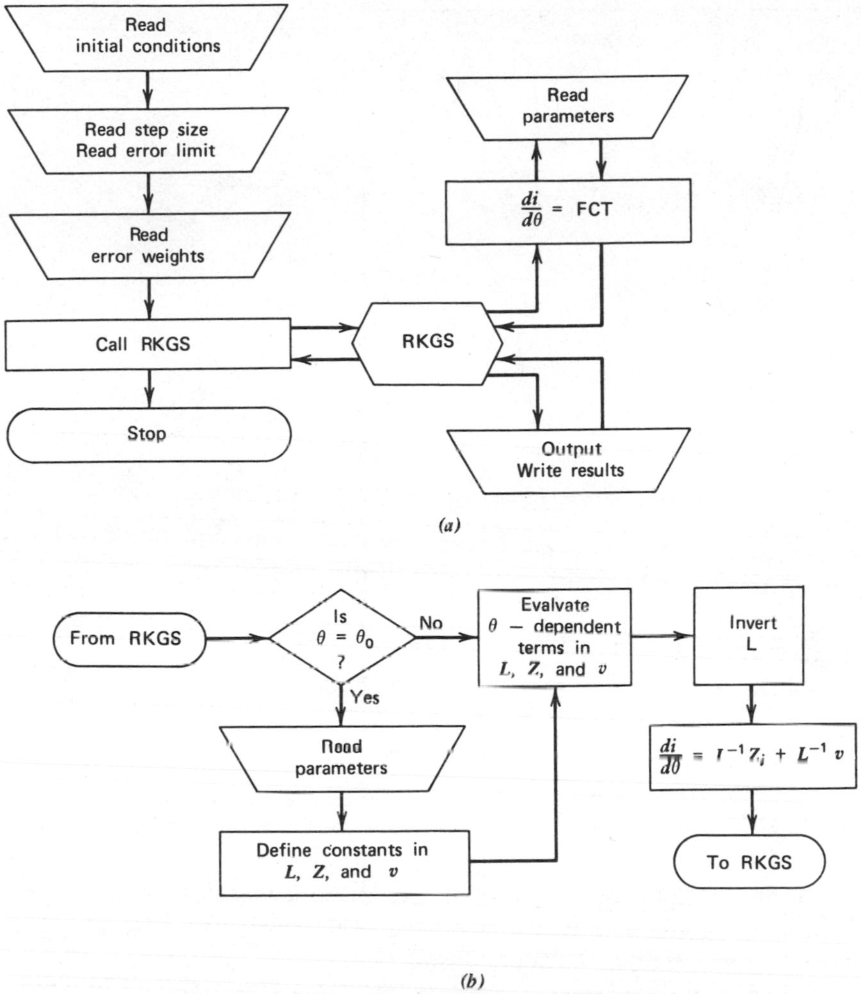

Figure 9.11. Flow chart for calculating the phase currents by the RKGS subroutine.

Although we do not intend to consider it here, we wish to point out that the application of the time-domain formulation to synchronous machine transients is extensively available in the literature. (See, for instance Reference 7.)

Example 9.3. Dynamics of induction motors
In our discussions in Chapter 6, we considered only the steady-state behavior of the induction motor. However, induction motors undergo transients because of switching, plugging, overspeeding, reversing, and sudden applications of load, as well as at the time of starting. Thus, it is helpful that we understand the dynamics of induction motors. While it is not practicable to examine most of the

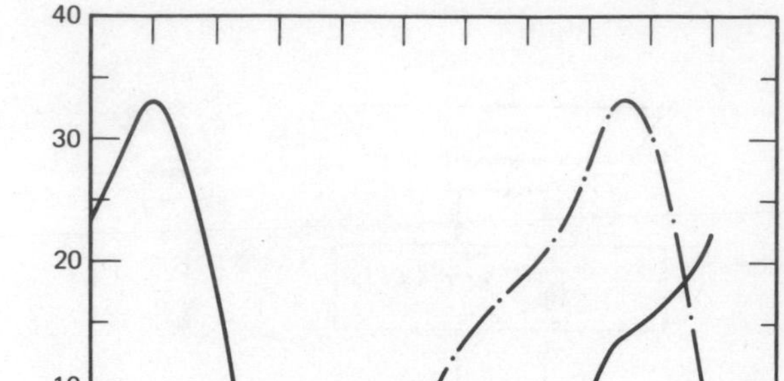

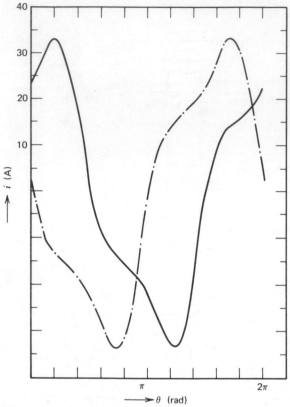

Figure 9.12. Computed phase currents. ——·——, current in phase a; ——, current in phase b.

instances when transients occur, we shall present now a general approach that can be adapted for many practical situations. This approach is based on the time-domain formulation of the equations of motion.

Proceeding as in Chapter 4, consider the induction motor as a multiwinding electromechanical energy converter. Using the notation of Section 4.7, the equations of motion may be written in matrix form as follows:

$$\mathbf{v} = \mathbf{ri} + \frac{d}{dt}(\mathbf{Li}) \tag{9.100}$$

$$T_e = \frac{P}{2}\tilde{\mathbf{i}}\frac{\partial \mathbf{L}}{\partial \theta}\mathbf{i} = J\dot{\omega}_m + b\omega_m + c \tag{9.101}$$

where $\tilde{\mathbf{i}}$ denotes the transpose of $\mathbf{i}$ and other symbols are defined later. In order to solve for the transient torque, for instance, the stator and rotor currents must be known. These are obtained from (9.100) and substituted in (9.101).

Because the above equations are nonlinear differential equations with time-varying coefficients, they are solved numerically as in the last example. For this purpose, it is convenient to express the equations of motion in the state-variables form. For (9.100) and (9.101), therefore, we have

$$\frac{d\mathbf{i}}{dt}=-\mathbf{L}^{-1}\left(\mathbf{r}+\frac{d\mathbf{L}}{dt}\right)\mathbf{i}+\mathbf{L}^{-1}\boldsymbol{v} \tag{9.102}$$

$$\dot{\omega}_m=\frac{1}{2J}\left(P\tilde{\mathbf{i}}\frac{\partial\mathbf{L}}{\partial\theta}\mathbf{i}\right)-\frac{b}{J}\omega_m-\frac{c}{J} \tag{9.103}$$

which may be combined to obtain

$$\dot{\mathbf{x}}=\mathbf{A}\mathbf{x}+\mathbf{B}\mathbf{v}' \tag{9.104}$$

where the various vectors are

$$\mathbf{x}=\begin{bmatrix}\mathbf{i}\\ \omega_m\\ \theta\end{bmatrix} \tag{9.105}$$

$$\mathbf{v}'=\begin{bmatrix}\mathbf{v}\\ 1\\ 0\end{bmatrix}$$

$$\mathbf{B}=\begin{bmatrix}\mathbf{L}^{-1} & 0 & 0\\ 0 & -\frac{c}{J} & 0\\ 0 & 0 & 0\end{bmatrix} \tag{9.106}$$

and

$$\mathbf{A}=\begin{bmatrix}-\mathbf{L}^{-1}\left(\mathbf{r}+\frac{d\mathbf{L}}{dt}\right) & 0 & 0\\ \frac{P}{2J}\left[F(\mathbf{i})\right] & -\frac{b}{J} & 0\\ 0 & 0 & 0\end{bmatrix} \tag{9.107}$$

with

$$\dot{\theta}=\frac{P}{2}\omega_m \quad\text{and}\quad F(\mathbf{i})=\dot{\theta}\,\frac{\partial\mathbf{L}}{\partial\theta}\mathbf{i}$$

Thus, the problem of transients during the start-up of an induction motor is reduced to the state equation with time-varying coefficients. This equation can

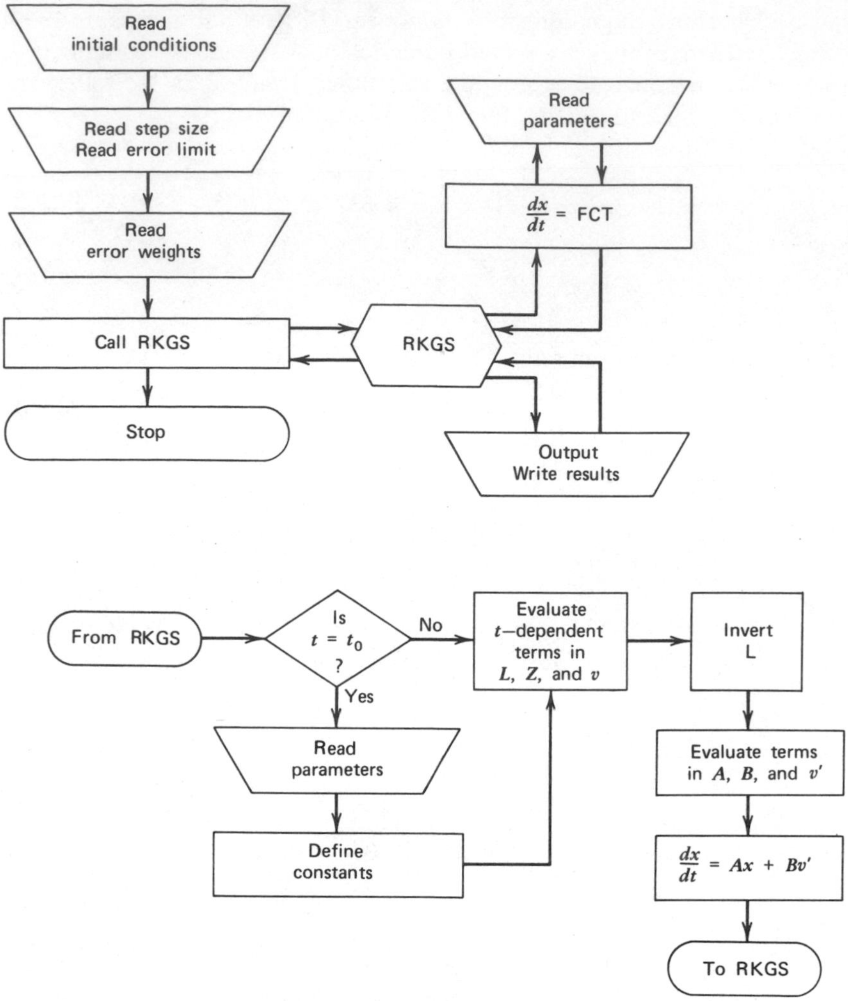

Figure 9.13. A flow chart showing steps in computation.

be solved by the Runge-Kutta-Gauss-Seidel (RKGS)–subroutine, as discussed in the last example. A slightly modified flow chart to be used with Figure 9.11*a* is shown in Figure 9.13. To illustrate the procedure, we consider a 30 hp, 220-V, four pole, 60-Hz, Y-connected motor, having a per-phase equivalent circuit shown in Figure 9.14. The total load torque on the motor is $(0.06\,\dot{\omega}_m + 0.03\,\omega_m + 6)$ N-m, where ω_m = rotor speed in rad/s. The acceleration characteristics are to be studied when the stator terminals are suddenly connected to a 220-V, three-phase supply.

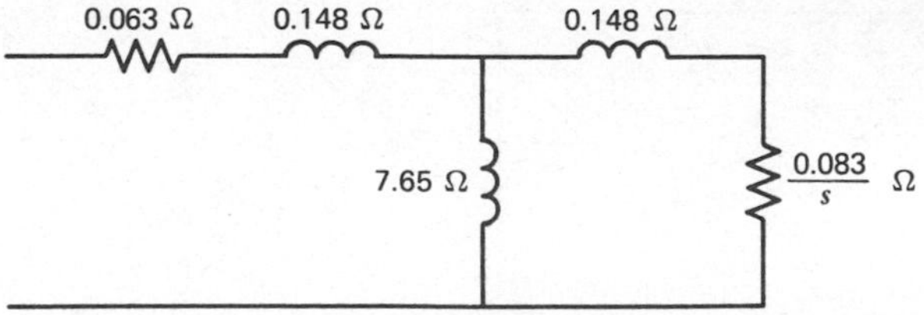

Figure 9.14. An approximate equivalent circuit of an induction motor.

We may rewrite (9.10) and (9.10) in expanded forms as

$$\begin{bmatrix} \mathbf{v}^s \\ \mathbf{v}^r \end{bmatrix} = \begin{bmatrix} \mathbf{r}^s + p\mathbf{L}^{ss} & p\mathbf{L}^{sr} \\ p\mathbf{L}^{rs} & \mathbf{r}^r + p\mathbf{L}^{rr} \end{bmatrix} \begin{bmatrix} \mathbf{i}^s \\ \mathbf{i}^r \end{bmatrix} \tag{9.108}$$

$$T_e = \frac{P}{2} [\tilde{\mathbf{i}}^s \tilde{\mathbf{i}}^r] \frac{\partial}{\partial \theta} \begin{bmatrix} \mathbf{L}^{ss} & \mathbf{L}^{sr} \\ \mathbf{L}^{rs} & \mathbf{L}^{rr} \end{bmatrix} \begin{bmatrix} \mathbf{i}^s \\ \mathbf{i}^r \end{bmatrix} \tag{9.109}$$

where $p = d/dt$ and the superscripts s and r correspond to stator and rotor quantities, respectively. For the case under consideration, the various submatrices are

$$\mathbf{v}^s = \begin{bmatrix} 180\cos 377t \\ 180\cos(377t - \phi) \\ 180\cos(377t + \phi) \end{bmatrix}; \mathbf{v}^r = \begin{bmatrix} 0 \\ 0 \\ 0 \end{bmatrix} \tag{9.110}$$

with $\phi = 2\pi/3$.

$$\mathbf{r}^s = \begin{bmatrix} .063 & 0 & 0 \\ 0 & 0.063 & 0 \\ 0 & 0 & 0.063 \end{bmatrix} \text{ohm}; \mathbf{r}^r = \begin{bmatrix} .083 & 0 & 0 \\ 0 & 0.083 & 0 \\ 0 & 0 & 0.083 \end{bmatrix} \text{ohm} \tag{9.111}$$

$$\mathbf{L}^{ss} = \mathbf{L}^{rr} = \begin{bmatrix} 20.3925 & -10.0 & -10.0 \\ -10.0 & 20.3925 & -10.0 \\ -10.0 & -10.0 & 20.3925 \end{bmatrix} \text{mH} \tag{9.112}$$

$$\mathbf{L}^{sr} = \begin{bmatrix} 20\cos\theta & 20\cos(\theta+\phi) & 20\cos(\theta-\phi) \\ 20\cos(\theta-\phi) & 20\cos\theta & 20\cos(\theta+\phi) \\ 20\cos(\theta+\phi) & 20\cos(\theta-\phi) & 20\cos\theta \end{bmatrix} \text{mH} \tag{9.113}$$

with $\phi = 2\pi/3$.

And $\tilde{\mathbf{L}}^{rs} = \mathbf{L}^{sr}$ – transposed. In (9.101) $P = 2$. The load torque, as mentioned earlier, is given by

$$T_L = 0.06\,\dot{\omega}_m + 0.03\,\omega_m + 6 \text{ N-m} \tag{9.114}$$

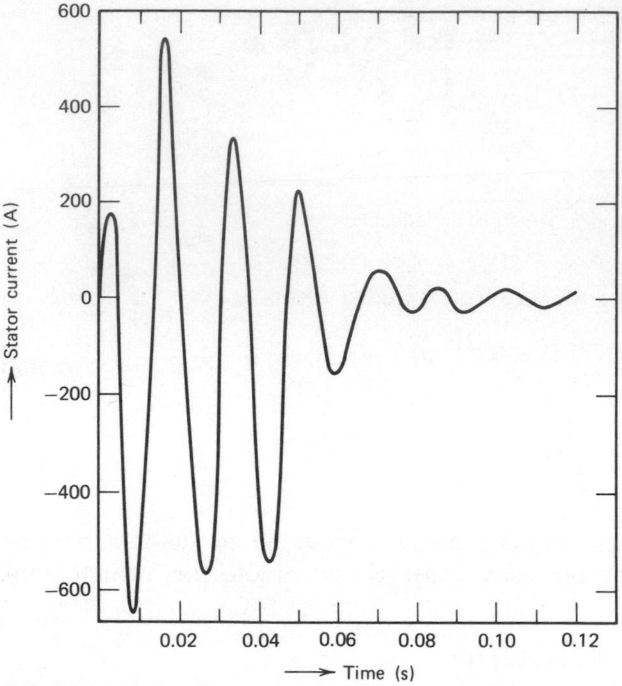

Figure 9.15. $i(t)$ during speed buildup.

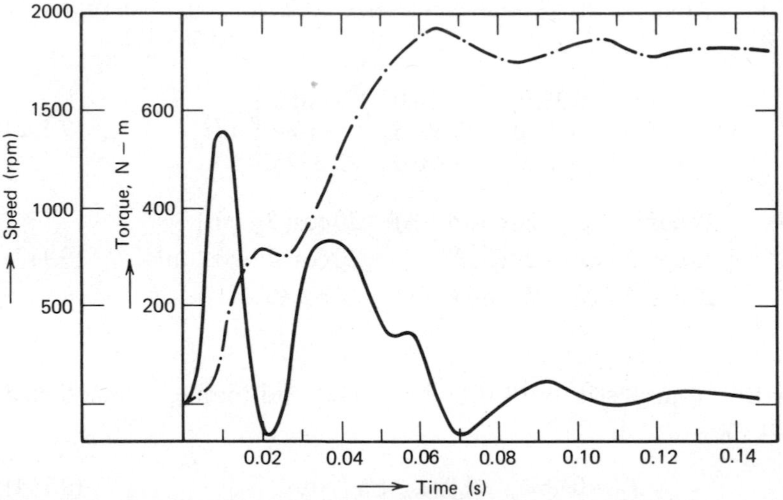

Figure 9.16. Speed and torque buildup. ——, torque; ——.——., speed.

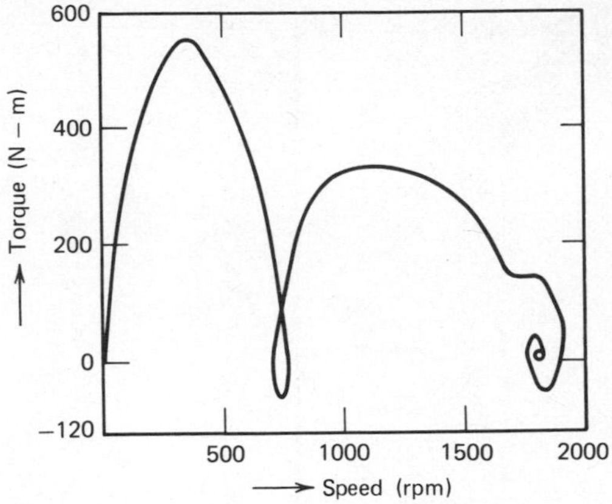

Figure 9.17. Transient torque/speed characteristics.

For the motor under consideration, with the numerical values of the inductances given and the load torque specified by (9.114), the vector $\mathbf{x}(t)$ can be numerically evaluated. The initial condition, of course, is $\mathbf{x}(0)=0$ because the motor is assumed to start from rest. Two of the components of the vector $\mathbf{x}(t)$ are $\mathbf{i}$ and ω_m, and, knowing $\mathbf{i}$, the torque is evaluated from (9.109). The computed $i(t)$, $T_e(t)$, and $\omega_m(t)$ are shown in Figures 9.15 and 9.16 and the transient torque /speed characteristic is shown in Figure 9.17. Notice that the dynamic torque/speed characteristics are strikingly different from the typical steady-state characteristics shown in Chapter 6.

Example 9.4. Dynamics of a dc motor

As the last example of the application of the time-domain formulation we consider the start-up of a separately excited dc motor on load, shown in Figure 9.18. From (9.45), the equations of motion in a slightly different nomenclature

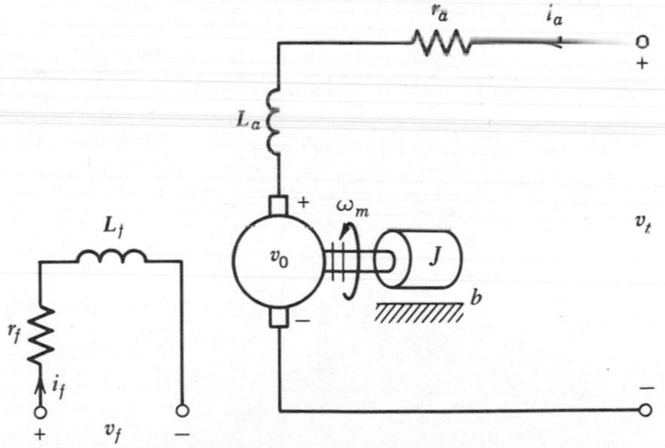

Figure 9.18. A separately excited dc motor.

```
****CONTINUOUS SYSTEM MODELING PROGRAM****

*** VERSION 1.3 ***

INITIAL
     IFI=16.7832
     IAI=0
     OMI=0
CONSTANT RA=0.0130,LA=0.01,RF=1.43,LF=0.167,J=0.21,Q=1.074E-6,C=2.493
     K=0.004
     VF=
     VT=24
DYNAMIC
     IAD=(VT-(K*OM*IF)-(RA*IA))/LA
     OMD=((K*IF*IA)-(Q*OM*OM)-C)/J
     IFD=(VF-(RF*IF))/LF
     T=K*IF*IA
     W=K*OM*IF
     OM=INTGRL(OMI,OMD)
     IA=INTGRL(IAI,IAD)
     IF=INTGRL(IFI,IFD)
     PRINT OM,IF,IA,VF,T,W
     TIME PRDEL=.1,FINTIM=20,DELT=.1
     PRTPLT OM,IF,IA,VF,T,W
END
STOP
OUTPUT VARIABLE SEQUENCE
IFI    IAI    OMI    K    VF    VT    OMD    OM    IAD    IA
IFD    IF     T      W
 OUTPUTS    INPUTS     PARAMS    INTEGS + MEM BLKS   FORTRAN   DATA CDS
 18(500)    50(1400)   10(400)     3+   0=  3(300)   15(600)       5
```

Figure 9.19. Program for computing dc motor transients.

are

$$L_a \frac{di_a}{dt} + r_a i_a + k\omega_m i_f = v_t \tag{9.115}$$

$$L_f \frac{di_f}{dt} + r_f i_f = v_f \tag{9.116}$$

$$J\dot{\omega}_m + C_2\omega_m + C_o = T_e = k i_a i_f \tag{9.117}$$

The numerical values of the parameters and constants for the motor under consideration are: $r_a = 0.013$ ohm; $L_a = 0.01$ H; $r_f = 1.43$ ohm; $L_f = 0.167$ H; $J = 0.21$ kg-m^2; $C_2 = 1.074 \times 10^{-6}$ N-m-s^2 (specified as Q in the computer program, Figure 9.19); $C_o = 2.493$ N-m (specified as C in the program); $k = 0.004$ N-m/A^2; $v_f = 12$ V; and $v_t = 24$ V.

The motor start-up characteristics are to be studied with v_f and v_t applied at the terminals at $t=0$. Notice that the left-hand side of (9.117) represents the load on the motor. Clearly, (9.115) to (9.117) is a set of nonlinear simultaneous differential equations, and only numerical solutions are practicable. For this

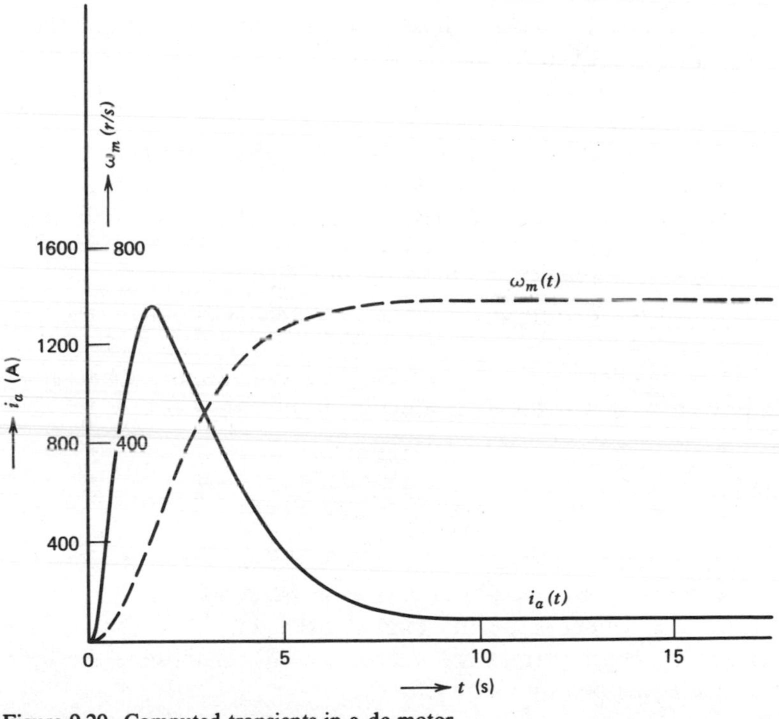

Figure 9.20. Computed transients in a dc motor.

purpose we use the Continuous System Modelling Program (CSMP) shown in Figure 9.19. The speed buildup and the variation of the armature current are shown in Figure 9.20, which may be said to constitute the solution of the equations of motion. We have plotted only $\omega_m(t)$ and $i_a(t)$, but the program shown in Figure 9.19 also gives the field-current, field-voltage, torque, and the back emf.

In conclusion, over the last 40 years or so, the generalized machine theory has offered an extremely useful approach to the study of electric machines. Almost all types of machines have been analyzed through the use of some type of transformation, such as the *dq*-transformation, symmetrical components transformation, and several others. In applying such transformations, it is assumed that the machine has some degree of geometric symmetry, which permits the use of various symmetric and cyclic-symmetric impedance matrices. For the transformation method to be applicable, restrictions have to be imposed which are often unrealistic. With the availability of efficient scientific subroutine packages for computers, the time-domain formulation offers an effective method of solving the dynamic and steady-state problems in electric machines.

REFERENCES

1. S. A. Nasar, *Electromagnetic Energy Conversion Devices and Systems*, Prentice-Hall, Inc., Englewood Cliffs, N.J., 1970.
2. D. C. White and H. H. Woodson, *Electromechanical Energy Conversion*, John Wiley & Sons, Inc., 1959.
3. D. O'Kelly and S. Simmons, *Generalized Electrical Machine Theory*, McGraw-Hill Book Company, New York, 1968.
4. H. K. Messerle, *Dynamic Circuit Theory*, Pergamon Press, Inc., Elmsford, N.Y., 1968.
5. G. Kron, *Tensors for Circuits*, Dover Publications, Inc., New York, 1959.
6. B. Adkins, *The General Theory of Electrical Machines*, Chapman and Hall, London, 1957.
7. R. E. Fite, "Transient Performance of an Isolated Synchronous Generator," MS Thesis, University of Kentucky, Lexington, Kentucky, 1974.

PROBLEMS

9.1. Proceeding in a manner similar to that for a salient rotor machine, by resolving the mmf's along the *d*- and *q*-axis, derive (9.40).

9.2. The *L*-matrix of a salient stator machine is given by (9.42). The stator is unexcited and the rotor current is $i^r = I^r \sin(\omega t + \delta)$. Determine the time-average torque developed by the machine.

9.3. Apply the appropriate constraints to (9.48) to obtain the equations for the steady-state operation of the machine as (a) a generator and (b) a motor.

9.4. From (9.20) to (9.22) obtain the L-matrix of a round-rotor synchronous machine. Recall from Section 9.4 that we applied certain constraints to obtain the characteristics of an induction motor. Apply appropriate constraints to obtain the steady-state torque characteristics of the two-phase round-rotor synchronous machine.

9.5. Apply the dq-transformation to a two-phase induction motor and hence obtain equivalent circuit of the motor. Verify that the circuit is identical to that obtained otherwise. (See Figure 6.17.)

9.6. To obtain a single-phase operation, one of the phases of a two-phase induction motor may be left unexcited. From the dq-transformed equations of Problem 9.5, obtain the torque-speed characteristics of a single-phase induction motor.

9.7. Starting with (9.75) obtain the steady-state per-phase equivalent circuit of a balanced two-phase induction motor.

9.8. A rotating machine has two windings on the rotor and one on the stator. The stator is supplied with a 24 V direct current and the rotor rotates at 100π rad/s. With the following numerical values compute $i_1{}^r(\theta)$: $L_{00}{}^{ss} = 2.9$ H; $L_{11}{}^{rr} = L_{22}{}^{rr} = 1.95$ Hh; $L_{01}{}^{sr} = (-1.49 \sin\theta - 0.665 \sin 3\theta)$ H; $L_{02}{}^{sr} = 1.665 \cos\theta$ H; $R_0{}^s = 30\ \Omega$; $R_1{}^r = R_2{}^r = 15\ \Omega$; $R_L = 1000\ \Omega$; $\omega_m = 100\pi$ rad/s; $v_o{}^s = 24$ V; $i_o{}^s(0) = 0.8$ A; $i_1{}^r(0) = 0.0124$ A; and $i_2{}^r(0) = 0.025$ A, where R_L = load resistance and the subscripts 0, 1, and 2 identify the windings.

9.9. Consider an unbalanced three-phase induction motor for which $\mathbf{v}^s$, $\mathbf{v}^r$, $\mathbf{r}^s$, $\mathbf{r}^r$, and $\mathbf{L}^{rr}$ are the same as in Example 9.3. However $\mathbf{L}^{ss}$ and $\mathbf{L}^{sr}$ are as follows:

$$\mathbf{L}^{ss} = \begin{bmatrix} 20.3925 & -20\cos 40 & -20\cos 70 \\ -20\cos 40 & 20.3925 & -20\cos 70 \\ -20\cos 70 & -20\cos 70 & 20.3925 \end{bmatrix} \text{ millihenry}$$

$$\mathbf{L}^{sr} = \begin{bmatrix} 20\cos\theta & 20\cos(\theta - 140) & 20\cos(\theta + 110) \\ 20\cos(\theta + 120) & 20\cos(\theta - 20) & 20\cos(\theta - 130) \\ 20\cos(\theta - 120) & 20\cos(\theta + 100) & 20\cos(\theta - 10) \end{bmatrix} \text{ millihenry}$$

Compute the dynamic start-up characteristics of this motor. Verify that because of a slight unbalance in the stator windings the motor undergoes severe transients as compared with a balanced motor.

9.10. Repeat the problem of Example 9.4 for $r_a = 0.0045$ ohm. Plot the oscillations of $i_a(t)$.

Appendix 1
Definition Of Symbols And Units

A. STANDARD INTERNATIONAL UNITS

SYMBOL	*DESCRIPTION*	*SI UNIT*
A	Magnetic vector potential	weber/meter
B	Magnetic flux density	tesla (T)
C	Capacitance	farad (F)
D	Electric flux density	coloumb/meter
D	Viscous damping coefficient	newton-meter-second
E	Electric field intensity	volt/meter
E,e	Electromotive force (emf)	volt (V)
f	Frequency	hertz
F, F	Force	newton
g	Air-gap length	meter
H	Magnetic field intensity	ampere/meter
hp	Horsepower (one hp = 746 W)	horsepower
I,i	Current	ampere (A)
J	Current density	ampere/(meter)^2
J	Moment of inertia	Kg-(meter)^2
L	Self inductance	henry
l	Length	meter
M	Mutual inductance	henry
M	Magnetic scalar potential (mmf)	ampere (ampere-turn)

N	Turns	—
P,p	Power, number of poles	watt (W)
P	Permeance (magnetic)	henry
Q	Electrical charge	coloumb
Q	Quality factor	—
R,r	Electric resistance	ohm
R	Magnetic reluctance	henry^{-1}
$\mathbf{S}$	Surface area vector	$(\text{meter})^2$
T	Torque	newton-meter
t	Time	second(s)
$\mathbf{U}$	Velocity vector	meter/second
V,v	Electric potential (or voltage)	volt (V)
v	Volume	$(\text{meter})^3$
W	Energy	joule or wattsec
σ	Electrical conductivity	$(\text{ohm-meter})^{-1}$
ϵ_o	Dielectric constant (permittivity of free space)	farad/meter
θ	Angular measure (radian)	—
λ	Magnetic flux linkage	weber
μ	Magnetic permeability	henry/meter
μ_o	Magnetic constant (permeability of free space)	henry/meter
μ_R	Relative magnetic permeability	—
ρ	Electrical resistivity	ohm-meter
Φ,ϕ	Magnetic flux	weber (Wb)
ω	Angular velocity	radian/second

B. ALTERNATE UNITS

Two additional systems of units are in relatively common usage in certain industrial and scientific applications. These are the English and CGS systems.

The relationships among these systems of units and the SI System are best described—for the purposes of electromagnetic analysis—by means of the equations relating the two fundamental magnetic parameters, (2.9) and (2.10) of Chapter 2. Equation (2.10) is repeated below in appropriate representation for the three systems

$$\mathbf{B} = \mu_o \mu_R \mathbf{H} \text{ (SI and English)}$$

$$\mathbf{B} = \mu_R \mathbf{H} \text{ (CGS)}$$

where units and numerical values are as follows:

	SI	English	CGS
μ_o	$4\pi \times 10^{-7}$	3.19	1
B	tesla	lines/(inch)^2	gauss
H	ampere/meter	ampere/inch	oersted

C. UNIT CONVERSION

SYMBOL	*DESCRIPTION*	one: *(SI UNIT)*	is equal to: *(ENGLISH UNIT)*	*(CGS UNIT)*
B	Magnetic flux density	tesla (= 1 weber/m^2)	6.452×10^4 lines/in^2	10^4 gauss
H	Magnetic field intensity	ampere/m	0.0254 ampere/in	0.004π oersted
ϕ	Magnetic flux	weber	10^8 lines	10^8 maxwells
D	Viscous damping coef.	newton-m-sec	.73756 lb-ft-sec	10^7 dyne-cm-sec
F	Force	newton	0.2248 pounds	10^5 dynes
J	Inertia	kg-m^2	23.73 lb-ft^2	10^7 g-cm^2
T	Torque	newton-m	0.73756 ft-pound	10^7 dyne-cm
W	Energy	joule	1 watt-s	10^7 ergs

Appendix 2
Magnet Wire Table For Single Film Coated Round Wire

Awg Size	Bare Wire Diameter Nominal (Inches)	FILM ADDITIONS (Inches)		OVERALL DIAMETER (Inches)			WEIGHT AT 20°C-68°F		RESISTANCE AT 20°C-68°F		Wires/In. Nom.	Awg Size
		Min.	Max.	Min.	Nom.	Max.	Lbs./M Ft. Nom.	Ft./Lb. Nom.	Ohms/M Ft. Nom.	Ohms./Lb. Nom.		
8	.1285	.0016	.0026	.1288	.1306	.1324	50.20	19.92	.6281	.01251	7.66	8
9	.1144	.0016	.0026	.1149	.1165	.1181	39.81	25.12	.7925	.01991	8.58	9
10	.1019	.0015	.0025	.1024	.1039	.1054	31.59	31.66	.9988	.03162	9.62	10
11	.0907	.0015	.0025	.0913	.0927	.0941	25.04	39.94	1.26	.05032	10.8	11
12	.0808	.0014	.0024	.0814	.0827	.0840	19.92	50.20	1.59	.07982	12.1	12
13	.0720	.0014	.0023	.0727	.0738	.0750	15.81	63.25	2.00	.1265	13.5	13
14	.0641	.0014	.0023	.0649	.0659	.0670	12.49	80.06	2.52	.2018	15.2	14
15	.0571	.0013	.0022	.0578	.0588	.0599	9.948	100.5	3.18	.3196	17.0	15
16	.0508	.0012	.0021	.0515	.0525	.0534	7.880	126.9	4.02	.5101	19.0	16
17	.0453	.0012	.0020	.0460	.0469	.0478	6.269	159.5	5.05	.8055	21.3	17
18	.0403	.0011	.0019	.0410	.0418	.0426	4.970	201.2	6.39	1.286	23.9	18
19	.0359	.0011	.0019	.0366	.0374	.0382	3.943	253.6	8.05	2.041	26.7	19
20	.0320	.0010	.0018	.0327	.0334	.0341	3.138	318.7	10.1	3.219	29.9	20
21	.0285	.0010	.0018	.0292	.0299	.0306	2.492	401.2	12.8	5.135	33.4	21
22	.0253	.0010	.0017	.0260	.0267	.0273	1.969	507.9	16.2	8.228	37.5	22
23	.0226	.0009	.0016	.0233	.0238	.0244	1.572	636.1	20.3	12.91	42.0	23
24	.0201	.0009	.0015	.0208	.0213	.0218	1.240	806.5	25.7	20.73	46.9	24
25	.0179	.0009	.0014	.0186	.0191	.0195	.988	1012	32.4	32.79	52.4	25
26	.0159	.0008	.0013	.0165	.0169	.0174	.779	1284	41.0	52.64	59.2	26
27	.0142	.0008	.0013	.0149	.0153	.0156	.623	1605	51.4	82.50	65.4	27
28	.0126	.0007	.0012	.0132	.0136	.0139	.491	2037	65.3	133.0	73.5	28
29	.0113	.0007	.0012	.0119	.0122	.0126	.395	2532	81.2	205.6	82.0	29
30	.0100	.0006	.0011	.0105	.0109	.0112	.310	3226	104	335.5	91.7	30
31	.0089	.0006	.0011	.0094	.0097	.0100	.246	4065	131	532.5	103	31
32	.0080	.0006	.0010	.0085	.0088	.0091	.199	5025	162	814.1	114	32
33	.0071	.0005	.0009	.0075	.0078	.0081	.157	6394	206	1317	128	33
34	.0063	.0005	.0008	.0067	.0070	.0072	.123	8130	261	2122	143	34
35	.0056	.0004	.0007	.0059	.0062	.0064	.0977	10235	331	3388	161	35
36	.0050	.0004	.0007	.0053	.0056	.0058	.0783	12771	415	5300	179	36
37	.0045	.0003	.0006	.0047	.0050	.0052	.0632	15823	512	8101	200	37
38	.0040	.0003	.0006	.0042	.0045	.0047	.0501	19960	648	12934	222	38
39	.0035	.0002	.0005	.0036	.0039	.0041	.0383	26110	847	22115	256	39
40	.0031	.0002	.0005	.0032	.0035	.0037	.0301	33222	1080	35880	286	40
41	.0028	.0002	.0004	.0029	.0031	.0033	.0244	40984	1320	54099	323	41
42	.0025	.0002	.0004	.0026	.0028	.0030	.0195	51282	1660	85128	357	42
43	.0022	.0002	.0003	.0023	.0025	.0026	.0153	65360	2140	139870	400	43
44	.0020	.0001	.0003	.0020	.0022	.0024	.0124	80645	2590	208870	455	44

Appendix 3
Computer Techniques In The Analysis Of Electromechanical Systems And Electric Machines

Analog and digital computer techniques have been applied to the analysis and design of transformers, rotating electrical machines, and control systems from the earliest days of the "computer era." Computer techniques are used today in every aspect of the study of machines, transformers, and other electromechanical systems. This appendix is included to acquaint the reader with a very few of the many computer techniques and specific programs used for this purpose.

A. MAGNETIC CHARACTERISTICS

A principal impediment to accurate analysis of electromechanical systems is the nonlinear relationship of the magnetic circuit. Computer methods offer an invaluable alternative to the graphical, analog, and manual techniques used in early machine analysis. Some approaches for handling the magnetic characteristics include:

a. An actual B versus H (or ϕ versus NI or E versus NI) characteristic can be stored in a data file.
b. The curve can be represented by two or more straight-line segments. Figure III-1 is a typical flux versus mmf characteristic for a starter motor. Two

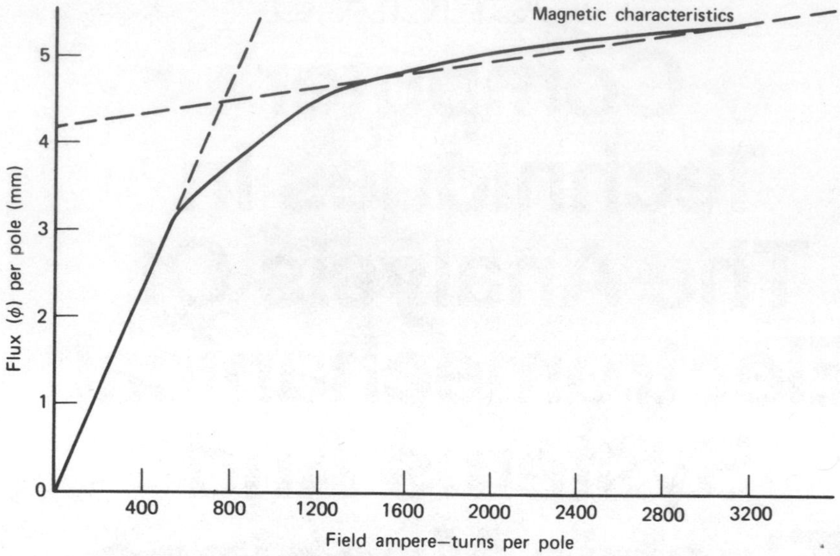

Figure III.1. Flux versus ampere-turns curve and straight-line approximation.

straight line segments are shown for the major slopes of the curve. The equations for these segments are

$$\phi = 6.786 \times 10^{-6}(\mathrm{NI})_f \text{webers}, \ (\mathrm{NI})_f < 700 \tag{5.36}$$

$$\phi = 4.2 \times 10^{-3} + 4.1 \times 10^{-7}(\mathrm{NI})_f \text{webers}, \ (\mathrm{NI})_f > 700$$

A third segment could be introduced to more accurately depict the curve in the region of the knee of ϕ versus NI characteristic.

c. There are various methods for representing a curve, such as Figure III-1, by means of a power series in NI (or H). In general, a minimum of three terms is required for such a representation. In many computer systems that may be available to the reader, software programs for regression analysis can give the power series expressions for specific B versus H characteristics.

d. Many analytical expressions for the B versus H characteristic have been proposed. One very useful analytical expression that has been used in finite-element methods of magnetic circuit analysis is[1]

$$H = \left(k_1 e^{k_2 B^2} + k_3\right) B$$

where k_1, k_2, and k_3 are determined by causing (5.38) to pass through three experimental points on a measured B versus H characteristic.

e. Finite elements. There are many computer programs for the general analysis of magnetic circuits used in electromechanical devices. Principle use of the finite element technique is the derivation and plotting of equipotential lines and flux lines, calculation of energy and inductance, and force calculation. One program available on many computer user systems is AOS/MAGNETIC.[2]

B. TRANSFORMERS AND INDUCTORS

A series of transformer and inductor design software programs are available on many computer user systems.[3]

1. TRANS—a general transformer design program.
2. INDUCTOR—inductor design program.
3. TOPT—a transformer optimization technique.
4. FERRO—a design program for ferro-resonant transformers.

REFERENCES

1. J. R. Brauer, "Simple Equations for the Magnetization and Reluctivity Curves of Steel," *IEEE Transactions on Magnetics*, New York, January, 1973.
2. A. O. Smith, Brochure, "Engineering Consulting Service," MKT-110577, May 1977.
3. Available from Optimized Program Services, Inc., Berea, OH.

Index